World Regional Geography Book Series

What does Finland mean to a Finn, Sichuan to a Sichuanian, and California to a Californian? How are physical and human geographical factors reflected in their present-day inhabitants? And how are these factors interrelated? How does history, culture, socio-economy, language and demography impact and characterize and identify an average person in such regions today? How does that determine her or his well-being, behaviour, ambitions and perspectives for the future? These are the type of questions that are central to The World Regional Geography Book Series, where physically and socially coherent regions are being characterized by their roots and future perspectives described through a wide variety of scientific disciplines. The Book Series presents a dynamic overall and in-depth picture of specific regions and their people. In times of globalization renewed interest emerges for the region as an entity, its people, its landscapes and their roots. Books in this Series will also provide insight in how people from different regions in the world will anticipate on and adapt to global challenges as climate change and to supra-regional mitigation measures. This, in turn, will contribute to the ambitions of the International Year of Global Understanding to link the local with the global, to be proclaimed by the United Nations as a UN-Year for 2016, as initiated by the International Geographical Union. Submissions to the Book Series are also invited on the theme 'The Geography of…', with a relevant subtitle of the authors/editors choice. Proposals for the series will be considered by the Series Editor and International Editorial Board. An author/editor questionnaire and instructions for authors can be obtained from the Publisher.

This book series is published in cooperation with the International Geographical Union (IGU). The IGU is an international, non-governmental, professional organization devoted to the development of the discipline of Geography. The purposes of the IGU are primarily to promote Geography through initiating and coordinating geographical research and teaching in all countries of the world.

Raquib Ahmed • Abdullah Al-Maruf
J. Craig Jenkins
Editors

Transforming Bangladesh

Geography, People, Economy and Environment

Springer

Editors
Raquib Ahmed
Fareast International University
Dhaka, Bangladesh

Department of Geography and
Environmental Studies
University of Rajshahi
Rajshahi, Bangladesh

J. Craig Jenkins
Department of Sociology
The Ohio State University
Columbus, OH, USA

Abdullah Al-Maruf
Department of Geography and
Environmental Studies
University of Rajshahi
Rajshahi, Bangladesh

ISSN 2363-9083 ISSN 2363-9091 (electronic)
World Regional Geography Book Series
ISBN 978-3-031-45092-1 ISBN 978-3-031-45093-8 (eBook)
https://doi.org/10.1007/978-3-031-45093-8

Picture Credit: Anton Balazh - Fotolia.com

This Springer imprint is published by the registered company Springer Nature Switzerland AG
The registered company address is: Gewerbestrasse 11, 6330 Cham, Switzerland

Paper in this product is recyclable.

Acknowledgement

The idea of editing the book first surfaced at the Pädagogische Hochschule Freiburg in Germany after a long discussion with Professor Gregor C Falk who happened to be a good authority on Bangladesh geography. Bangladesh experienced rapid and extensive changes in all aspects of its development all through colonial and post-colonial era – particularly in the last 52 years. During many research, we came across a perpetual shortage of up-to-date information on transformational process of Bangladesh's economy, environment and physical changes. Later on, the idea was solidified at a meeting in München after a long discussion, where Professor Boris Braun of the University of Cologne, Professor Craig Jenkins of the Ohio State University and myself were present and extensively examined the possible coverage of the book. We all felt the necessity of writing a book or editing a book on Bangladesh creating four windows, viz. geography, people, economy and environment, composed of 19 chapters. Selected authors are from North America, Europe and Asia, but all of them are strongly fluent about Bangladesh. Deutscher Akademischer Austauschdienst (German Academic Exchange Program or DAAD) was instrumental behind all through which has made us deeply obliged. Also we all are grateful to the University of Cologne and Pädagogische Hochschule Freiburg for hosting our activities there. We owe a lot to the publishers of the materials used in the chapters as reference by the authors of the book as well as all the reviewers who took much care to help us improve the chapters. But the larger share of our gratefulness is due to the authors who are the main part of the activity. Success of this book depends on the readers if they are benefited from it and we welcome their comments and observations of the chapters of the book. We would like to express thanks to the users in advance with a hope that the book will provide them some information about Bangladesh and its transformation through time. University of Rajshahi acted from behind which was useful for the book and for the editors to complete the editing part with full dedication and we are grateful to them. Finally, warm thanks are for the publisher, Springer, who helped the book to come into a book form.

Raquib Ahmed

Craig Jenkins

Abdullah Al-Maruf

Contents

Contributors

Fazle Rabbi Sadeque Ahmed Palli Karma-Sahayak Foundation (PKSF), PKSF Bhaban, Dhaka, Bangladesh

Raquib Ahmed Fareast International University, Dhaka, Bangladesh
Department of Geography and Environmental Studies, University of Rajshahi, Rajshahi, Bangladesh

Saleh Ahmed School of Public Service, Boise State University, Boise, ID, USA

Shajeda Aktar Department of Public Administration, University of Rajshahi, Rajshahi, Bangladesh

Md. Yusuf Ali Bangladesh Agricultural Research Institute, Rajshahi, Bangladesh

Abdullah Al-Maruf Department of Geography and Environmental Studies, University of Rajshahi, Rajshahi, Bangladesh

Rejaul Karim Bakshi Department of Economics, University of Rajshahi, Rajshahi, Bangladesh

María Gabriela Costa Faculty of Environmental and Natural Resources, Albert Ludwig Freiburg University, Freiburg im Breisgau, Germany
Facultad Jurídica, Social y Administrativa, Universidad Nacional de Loja, Loja, Ecuador

Gregor C. Falk Institute for Geography and Geography Education, University of Education Freiburg, Kunzenweg, Freiburg, Germany

A. K. M. Mahmudul Haque Department of Political Science, University of Rajshahi, Rajshahi, Bangladesh

Shahadat Hossain University of Duhok, Duhok, Iraq

Ferdous Farhana Huq Department of Urban & Regional Planning, Rajshahi University of Engineering & Technology (RUET), Rajshahi, Bangladesh

Saleemul Huq International Centre for Climate Change and Development, Independent University, Bangladesh, Dhaka, Bangladesh

M. Gulam Hussain Feed the Future Fish Innovation Lab, Mississippi State University, Mississippi State, MS, USA

Md. Saidul Islam Division of Sociology, Nanyang Technological University Singapore (NTU), Singapore, Singapore

Rabiul Islam Department of Social Work, University of Rajshahi, Rajshahi, Bangladesh
Macquarie School of Social Sciences, Macquarie University, Sydney, NSW, Australia

Chowdhury Sarwar Jahan Department of Geology and Mining, University of Rajshahi, Rajshahi, Bangladesh

J. Craig Jenkins Department of Sociology, The Ohio State University, Columbus, OH, USA

Md. Kamruzzaman Department of Civil Engineering, Rajshahi University of Engineering & Technology (RUET), Rajshahi, Bangladesh

Nazmus Sadat Khan Economist, The World Bank, Dhaka, Bangladesh

Shahrin Mannan International Centre for Climate Change and Development, Independent University, Bangladesh, Dhaka, Bangladesh

Quamrul Hasan Mazumder Department of Geology and Mining, University of Rajshahi, Rajshahi, Bangladesh

Craig A. Meisner Retired Agricultural Researcher from IFPRI, CIMMYT, FAO, Chapel Hill, NC, USA

Jahan Box Moral Department of Geography and Environmental Studies, University of Rajshahi, Rajshahi, Bangladesh

Md. Ferozur Rahaman Institute of Environmental Science, University of Rajshahi, Rajshahi, Bangladesh

M. Feisal Rahman International Centre for Climate Change and Development, Independent University, Bangladesh, Dhaka, Bangladesh

Tasnim Nazira Rida Department of Philosophy, University of Rajshahi, Rajshahi, Bangladesh

Chandan Roy Department of Geography and Environmental Studies, University of Rajshahi, Rajshahi, Bangladesh

Part I

Geographic Transformations

1 Bangladesh in Transformation: Geography, People, Economy, and Environment

Raquib Ahmed, Abdullah Al-Maruf, and J. Craig Jenkins

Abstract

Bangladesh is transforming from a rural subsistence agricultural economic system to a mixed economic partner that complements global processes. It exemplifies contemporary changes in the global south where mismatches often develop in the processes linking resources – activities – environment – sustainability, partially due to insufficient baseline knowledge. This book uses an interdisciplinary scientific approach to understand the real-time problems associated with these transformations and to identify possible solutions. Bangladesh is approaching a vulnerable situation of environmental degradation and raising questions of sustainability in resource exploitation as faster growth in GDP is depleting natural resources, creating soil degradation and resource scarcities along with severe pollution problems. These changes are creating significant economic growth alongside great unevenness in natural resource availability, a prime example being the rapid growth of fresh and saltwater aquaculture that exists alongside a collapse in the number of aquatic species and biodiversity. Parallel problems are manifesting in other areas. These problems are also compounded by the long-term challenges associated with global climate change with more volatile rainfall, increased storm intensity, and sea-level rise. This book should help academics, policymakers, and activists engaged in environmental studies and development activities to better understand these problems and develop solutions. Our goal is to open up windows to investigate these problems further and to widen the knowledge frontier.

Keywords

Bangladesh · Geography · Environmental degradation · Transformation

1.1 Context

Bangladesh stands on the earth's largest river delta formed by a geo-tectonic process and successively modified through the deposition of sediments by the large rivers viz. the Ganges, the Brahmaputra and, to some extent, the river Meghna. Although these processes started during the Tertiary Period in geological history, major surface modifications took place during the Pleistocene time, and have been steadily modified by recurrent river-floods during the Quaternary Period. Hence, this delta received its name the "Bengal Flood Plain". Broadly speaking, the surface elevations range up to about 10 meters, which makes most of this plain suitable for easy cultivation (Wadia 1926). Due to its homogenous and level surface, except for south-eastern mountainous regions formed during the Tertiary Period and smaller isolated parts of the north-east, the entire region contains fertile soil that is recharged annually by the rivers and distributaries and which has attracted people of different races and ethnicities from northern India, the Arabian Peninsula, and from central Asia. People from south-east Asian areas and even from the north side of the Himalayan range have also come and settled in the southeastern part of Bangladesh. Nature's influence from earlier times is today overwhelmed by the human impact of the late *Holocene epoch*, particularly that of the last 100 years, which might be termed as the *Anthropocene* (LeCain 2015) to reflect the

R. Ahmed (✉)
Fareast International University, Dhaka, Bangladesh

Department of Geography and Environmental Studies, University of Rajshahi, Rajshahi, Bangladesh
e-mail: vc@fiu.edu.bd

A. Al-Maruf
Department of Geography and Environmental Studies, University of Rajshahi, Rajshahi, Bangladesh

J. C. Jenkins
Department of Sociology, The Ohio State University, Columbus, OH, USA

R. Ahmed et al. (eds.), *Transforming Bangladesh*, World Regional Geography Book Series,
https://doi.org/10.1007/978-3-031-45093-8_1

transformative impact of humans on the ecosystem. The influence of different people-times and the different needs of society is reflected in changing geographical spaces. These changes are more dramatic and thoroughgoing during the last 50 years with the country's rapid population growth, urbanization, and global change. Increasing development and global change have brought about a cultural transformation in consumption patterns that parallel the major changes in the land and the environment, creating a scope for conflict between natural processes and human influence, posing new environmental sustainability questions.

Downstream geomorphological changes are highly visible in Bangladesh, particularly in the flood plain and the lower delta region along the coast. One can quickly identify how the natural system has created extensive modifications on the surface. For example, an estimated 1.2 billion tons of sediment yearly is carried downriver in the Ganges-Brahmaputra-Meghna river system, much of which eventually reaches the coastal area where new islands are formed. Over the past two decades, many new islands have formed that now hold settlements and major agricultural activities. An estimated 5 km^2 of new land is formed yearly in the lower delta area. Regular plain land floods and hilly flash floods change the landscape almost every year. At the same time, these river courses, the extent of flooding, and extreme climate events such as coastal cyclones are also influenced by human-induced climate change (USGS-Bangladesh Gas Assessment Team 2001; Kuehl et al. 2005).

1.2 Processes of Transformation

Transformation takes place in many ways, such as natural processes (fluvial, biological, hill slope, and marine) and economic processes (agriculture, deforestation, industrial, and development processes). Natural processes work behind, and the socioeconomic methods work on the imminently visible surface.

The most dominant and visible economic processes come from historical change in agriculture, which still employs almost half of the population but contributes only about 15% to the national GDP. This cost-benefit pattern looks unviable, yet it bears importance as agriculture is considered to be a subsistence type of activity in the region. An insignificant range of cash crops, viz. jute, tobacco, and tea, are exported but rice is the dominant crop and it remains central to the country's society and economy. Bangladesh has always been a food-deficit region due to the excessive food demand growth. However, during the last 5 years, the country has attained self-sufficiency in rice production – which is the staple food for all in Bangladesh. This is a vital point in food security in the absence of food-intake diversification. Thus, rice production plays a key role in political agendas and social stability. Rice production is going through a transformation from traditional to scientific and mechanized processes using massive inputs of high-yield varieties of seeds, extensive irrigation, the introduction of genetically modified breeds, modern farm mechanization and management, and inputs of extensive use of fertilizer and pesticides (Al-Maruf et al. 2022). The rapid increase in the production of food crops has brought the land under intensive use and exposes the soil to potential degradation. In parts of the southwestern coastal zone, extensive pumping of irrigation water combined with sea-level rise and land subsidence has created soil salinization, undermining agricultural productivity and limiting access to freshwater (Dasgupta et al. 2014). The introduction of saltwater shrimp farming has created a highly profitable export industry but also set off intense and sometimes violent conflicts over land development, aggravating soil salinization and reducing vegetative cover and storm resilience (Al-Maruf 2017). Alongside these challenges, the broader picture is the major agricultural growth. Overall evidence suggests that soil productivity is likely to continue to increase with improved scientific innovations. The agricultural practices are broadly similar across the country, but differences exist in the combinations of crops and depending on topography, rain, and soil. A significant part of the land has been brought from a single rice crop per year to triple crops per year. Annual growth in agriculture has risen from less than 2% during the 1970s–1990s to 3.5% in the last two decades (World Bank 2022a). Where possible, rice is supplemented by higher-value specialty crops and, in many coastal areas, pond-based freshwater fish production and other aquaculture.

The only difference exists in the south-eastern hills where cropping is only on small tracts of plain lands in the valleys and on slopes, but this is not too productive. The sedentary agricultural method is traditionally practiced by the people living in hilly areas. It is believed to be environmentally damaging as the land is acquired by clearing forest either by cutting or by burning, causing deforestation. There is a debate whether really such a small tribal population could bring such extensive deforestation or this was related to proportionate national economic growth and increasing demand for plain land. The hilly regions are threatened by landslides, especially in new settlements in more vulnerable areas.

During the last 100 years, there has been extensive deforestation. Records show that a century ago about 25% of the country's land area was covered by natural forest, which has been reduced to only about 12% by recent government estimates. Other research and satellite images suggest that current forest coverage is likely closer to 10% of the total land. Forest in the entire Hill Tracts area has been greatly affected. Some parts of deforested land have been brought under reforestation programs. A drawback of these programs is a less diversified forest, which is less likely to support diverse species even if it may improve human well-being. Parallel to that, there has been a significant amount of social forestry during the last 50 years mainly in the north-western part of the country. The cumulative amount of vegetation bio-mass puts Bangladesh on the positive side of CO_2 emission and carbon sink measurement.

At the same time, Bangladesh has developed rapidly over the past several decades. In 1971 when the country was formed, it was one of the poorest nations in the world. By 2015, it became a lower-middle-income country. Between 1991 and 2016, poverty declined from 43.5% of the population to 14.3%, making it one of the most rapidly developing countries in the world (World Bank 2022a). The World Bank forecasts that by 2030, poverty will have disappeared completely and the Bangladesh government predicts that the country will be at upper middle-income level. While these may be overly optimistic, they indicate the depth of the current transformations that are occurring in the economy and the social system. At the same time, it is important to note that such rapid growth will involve increased consumption of natural resources and major impacts on the ecosystem. Bangladesh will also have to recover fully from the COVID-19 pandemic and develop a long-term strategy for pandemic management.

The history of social and economic segregation in the Bengal region goes back to 1905, when the area was divided into two provinces, known as the Partition of Bengal. The socio-cultural and economic hub of the entire region was based on Calcutta, which allowed to grow an elite class of people to exert a strong influence on the economy and society. The colonial rulers divided Bengal, and a new province was created called East Bengal comprising present Bangladesh, all Assam province of India located in its north-eastern part. That incident allowed people of the Bangladesh region gave a sentiment of a separate cultural entity. This used to be the initial kernel of Bangladesh's emergence, finalized in 1971. Bangladesh region started getting new Dhaka-based administrative setup and economic growth center. But fear of losing benefit, the Calcutta elites opposed it and pressured the British ruler to reverse the policy to reunite it in 1911. Two World Wars greatly impacted the economy as the British rulers forced them to withdraw all resources from the Bengal region and kept them reserved for the war, particularly food grains, which was the root cause of the infamous Bengal Famine of 1943. Although it was for the soldiers' movement, the establishment and expansion of railway and road communication gave a foundation of economic growth through selected heavy industrialization around Calcutta. The Eastern part, i.e., the Bangladesh region was deprived of benefits and allowed a sentiment of urge for a new land. Thus a contemporary society came up with a new vision and need, a situation which was picked up by the Pakistan movement and separated the Bangladesh region from India to join with Pakistan under a new political system. This was the starting of regionalization in South Asia, an economic potential still untapped by both Bangladesh and India (Ahmed 2018).

The political transformation since decolonization in 1947 has been significant through Bangladesh's unification with Pakistan. Bangladesh with Pakistan (got independence in 1947) and was with India (before 1947 as a British colony) transformed through various economic and social changes. So, changes in present Bangladesh carry a legacy of colonial rule as part of India and uncomfortable attachment with Pakistan. During the colonial era this region suffered famine of 1943, a result of the World War II that killed an estimated population of 2.1–3.8 million (Sen 1976) and shattered the economic system. Later on a communal riot during 1946–1947 left thousands of Hindus and Muslims dead, and a mass migration was resulted. These two aspects changed demography and the region had to start its economy almost from scratch due to low industrialization. The whole economy and administrative infrastructure was designed to serve the imperial British ruler. When the British rule left South Asia in a vacuum of economy that paved a way for a new economic system. Bangladesh region started new pattern of industrialization, urbanization, and rural-to-urban migration reflected a primate urban character. It was primarily an agrarian economy based on limited uni-polar urban growth that created a new urban-based industrialization and a class of people. However, the economic growth was slow despite a continuous rural-to-urban migration where Dhaka (previously called Decca) was the focus of all cultural and economic development.

Culturally, Bangladesh's society is significantly integrated with the South Asian community and culture to a broad extent. But politically, it maintained a very distinct identity and discourse which finally paved way for its separation from Pakistan in 1971. Since then, it has maintained a distinct policy of foreign relations which has been generally a leaning toward the west with emphasis on a liberal market economic system and combining this with a neutral stance toward global power polarization. Its international trade reflects these policies, and exports are oriented mostly toward North America and Europe. But China and India dominate in import sectors. On the other hand, major financial aid comes from Japan and Europe. Bangladesh also draws in major remittances with over 2 million people in the Middle East and at least another million in Europe, North America, and the rest of the world. These remittances make up the largest source of foreign currency in Bangladesh and at times have constituted over 10% of GDP (2012) and most recently 6–7% (World Bank 2022b).

Changes in global consumption patterns have opened up windows for Bangladesh in two critical sectors: ready-made garments and shrimp farming. Garments industries have transformed the economy, creating major and permanent change in women's employment and thereby boosting women's status, earning foreign currency and constituting the second largest share of national GDP. Extensive work in the garments sector has transformed women's status, raising educational standards, lowering fertility, and contributing to women's empowerment. Further industrialization in other industries, especially shrimp processing, shipbuilding, cement, medicine and pharmaceuticals, and fertilizers has continued to grow. This growth is highly dependent on foreign investment, which comes to the sectors which have global demand and at the same time potential to reduce production costs through cheap labor. Since the 1990s, foreign direct investment has grown rapidly but in the last decade declined (World Bank 2022c). In part this reflects a mismatch with the government's industrialization policies and barriers in terms of business costs, corruption, insufficient skilled labor, and the like. Nonetheless, Bangladesh has a better Global Competitive Index (GCI) than many of its South Asian neighbors in many sectors. This is also related to weak regionalization of trade and investment within South Asia, which weakens growth prospects for Bangladesh and other neighbors.

As noted, Bangladesh is rapidly transforming from heavy dependency on a subsistence agrarian economic system to a diversified globally integrated economy with a still limited industrial setup. The only dominant industry before independence in 1971 used to be jute which lost its world market and collapsed despite governmental attempts to bail out this industry. At independence, food shortages were extensive due to heavy dependence on imports. Initially, Bangladesh developed strong relations with the Soviet Union and India, which helped with the reconstruction of basic physical infrastructure. However, after a series of military governments and a major financial crisis in the 1980s, civilian government was restored in 1990 and neo-liberal economic policies were gradually adopted, opening up the economy to foreign investment, greater export trade with Europe and the United States, and the rapid growth of these new export sectors. While the military continues to be an important political institution, these changes promise to continue into the future.

The world passed through a massive communication revolution that resulted in the internet's introduction in the 1980s, which developed into a global web by the 1990s. In the next two decades, the internet has become one of the strongest forces for social and economic change, breaking the barrier of distance and giving mankind an upper hand over nature. These developments were possible because of the laying of intercontinental submarine cables that made possible the shift from weather-dependent micro-wave to light-based faster communication. The other major communication change has been the adoption of mobile phones, which have now become widespread except among the older, typically rural generation. Almost all of the country now has access to at least 3G service and with it internet connectivity and the possibility of using smartphones, which has made possible greater internet access, mobile banking, payments systems, and access to commodity markets. Rural farmers now use their mobile phones to keep track of commodity markets. Families use the internet to order medicines and medical services. Rural access is being expanded by the building of access in the rural *parishad* (council) offices. Significant barriers to mobile phone use and internet access persist, especially among the older rural generation, but this is declining.

Growth of population and rapid urbanization poses major challenges to the country's environmental sustainability and social stability. Currently, about 38% of the population lives in urban areas (World Bank 2022d) and Dhaka has become a primate megacity with a population of over 20 million and containing a third of the country's total urban population. One consequence has been an outstripping of the transportation system, which has failed to keep pace with this dramatic growth, making Dhaka one of the most traffic-congested cities in the world. It has also created serious environmental degradation with urban pollution, many shanty-town dwellers living without access to urban services, and significant exposure to flooding and poor sanitation. If projections are made, half of the country's population will be urban by 2030. This will magnify these problems and also have a significant impact on agricultural lands, which are currently being converted to urban settlement at the rate of about 1% per year. Despite some growth of urban agriculture, this poses a potential threat to food security. At the same time, urbanization and the growth of the urban middle class is creating changes in lifestyles and consumption which are making the society more internationally interactive and opening up the promise of a stronger civil society and democratic politics.

Land use transformation is based on the competitive economic returns of land, often without consideration of environmental and sustainability consequences. One highly visible example is the transformation of rice fields into commercial shrimp farming in the southwestern coastal zone, which has created a prosperous export industry but contributed to deforestation, lower resilience to storm surges, and extensive salinity intrusion. This has also changed the occupational and livelihood patterns of the population with less demand for field labor and pressure to migrate either seasonally or permanently to the city. Other dynamics are also influencing changes in livelihoods and sustainability such as the increasing intensity of storms, increased droughts and heat waves associated with global warming and temperature volatility, industrialization in the eastern part of the country, decreasing water bodies almost everywhere, top-soil degradation, greater river-bank erosion, and a faster growth in urbanization.

1.3 Consequences of Transformation

While the world is experiencing changes under the impact of globalization, the driving dynamics and the patterns are visible by a divide in the global north and global south. The transformation is faster in the south than in the north. This is particularly vivid where the relation between population-resource-climate change-induced impacts is significant. Bangladesh is a geographical region that typically reflects various aspects of its Geography, People, Economic, and Environment, which are the four windows of the book Bangladesh belongs to the South Asian region and has commonness to all the countries in the area in terms of geography, culture, and economic practices reflecting the region's ideas. This is particularly applicable to the environmental consequences of economic activities. Focus on the environment is a significant perspective in the book. Sustainability in economic growth is intricately related to the ecological use of resources. Faster economic growth and self-sufficiency in food production have led to greater dependency on technology and resources, resulting in agricultural practices and diversification of economic sectors both for domestic need and export-oriented sectors under a liberal market financial system. This has linked Bangladesh in the global economy in and, successively, to local transformations that are linked to this global strategy. These transformations are visible in almost every economic sector, social arena, physical aspects of the built environment, and ultimately the larger ecosystem.

Bangladesh has long been threatened by floods, drought, regular visit of coastal cyclones, and the new threat of thunderstorms. All these used to claim a high number of lives, but this is reduced because of better early warning and preparation. At the same time, the threats from climate change are intensifying and are likely to present greater challenge in the future. The phenomena of threat-challenge and success in economy and society are a game of failures and achievements through which the land is transforming, leaving traces of different activities, mainly due to success in education, social awareness, health, and disaster management, having a significant impact on economic activities. The most vulnerable groups are poverty-driven people who are forced to adapt to new situations or to migrate for survival. In many respects, major portions of the community are coping and adapting well to these challenges and promise to continue to do so in the immediate future. Increased community participation, greater mass awareness about risks and adaptive measures, improvements in health and education, and gender participation have played critical roles in these developments. Government initiatives in these sectors have always been existed, but extensive help from non-governmental organizations or NGOs has played a vital role. Bangladesh at one stage has become a role model of NGOs' participation in national development through social and economic change, particularly in the sectors of poverty alleviation with the help of micro-financing and the spread of education through informal and sub-formal schools. Two of the world's most notable NGOs working in Bangladesh are the BRAC and Grameen. Both have contributed greatly to improvements in human well-being. The increased participation of women in economic growth has resulted in reduced gender discrimination and women's empowerment, especially in local-level decision-making.

Physical dynamics create a foundation of the other processes as a kind of deterministic constraint where nature impacts human actions through limiting resources and natural properties. Social and economic dynamics entail possibilistic approaches where people can modify things as needed and *make change* in the sense of positive adaptation. Over time, the human impact is becoming more important and opening the possibility of durable improvements. Perhaps inappropriate decisions have brought conflicts and created many environmental challenges we have discussed. An additional dynamic is the actions of the state and the people in addressing the weaknesses in these transformation processes. A major advantage Bangladesh has is a relatively young population that has acquired greater education and human capital and is therefore more adaptive to challenges. This demographic dividend, especially the human capital aspect of it, should not be underestimated. Further, the state has adopted adaptive planning and, working with the NGOs and civic groups, has made many improvements to address these problems. The free market system under globalization creates possibilities for future growth and adaptations. The successful utilization of these potentials depends on policy and the drive of social ingenuity.

1.4 Addressing Strategies

Bangladesh has maintained a set of policies, strategies, and programs to address the issues that we have discussed and direct the transformation toward sustainability. These are crystallized in 5-year national strategic plans for the nation's overall development. In addition, Bangladesh has sectoral plans for agriculture, urban development, infrastructure, environmental management, forestry, disaster management, climate change adaptation, etc. These are indicative planning tools, in other words, ways to refine our thinking and understand complex interactions. The government's target is to fulfill the Social Development Goal targets by 2030, turning Bangladesh into a middle-income country. Past success with the Millennium Development Goals adopted in 2000 suggests a strong foundation for such qualitative change.

The success of these plans, policies, and strategies is in question in the continuously changing global system. For example, in the dialogues on climate change, Bangladesh's position is not sufficiently influential due to the global power polarization. New developments, such as the Russia-Ukraine war and possible global food scarcity, may intervene. The challenges of global warming may well displace a large number of people, creating diverse challenges at the national level. Urban plans exist for solving the urban challenges discussed above. Still, these plans do not work well if there is continuous massive rural-to-urban migration, especially one centered in Dhaka city. The clearest success story in Bangladesh lies in agriculture where food self-sufficiency has been achieved but future challenges of rapid population growth, exhaustion of soil, and new phenomenon such as the influx of 1.2 million Burmese people to Bangladesh living in political asylum pose new challenges to future food security. It is possible to imagine other scenarios that create new sustainability challenges in the future.

1.5 Book Structure

Following the above themes, the book is organized into 12 parts progressively linked to one another, explaining the context, process, consequences, and the way forward. These are as follows (Table 1.1).

Table 1.1 Overall structure of the book

Themes	Parts
Context	Geo-physical dynamics
	Changes in bio-geographical aspects
	Land resources, land Use patterns, and rural settlements
Processes of transformation	Society in transition
	Food production and food security
	Urbanization: a future challenge
	Trade, communication, and industries
	Bangladesh in the international context
Impact	Natural hazards and changes in livelihood patterns
	Environmental problems and policies for sustainable living
Addressing strategies	Bangladesh in times of climate change
	Development policies, NGOs, and the future of Bangladesh

Transforming Bangladesh aims to offer knowledge about the transformation process as a concept through an actual situation visible in Bangladesh and using a range of problem-oriented discussions. Therefore, the chapter headings are selected to represent various issues related to the book's central message. The economic and social transformation are possible if informed, and scientifically based judgments are brought to bear in a timely fashion. Our hope is that this volume provides insights that enhance adaptive resilience and pave the way to future sustainability.

References

Ahmed R (2018) Regionalization process and the economic integration in South Asia

Al-Maruf A (2017) Enhancing disaster resilience through human capital: prospects for adaptation to cyclones in coastal Bangladesh. Doctoral dissertation, Universität zu Köln

Al-Maruf A, Pervez AK, Sarker PK, Rahman MS, Ruiz-Menjivar J (2022) Exploring the factors of farmers' rural–urban migration decisions in Bangladesh. Agriculture 12(5):722

Dasgupta S, Hossain MM, Huq M, Wheeler D (2014) Climate change, soil salinity and the economics of high-yield rice production in coastal Bangladesh. Policy research working paper 7140. World Bank Research Group

Kuehl S, Allison M, Goodbred S, Kudras H (2005) River deltas – concepts, models, and examples. Society for Sedimentary Geology Special Publication, no. 83, pp 413–434

LeCain TJ (2015) Against the anthropocene: a neo-materialist perspective. Int J Hist Cult Mod 3(1):1–28

Sen A (1976) Famines as failures of exchange entitlements. Economic and Political Weekly 11 (31/33)

USGS-Bangladesh Gas Assessment Team (2001) US Geological Survey – PetroBangla Cooperative Assessment of Undiscovered Natural Gas Resources of Bangladesh, US Geological Survey Bulletin 2208 – A, pp 2–8

Wadia DM (1926) Geology of India. Macmillan, London, pp 30–48

World Bank (2022a) The World Bank in Bangladesh. https://www.worldbank.org/en/country/bangladesh/overview. Accessed 14 June 2022

World Bank (2022b) Personal remittances, received (% of GDP) – Bangladesh. https://data.worldbank.org/indicator/BX.TRF.PWKR.DT.GD.ZS?locations=BD. Accessed 14 June 2022

World Bank (2022c) Foreign direct investment, net inflows (% of GDP) – Bangladesh. https://data.worldbank.org/indicator/BX.KLT.DINV.WD.GD.ZS?locations=BD. Accessed 14 June 2022

World Bank (2022d) Urban population (% of total population) – Bangladesh. https://data.worldbank.org/indicator/SP.URB.TOTL.IN.ZS?locations=BD. Accessed 14 June 2022

The Thirsty Pleistocene Landscape and Challenges for Water Resource Management Plan

2

Chowdhury Sarwar Jahan, Md. Ferozur Rahaman, and Quamrul Hasan Mazumder

Abstract

The United Nations reports that globally more than 700 million people live at present without safe water supply close to home. For many people across swathes of drought-prone landscape 'Barind Tract' in the northwestern part of Bangladesh, spending hours queuing for water, trekking to distant water sources, and coping with the effects of contaminated supplies are usual since 1980s have the severe problem: water scarcity. So the 'Barind' is the thirsty landscape where life is written in water. This elevated 'Pleistocene' Tract has undulating to rolling topography with gentle slopes. Very scarce and seasonal rainfall has been exacerbated by frequent droughts (moderate to high-risk), and hit paddy cultivation even in rainy season also forthcoming paddy in summer when groundwater irrigation is required. Soil moisture deficit starts in October and becomes prominent in December continuing till May. The aquifer system is mostly confined and does not get an appreciable vertical recharge through thick top surface 'Barind' clay. The Tract bounding rivers like the Padma, Mahananda, Purnabhaba and Atrai are shrinking and many of its tributaries are drying up for want of rainfall. Farmers' changing groundwater withdrawal habits are to blame as three decades ago Barind farmers harvested one crop per year, but now they grow three crops annually by groundwater irrigation. Increasing crop production made possible since the 1990s, and the issue of food security couldn't be ignored. Consequently, over-exploitation of groundwater against lesser recharge causes still in drastic depletion its level, and on hot summer days, it falls further. As a result, people face hardship for availability of drinking water, and the anxiety deepens as the irrigation machines cannot pump up water properly due to drop in groundwater level. Moreover, after the extended for days 2005 drought spell rainfall scarcity followed by shortage of water resource, and extension of irrigated agricultural practice had a major role in over-exploitation of groundwater. All these together put stress and threat for groundwater resource sustainability and throw challenges for its proper management plan which will be aggravated in the coming days. Finally, keen attention should be paid for an efficient irrigation management plan that essentially reduces pressure on groundwater resources, ensuring its sustainability as an adaptation measure.

Keywords

Water scarcity · Drought · Water management plan · Barind Tract · Bangladesh

2.1 Introduction

As an episodic but complex phenomenon, drought happens due to rainfall scarcity and creates a gap in demand and supply of the water management chain (Liu et al. 2016). The food productivity and its security are primarily hampered by this event resulting in food grain import for the societal need (Benson and Clay 2004; Scott and Sharma 2009). In the northwestern part of Bangladesh, the drought-prone area popularly known as 'Barind Tract'is the largest Pleistocene physiographic unit of the Bengal basin. The Tract has comparatively higher elevation (47 m AMSL) than the adjoining floodplains (11 m AMSL). Present study area includes Godagari and Tanore *Upazilas* of Rajshahi district; Nachole and Gomastapur *Upazilas* of Chapai Nawabganj district; and Porsha, Sapahar and Niamatpur *Upazilas* of Naogaon district (District and *Upazila* are the second and third lowest tiers of administrative units in Bangladesh) (Fig. 2.1) covering

C. S. Jahan (✉) · Q. H. Mazumder
Department of Geology and Mining, University of Rajshahi, Rajshahi, Bangladesh

Md. F. Rahaman
Institute of Environmental Science, University of Rajshahi, Rajshahi, Bangladesh

R. Ahmed et al. (eds.), *Transforming Bangladesh*, World Regional Geography Book Series,
https://doi.org/10.1007/978-3-031-45093-8_2

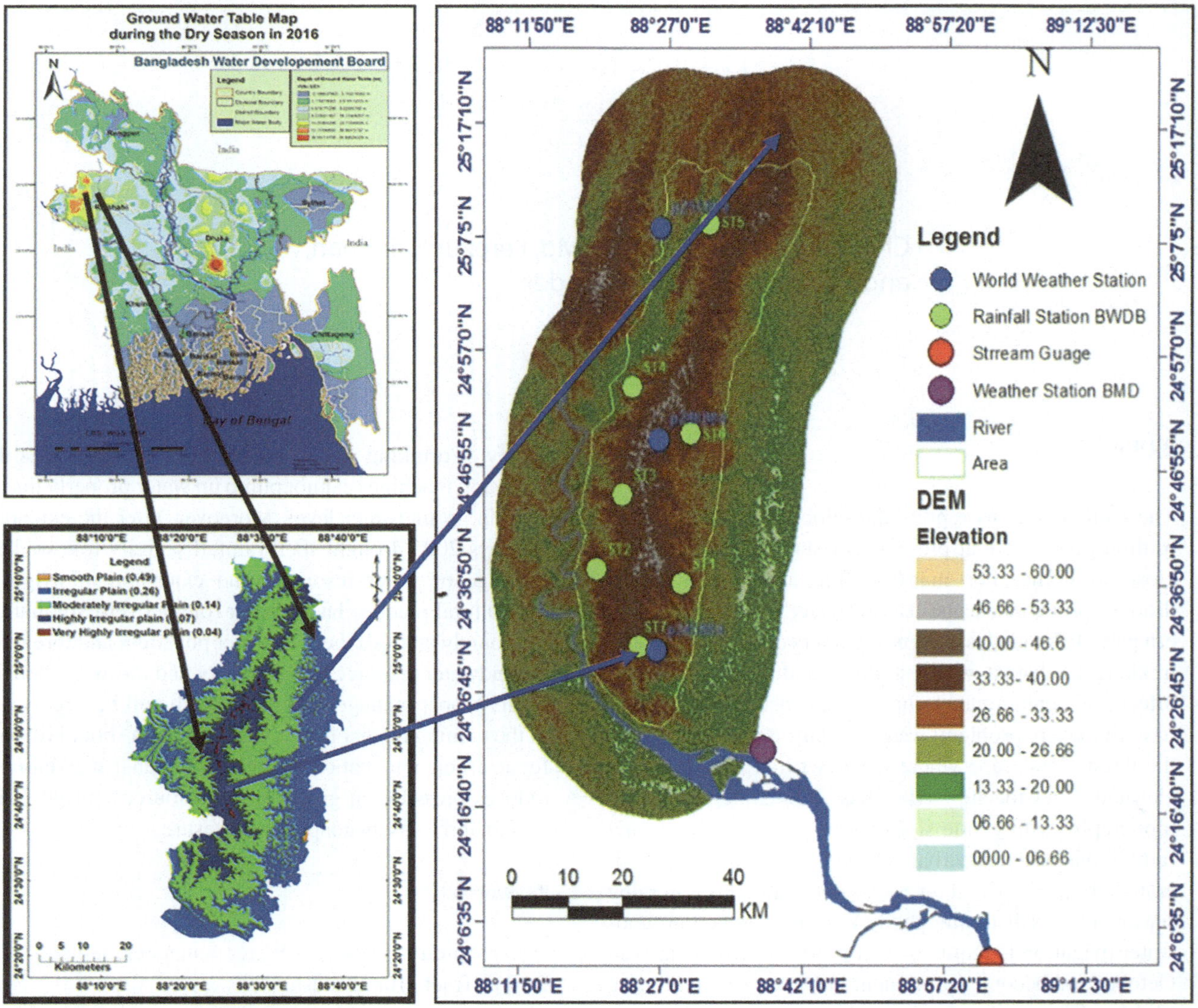

Fig. 2.1 Groundwater hot spots (Red and Yellow colour) of Bangladesh due to over-exploitations of groundwater. (Source: GWC of Bangladesh Water Development Board 2016) along with geomorphologic and elevation maps, and locations of meteorological and weather stations in Barind Tract

geographical area of 2586 km^2 with 1,417,928 inhabitants (BBS 2015).

The National Water Policy (NWPo 1999) of Bangladesh has declared drought as a major challenge to face water scarcity issues. It has caused loss of food production and starvation for many people leaving livelihood at risk, which will be aggravated in future. In order to ensure country's food security maintaining ecological balance in the Tract, the Barind Integrated Area Development Project (BIADP), later the Barind Multipurpose Development Authority (BMDA) under the Ministry of Agriculture (MoA), Government of Bangladesh (GoB) came forward to launch groundwater centric irrigation management system in the late 1980s. It has played a significant role in food production, overcoming constraints in surface water resource (Jahan et al. 2010), and has been practised to cultivate high-yield variety of rice (*Boro*) during dry pre-monsoon summer season (Mar–May). The late 1970s introduction of *Boro* rice production has necessitated entirely groundwater-based irrigation in the country by power pumps (PPs) during the summer season (Ravenscroft 2003). Presently 80% of rice cultivation depends on groundwater exploitation by low-capacity Shallow Tube Wells (STWs) (depth < 80 m bgl); and the rest by high-capacity Deep Tube Wells (DTWs) (depth > 80 m bgl). The BMDA's practice has boosted multi-cropping agriculture practice with increasing intensity from 117% (in pre-BIADP time) to more than 210% (national average: 180%) at present. The area has the potentiality for groundwater development of 8728 DTWs (0.056 m^3/sec capacity) (BMDA 2001), but at present, more than 23,000 DTWs have been operating by BMDA and private owners. Unfortunately, the increasing demand of exploited groundwater fails to replenish fully by nature recharge process even in rainy season. The problem has been becoming progressively more acute with the growing population and extension of irrigated agriculture here.

A number of research have been performed on hydrogeology (Ahmed and Burgess 1995; Islam and Kanumgoe 2005), groundwater occurrence and potentiality (Jahan and Ahmed 1997), climate modelling and drought risk assessment (Kamruzzaman 2014), impact of groundwater irrigation (Jahan et al. 2010), spatio-temporal analysis of climate variability and groundwater table (Kamruzzaman et al. 2016; Rahman et al. 2016a, b), long-term trend analysis of water table (Rahman et al. 2016a, b), morphometric analysis of watersheds and groundwater potentiality (Rahaman et al. 2017, 2019), etc. for the Barind Tract. Moreover, research work by Mustafa et al. (2017) has revealed that over-exploitation of groundwater due to agricultural activity is a major influencing factor for the declining trend of groundwater table (GWT), but a conclusion has been made without any analytical research findings.

No study on the effect of drought on groundwater resource management plan in light of analytical exploration has received significant scientific attention in Barind Tract and country's northwestern part (Alexander 1995; Shahid and Behrawan 2008). Considering the scenario of stress in groundwater resource, the present study has given emphasis to carry out on rainfall climatology (trend, variability and seasonality), drought risk (pattern and dynamics), groundwater scenario (temporal and depletion), land cover change, irrigation coverage, crop productivity, etc. by multi-step statistical approach of analytical exploration. The outcome will provide aid to policymakers to oversee the challenges for framing proper water resource management plan, and help local managers to take decisions for sustainable use of this resource as a demand of time.

2.2 Geography and Geology

The Tract – a landmass of elevated triangular wedge – has characteristic of dissected and undulatory nature and subdivided into: the Barind Tract (Pleistocene terraces) and the floodplains (Recent deposits) (Brammer 2012). The north-south orientated dome-shaped Tract (20–25 km wide in east-west direction) has edged by parallel rivers like the mighty Padma (Ganges) flowing in the south along with the rivers Mahananda and Purnabhaba in west, and that of Atrai, Sib-Barnai, etc. in the east of seasonal to minor seasonal nature. Topographically, the landform has an undulating to rolling nature, having low regional slopes towards SE and SW direction. The Tract enjoys characteristic subtropical monsoon climate having seasons like winter (Nov–Feb) – cool and dry with almost no rainfall; pre-monsoon (Mar–May) – hot, dry and rainless; and monsoon (Jun–Oct) – rainy days. The time-series average annual rainfall amount (1410 mm) is much less than that of the national average (2550 mm). The average monthly humidity ranges from 62% (Mar) to 87% (Jul) with mean annual of 78%, and monthly average temperature range is 10 °C (Jan)–33 °C (May).

The geoscientists thought that the Tract might have evolved due to tectonic upliftment and exists as an erosional geomorphic feature (Rashid et al. 2018a). Khondoker (1987) shows that it has elevated as a horst block at the close of Pleistocene. Nandy (1980) has shown a lineament along the river Padma (Ganga) that delineates the southern extremity of the surface possibly indicative of being fault-controlled. In some parts of the Tract, the boundary separating the Recent and Pleistocene sediments is sharp and quite straight (Klinski 1979). The evidence has strongly supported the structural origin of the Tract rather than only geomorphic origin. During the Quaternary block adjustment of different parts of the Bengal Basin, it has moved up and down due to the Himalayan upliftment. The structural interpretation has revealed some direct influence in building up the present morphological configuration (Rashid et al. 2018b). Displacement of lithologic units, undulating nature, higher thickness of Pleistocene 'Barind'clay and presence of Barind clay below the Recent Alluvium along river sides indicate the structural control in the origin of this elevated Tract.

Hydrogeology of the Tract has indicated the groundwater potentiality possibly related to fault-controlled grabens and adjacent horsts. The Recent and Pleistocene age sediments have comprised of surface lithology of clay-silt aquitard [infiltration rate: 1–2 mm/day (UNDP 1982) having thickness of 3.0–47.5 m] of semi-impervious nature. The aquifer system has revealed single to multiple layers (two–four) of Plio-Pleistocene age (thickness: 5.0–42.5 m) (Jahan et al. 2007) having suitability mostly for domestic water supply with limited scope in irrigation purposes (Pitman 1981; Jahan and Ahmed 1997).

2.3 Challenges of Water Resource Management Plan

2.3.1 Rainfall Climatology

The average rainfall amount for the period 1980–2017 is 1410 mm, about 44% less than the national average of 2550 mm. Here rainfall is very much seasonal with longer dry period, where more than 79% occurs during the months May–October, and that of rest 21% during November to April showing an alarming situation for the rain-fed agriculture even in rainy season when supplementary groundwater irrigation is required for rice crops. The rainfall variability during the summer season (37–41%) has decreased during the rainy season (24–31%). It has irregular to strongly irregular distribution along with high seasonal variability and has represented higher percentages of the annual total precipitation within very few days in rainy season and effect the time and degree of its concentration on yearly total precipitation. Thus, phenomenon has made potential cause of frequent droughts putting considerable stress on water resources (Zhang and Qian 2003; Luis et al. 2011). As the average seasonal and annual rainfall variability is high (more than 20%) (Oliver 1980) along with drought event, so it has made highly difficult for framing water management plan in the Tract.

The statistical approach like Mann-Kendall Test (Mann 1945; Sen 1968; Kendall 1975) reveals a general decreasing trend in the magnitude of seasonal rainfall pattern in winter, summer and rainy season at a rate of 0.94, 0.15 and 0.80 mm/year, respectively with an average annual average value of 0.45 mm/year. The rainfall pattern scenario has started in winter season in 1980, and that of summer season in 1990. So it may reveal that the irrigation system in winter and summer seasons, i.e., 7–8 months in *Boro* rice and *Rabi* cropping seasons respectively is fully groundwater dependent. The linear trend in annual rainfall in the Tract for the period of 1980–2017 is shown in Fig. 2.2.

The monthly water balance study (Fig. 2.3) has revealed that the soil moisture utilization started in October when soil has become dry, and deficit condition has prevailed from December until May. The highest deficit amount is 35% (503 mm) of the annual rainfall and that of the lowest surplus amount (32%) has indicated lower amount of available water for groundwater recharge for further use. Here the potential evaporation-transpiration value (P_{ET}) (Thornthwaite and Mathur 1957) is higher than the annual and monthly total rainfall amount, and has shown an increasing trend which is consistent with the increasing trend of annual average temperature, and might be responsible for increasing P_{ET}.

2.3.2 Drought Effect

The drought categorization based on average rainfall amount [calculated from Standardized Precipitation Index (SPI) as meteorological (SPI-6), hydrological (SPI-12) and agricultural (SPI-3) events] (Lloyd-Hughes and Saunder 2002) for different time steps is shown in Fig. 2.4. The total number of dry episodes is higher in the past decade (2001–2010). The number of mild and moderate drought events has been increasing rapidly in rainy and summer seasons with steady rising trends in winter season. In that decade (especially after 2005), the number of hydrological drought has increased along with meteorological and agriculture droughts. It results in frequent drought effect of mild to moderate categories due to rainfall variability, and extension of cultivated and irrigated agriculture in the Tract has inevitably resulted in over-exploitation of groundwater resulting hydrological drought situation, and effects will be intensified in future. The agricultural drought (moderate to extreme nature) has been prevailing frequently in annual basis in rainy season since 1980 having coincidence in occurrence with that of meteorological and hydrological drought conditions. As a result, it will hamper country's food security leading to rice price hike. The drought ranking based on frequency (Ramamasy and Bass 2007) has shown the highest risk in the Tract belongs to class **B** (moderate to high category) that must be addressed by adaptation measure. Frequencies of different drought events (SPI-3, SPI-6 and SPI-12) show variability within the

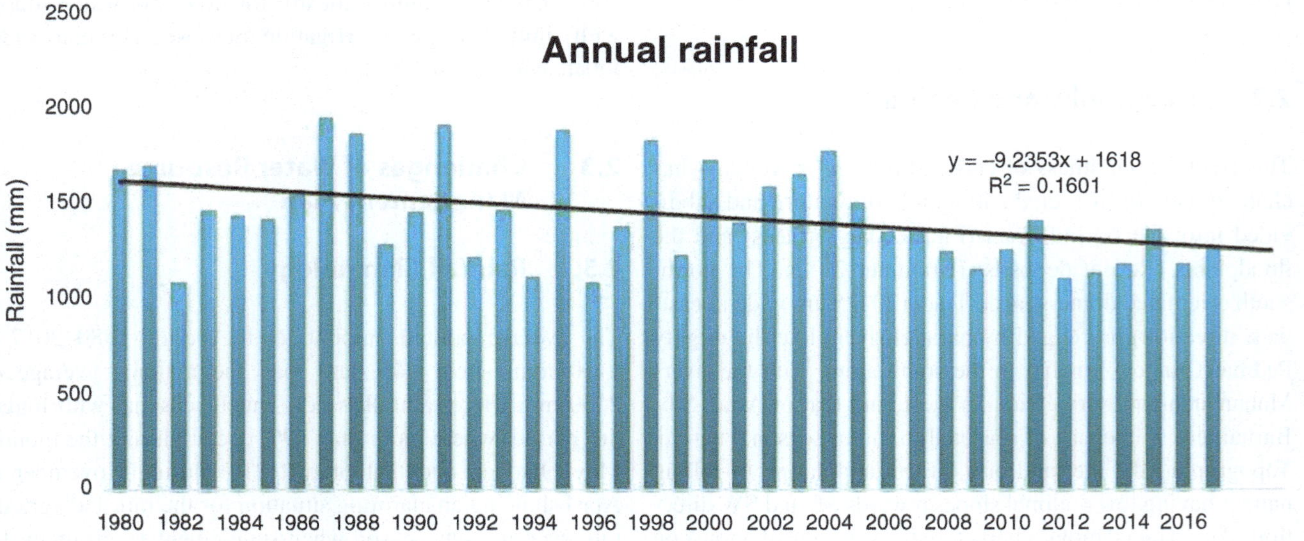

Fig. 2.2 Linear downward trend in annual rainfall for the period 1980–2017 in Barind Tract. (Data Source: Bangladesh Water Development Board – BWDB)

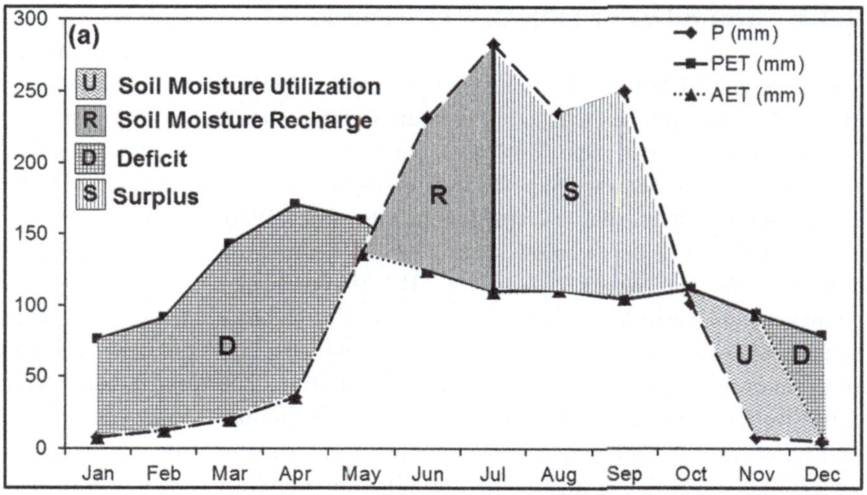

Fig. 2.3 Monthly water balance of Rajshahi meteorological station adjacent to Barind Tract. (Data Source: Bangladesh Meteorology Department- BMD)

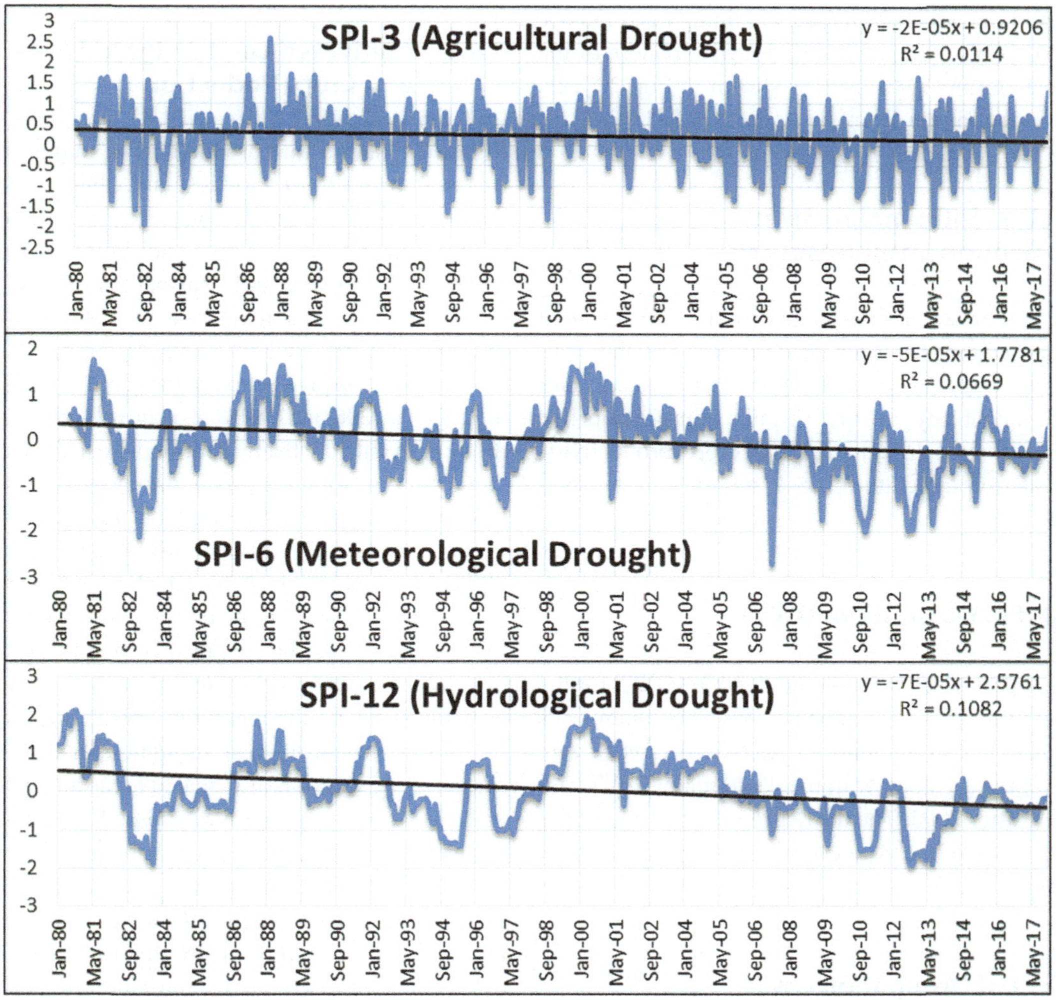

Fig. 2.4 Drought categorization based on average rainfall amount as calculated from Standardized Precipitation Index (SPI) for different time steps as meteorological (SPI-6), hydrological (SPI-12), and agricultural (SPI-3) events in Barind Tract

categories which may be caused due to variability in the duration of drought conditions, i.e., short, moderate and long. In general, when drought has started and continued for long time, frequencies of agricultural and hydrological drought events remained closer rather than that of meteorological drought.

2.3.3 Groundwater Table (GWT) Scenario

The long-run (1980–2017) time-series scenario of GWT shows that its depth has reached at maximum in the end of pre-monsoon summer (Apr–May) and that of minimum in the end of rainy season (Sep–Oct). The depth has been increasing in past decades (except recent times) since 1980, and has run at 7 m depth below suction level ['*Krishi Kaje Bhugorvostho Pani Bebosthapona Ain* 2017' (Groundwater Management Law in Agriculture 2017)] only during summer time, but has regained its original level due to augmentation of groundwater through recharge in rainy season. But since 2010, its depth runs below the suction level both in summer and rainy seasons, and people are not getting even drinking water. The gap between maximum and minimum depth of GWT has increased from 2 to 17 m since 1980 with higher value in recent time with 0.23 m/year declining rate of GWT (Rahaman 2018). The temporal (decade-wise) pattern of the depth of GWT during 1980–2017 is shown in Fig. 2.5 as a representative one for the Tract.

2.4 Factors Influencing Stress on Groundwater Resource

To frame a proper water management policy for drought adaptation, an attempt has been made to find out the challenges that the Tract has been facing for its groundwater resource unsustainability considering influencing factors like rainfall climatology, drought effect, crop productivity, demand of water for agriculture, etc. to minimize water scarcity (Fig. 2.4).

2.4.1 Rainfall Vs. Depth of GWT

The relationship between rainfall and depth of GWT for the time series of 1980–2017 has shown in Fig. 2.6, and reveals that the annual rainfall amount has decreased rapidly with rate of 2.76 mm/year, along with the depletion of GWT rapidly especially after 2005. The decreasing rate of rainfall has been playing as one of the most essential responsive roles for the depletion of GWT.

2.4.2 Effect of Drought on GWT

Relationship between the trends of annual average changes in depth of GWT, and that of rainfall pattern, temperature and droughts like meteorological, hydrological and agricultural events in the past couple of decades since 1980 is shown in Fig. 2.7. The decreasing trend of rainfall pattern, the increasing trend of temperature and rising trend in meteorological drought event are consistent with the trends of hydrological as well as agricultural droughts. The total number of dry episodes due to meteorological drought is highest in the past decade (2001–2010), and the number of mild and moderate drought events like meteorological, hydrological and agricultural time steps have been increasing rapidly in rainy and summer seasons along with steady rising trends in winter seasons in recent times. The annual depth of GWT is influenced by rainy season's monthly meteorological drought event. Moreover, after 2004 the depletion of GWT has continued during the rainy season even without any meteorological drought effect, due to huge withdrawal of groundwater for irrigation demand in extended agricultural land.

2.4.3 Cultivated Land, Irrigated Area and Crop Production

Figure 2.8 shows that productions of major crops like rice *Aman, Aus* and *Boro* varieties, and wheat since 1980 in the cultivated and irrigated land area by different pumping modes like DTWs, STWs, and Power Pumps (PPs) in Chapai Nawabganj and Rajshahi districts are steadily increasing with an increment rate of 126, 81, 143 and 251%; and 787, not irrigated, 347 and 297% respectively. On the other hand, crop production in the same period has increased at 158, 107, 206 and 289% respectively. So, meteorological drought event has been identified as the most influencing factor for developing stress in groundwater resource, which has ultimately invited hydrological and agricultural drought events successively. In the last decade (especially after 2005), the hydrological drought has increased along with agricultural as well as meteorological events, and reveals that drought effect has aggravated due to over-exploitation of groundwater for more cultivated land area and groundwater-based irrigation in dry summer season as well as during meteorological drought situation in the rainy season. So, extension of irrigated areas along with cropping patterns (mostly groundwater-based irrigated *Boro* rice cultivation in the summer season) and water management practices influence hydrological drought. Finally, it is revealed that meteorological drought due to rainfall scarcity and variability, followed by hydrological drought and extension of irrigated agricultural practice have consequenced huge withdrawal of groundwater in recent years. These factors have altogether put threat and challenges for sustainable management of water resource, and aggravated agricultural drought effect damaging crop productivity in this Landscape.

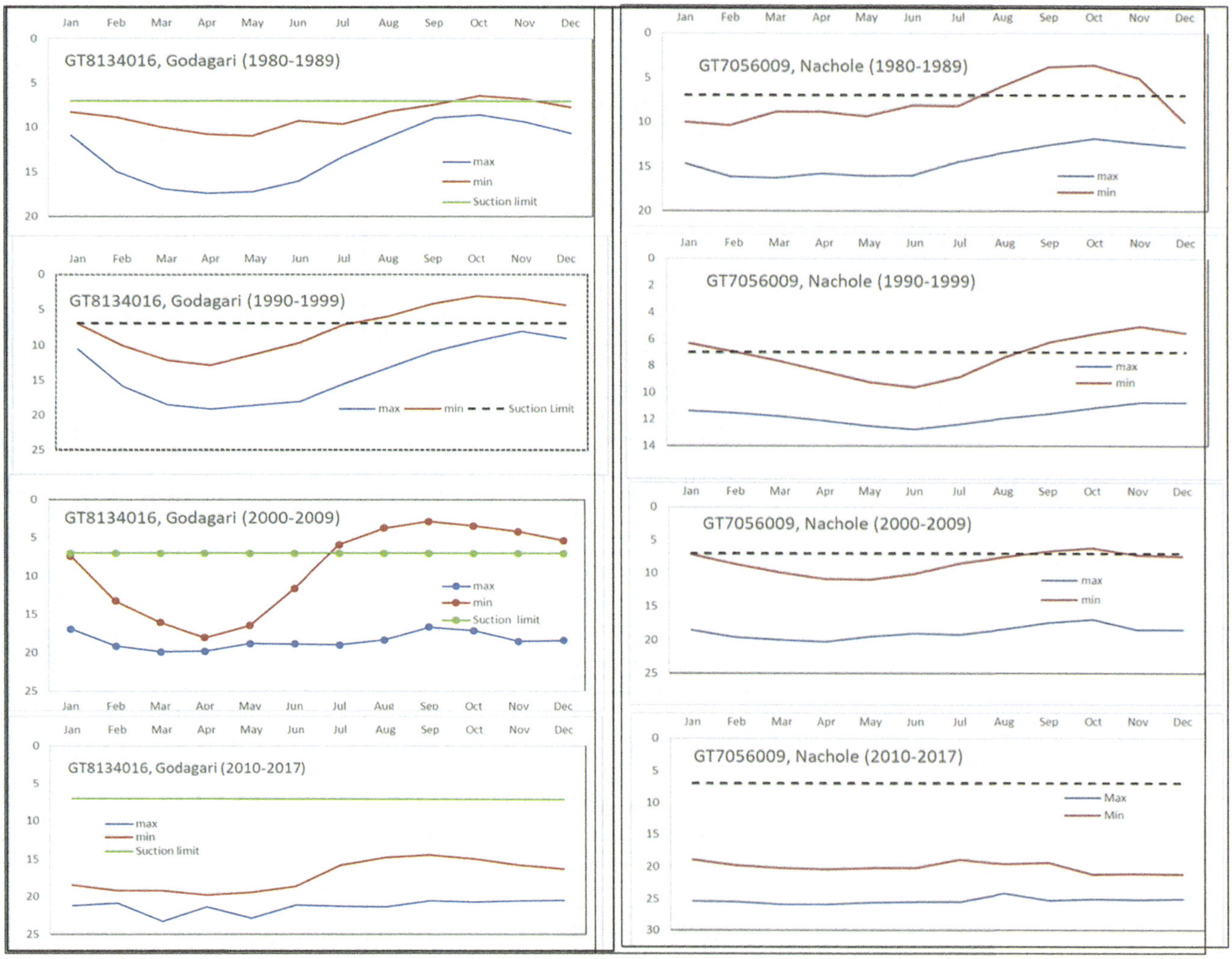

Fig. 2.5 Representative temporal pattern of decade-wise depth of Groundwater Table (GWT) (1980–2017) in Barind Tract. (Data Source: Bangladesh Water Development Board – BWDB)

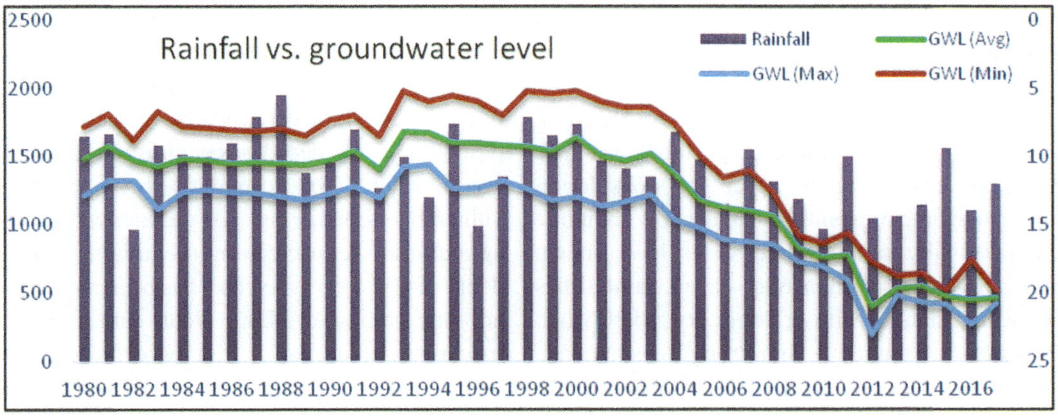

Fig. 2.6 Plot of rainfall (mm) vs. groundwater depth (m) in Barind Tract. (Data Source: Bangladesh Water Development Board – BWDB)

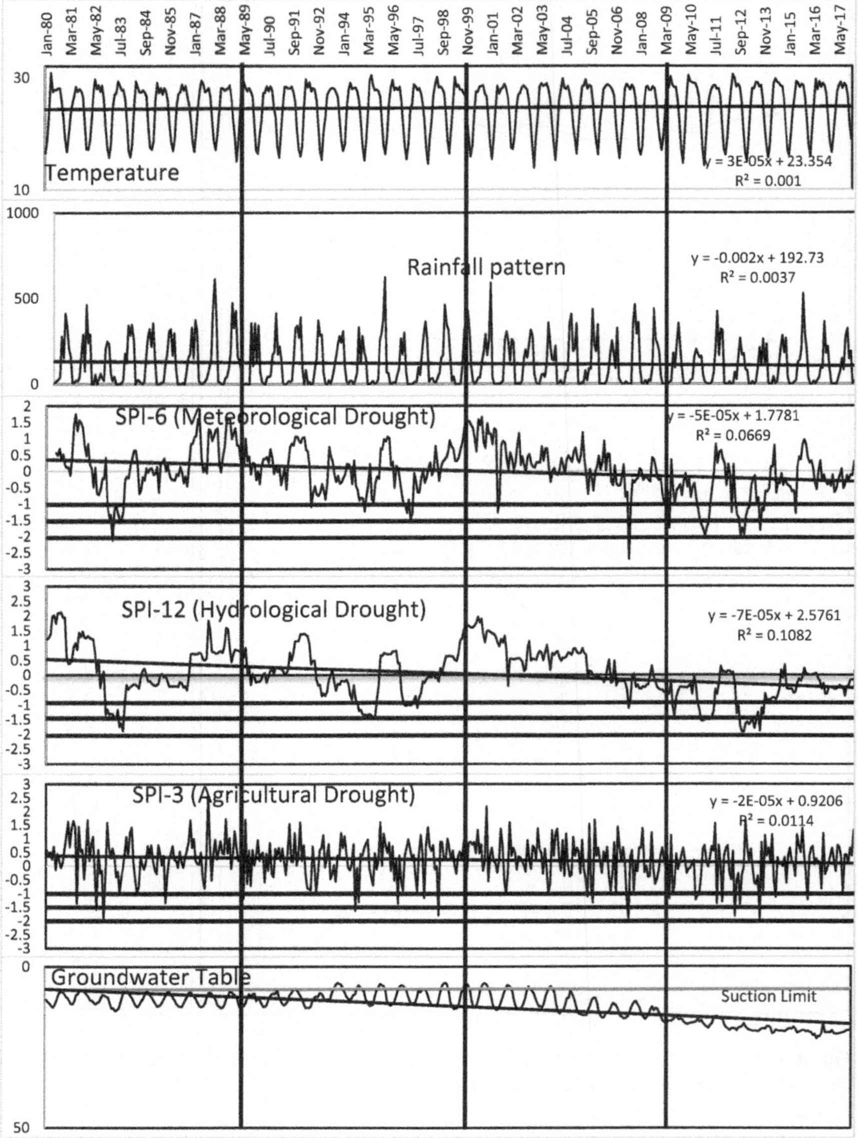

Fig. 2.7 Relationship between trends of annual changes in GWT depth, and rainfall pattern, temperature and meteorological, hydrological and agricultural drought events in the years since 1980 in Barind Tract. (Data Source: Bangladesh Water Development Board – BWDB)

2.5 Conclusion

The quest of nexus between drought (moderate to high-risk nature) and groundwater resource depletion in the Pleistocene landscape has provided valuable input for considering challenges in management planning. In general monthly, annually and seasonally rainfall time series have shown declining trends. Even in the monsoon season, rainfall has unusual seasonality nature with a long dry spell when precipitation occurs for few days, and drought events have become inevitable. The meteorological drought event (rainfall deficiency) directly relates to the insufficient groundwater recharge resulting hydrological drought event. Since 2005, hydrological drought has become acute at the end of rainy season (winter and summer seasons), and even during the rainy season's months. The land area suffers from high soil moisture stress for 7–8 months annually, and the demand for groundwater-dependent irrigation comes forward as a consequence for long period. Moreover, groundwater-based irrigated areas have been increasing rapidly since the past couple of decades resulting in over-exploitation of groundwater. The drought phenomenon has created vulnerable conditions for groundwater resources and agricultural sector creeping the area to water scarcity. The depth of GWT over the years has been increasing, and people are not getting drinking water and facing hardship for irrigation demand also. Finally, decreasing rainfall due to climate variability, frequent droughts, extension of irrigated and cultivated area, increasing crop production, etc. altogether act as influencing factors for rapid depletion of groundwater resource and throwing challenges for framing its management plan – a burning issue for this landscape. Thus, this outcome will be helpful for the concerned

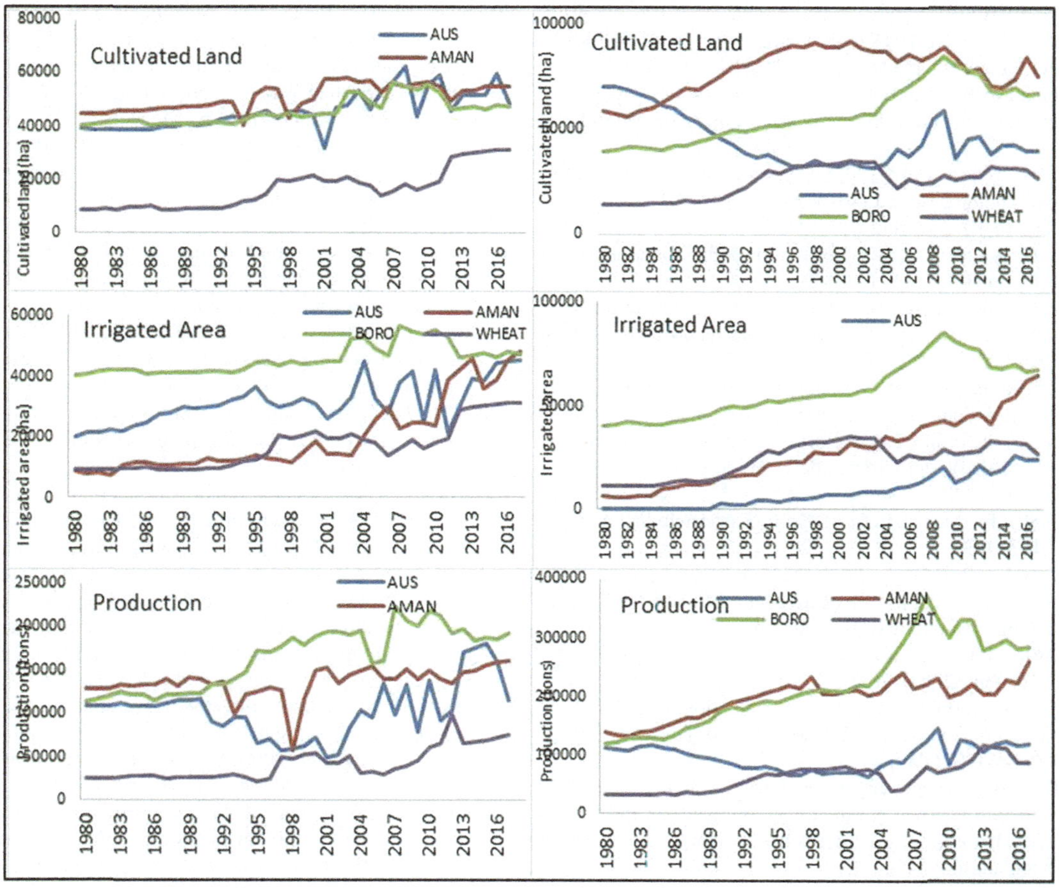

Fig. 2.8 Cultivated and irrigated areas in Chapai Nawabganj and Rajshahi districts along with productions of major crops like rice (*Aman, Aus* and *Boro*) and wheat for the period 1980–2017 in Barind Tract. [Data Source: Department of Agriculture Extension (DAE), Ministry of Agriculture (MoA), Govt. of Bangladesh]

authorities and local managers set proper planning for groundwater resource management for livelihood and make decisions for its sustainability agenda as a demand of time.

References

Ahmed K, Burgess W (1995) Bils and the Barind aquifer, Bangladesh. In: Brown AG (ed) Geomorphology and groundwater. Wiley, New York

Alexander D (1995) Changing perspectives on natural hazards in Bangladesh. Nat Hazards Obs J 10(1):1–2

BBS (Bangladesh Bureau of Statistics) (2015) Statistical Yearbook Bangladesh, 2015. Bangladesh Bureau of Statistics, GoB

Benson C, Clay EJ (2004) Understanding the economic and financial impacts of natural disasters. Disaster Risk Management series. no. 4. World Bank, Washington, DC. © World Bank. https://openknowledge.worldbank.org/handle/10986/15025 License: CC BY 3.0 IGO

BMDA (Barind Multipurpose Development Authority) (2001) Borelog fixtures of the BIADP area, Rajshahi

Brammer H (2012) The physical geography of Bangladesh. University Press, Dhaka

Islam MM, Kanumgoe P (2005) Natural recharge to sustainable yield from the barind aquifer: a tool in preparing effective management plan of groundwater resources. Water Sci Technol 52(12):251–258

Jahan CS, Ahmed M (1997) Flow of groundwater in the Barind area, Bangladesh: implication of structural framework. J Geol Soc India 50:743–752

Jahan CS, Islam MA, Mazumder QH, Asaduzzaman M, Islam MM, Islam MO, Sultana A (2007) Evaluation of depositional environment and aquifer condition in the Barind area, Bangladesh, using gamma ray well log data. J Geol Soc India 70:1070–1076

Jahan CS, Mazumder QH, Islam ATMM, Adham MI (2010) Impact of irrigation in Barind area, NW Bangladesh – an evaluation based on the meteorological parameters and fluctuation trend in groundwater table. J Geol Soc India 76:134–142

Kamruzzaman M (2014) Climate modeling, drought risk assessment and adaptation strategies in the western part of Bangladesh. Unpublished PhD thesis, Institute of Bangladesh Studies, University of Rajshahi, Bangladesh

Kamruzzaman M, Kabir ME, Rahman ATMS, Jahan CS, Mazumder QH, Rahman M (2016) Spatio-temporal analysis of climatic variables in the Western part of Bangladesh. Environ Dev Sustain J 20(2):569–588. https://doi.org/10.1007/s10668-016-9872-x

Kendall MG (1975) Rank correlation methods. Griffin, London

Khandoker RA (1987) Origin of elevated Barind-Madhupur area, Bengal Basin: result of neotectonic activities. Bangladesh J Geol 6:1–9

Klinski TS (1979) Hydrogeology of the Barind area, BWDB water supply paper 417, UNDTCD Groundwater Survey BGD/74/009, Groundwater Circle, Bangladesh Water Development Board
Krishi Kaje Bhugorvostho Pani Bebosthapona Ain (2017) (Groundwater Management Law in Agriculture, 2017). Ministry of Agriculture, GoB
Liu X, Zhu X, Pan Y, Li S, Liu Y (2016) Agricultural drought monitoring: progress, challenges, and prospects. Geogr Sci J 26:750–767. https://doi.org/10.1007/s11442-016-1297-9
Lloyd-Hughes B, Saunders MA (2002) A drought climatology for Europe. Int J Climatol 22:1571–1592
Luis M, González-Hidalgo JC, Brunetti M (2011) Precipitation concentration changes in Spain 1946-2005. J Nat Hazards Earth Syst Sci 11:1259–1265
Mann HB (1945) Nonparametric tests against trend. J Econ 13:245–259
Mustafa SMT, Abdollahi K, Verbeiren B, Huysmans M (2017) Identification of the influencing factors on groundwater drought and depletion in North-Western Bangladesh. Hydrogeol J. https://doi.org/10.1007/s10040-017-1547-7
Nandy DR (1980) Tectonic pattern in North-Eastern India. Indian J Earth Sci 7:103–107
NWPo (National Water Policy) (1999) National water policy 1999. Water Resources Planning Organization, Ministry of Water Resource, People's Republic of Bangladesh
Oliver JE (1980) Monthly precipitation distribution: a comparative index. J Prof Geogr 32:300–309
Pitman GTK (1981) Aquifer and recharges evaluation in Bangladesh. Technical note no. 8, UNDP/UNDTCD Groundwater Survey, BGD/74/009, BWDB, GWC
Rahaman MF (2018) Integrated water resource management for water stressed Barind Area, Northwestern Bangladesh, PhD thesis (Unpublished), Institute of Environmental Science, University of Rajshahi, Bangladesh
Rahaman MF, Jahan CS, Arefin R, Mazumder QH (2017) Morphometric analysis of major watersheds in Barind tract, Bangladesh: a remote sensing and GIS-based approach for water resource management. Hydrol J 5(6):86–95. https://doi.org/10.11648/j.hyd.20170506.12
Rahaman MF, Jahan CS, Arefin R, Mazumder QH (2019) Groundwater potentiality study in drought- prone Barind Tract, NW Bangladesh using remote sensing and GIS. J Groundw Sustain Dev 8:205–215. https://doi.org/10.1016/j.gsd.2018.11.006
Rahman ATMS, Jahan CS, Mazumder QH, Kamruzzaman M, Hossain A (2016a) Evaluation of spatio-temporal dynamics of water table in NW Bangladesh: an integrated approach of GIS and statistics. J Sustain Water Resour Manage 2(3):297–312
Rahman ATMS, Kamruzzaman M, Jahan CS, Mazumder QH (2016b) Long-term trend analysis of water table using 'MAKESENS' model and sustainability of groundwater resources in drought-prone Barind area, NW Bangladesh. J Geol Soc India 87(2):179–193
Ramamasy S, Baas S (2007) Climate variability and change: adaptation to drought in Bangladesh. A resource book and training guide. Institution for Rural Development, FAO, Rome
Rashid B, Islam MS, Islam B (2018a) Physiography of the Barind tract and its surrounding areas Bengal Basin, Bangladesh. Am J Earth Sci 5(1):1–9
Rashid B, Islam MS, Islam B (2018b) Structure and lineaments of the northwestern part of Bangladesh and evolution of the Barind tract. Am J Earth Sci 5(3):26–36
Ravenscroft P (2003) Overview of the hydrogeology of Bangladesh. In: Rahman AA, Ravenscroft P (eds) Groundwater resources and development in Bangladesh – background to the arsenic crisis, agricultural potential and the environment, Bangladesh Centre for advanced studies. University Press, Dhaka, pp 43–86
Scott CA, Sharma B (2009) Energy supply and the expansion of groundwater irrigation in the Indus-Ganges Basin. Int J River Basin Manage 7(1):1–6
Sen PK (1968) Estimates of the regression coefficient based on Kendall's tau. J Am Stat Assoc 63(324):1379–1389
Shahid S, Behrawan H (2008) Drought risk assessment in the western part of Bangladesh. Nat Hazards 46:391–413
Thornthwaite CW, Mather JR (1957) Instructions and tables for computing potential evapotranspiration and the water balance, vol. 10, issue 3. Publications in Climatology, Laboratory of Climatology, Drexel Institute of Technology, Centerton, pp 183–311
UNDP (United Nations Development Programme) (1982) Groundwater survey: the hydrogeological conditions of Bangladesh, United Nations Development Programme (UNDP), Technical report DP/UN/BGD-74-009/1, New York, 113p
Zhang LJ, Qian YF (2003) Annual distribution features of precipitation in China and their inter-annual variations. Acta Meteor Sin 17:146–163

Land Use Transformation in Active Delta

3

Raquib Ahmed, Gregor C. Falk, Jahan Box Moral, Chandan Roy, and Abdullah Al-Maruf

Abstract

Bangladesh's coast is vulnerable due to its increasing high man-land/natural resource ratio, fragile process of delta formation, and increasing threat from climate instability. The result is visible from faster land use changes, which is decided more often by competitive economic returns than its environmental sustainability. The transformation pattern varies from estuarine islands at the east to inter-tidal western lowlands. The delta-forming process, climate change threat, and human activity are some driving forces for socio-demographic and environmental future there. The chapter investigates land use transformation by focusing on how the land uses are changing in spatial context, changes in cropping patterns in economic contest, shrimp farms' expansion in environmental context, and the pattern of physical changes in future context, which synthesizes the general process of change. It is observed that decreasing cropland and increasing shrimp farming indicate an increased pressure on land and environment. At the same time overall gain in new lands shows a future possibility. These are certain challenges Bangladesh is expected to face for the future decades referring environment, livelihood, and the sustainable use of resources.

Keywords

Livelihoods · Migration · Shrimp cultivation · Vulnerability · Agriculture

R. Ahmed · J. B. Moral · C. Roy · A. Al-Maruf (✉)
Department of Geography and Environmental Studies, University of Rajshahi, Rajshahi, Bangladesh
e-mail: ammaruf4@gmail.com

G. C. Falk
Institute for Geography and Geography Education, University of Education Freiburg, Kunzenweg, Freiburg, Germany

3.1 Introduction

Bangladesh coast is unique in many ways – the coast includes diverse geomorphologic contexts, different human-nature interaction, and varied vulnerability of land toward environment degradation and social change. Obviously, potential climate change impact is pursued differently based on context. Under continuous geomorphologic processes, land use changes significantly reflecting human need and adjustment toward the system-induced result. Naturally, the questions appear: what are the major changes visible in the Bangladesh coast, who are the drivers, and how stakeholders adjust in the change system?

Western part of Bangladesh coast is under active delta growing process. It reflects visible changes in the nature of water discharge through desiccating rivers, pattern of river shifting, and changes in the nature of soil and salinity. Decreasing water discharge directly influences on agriculture through availability of irrigable water and responsible for increase in salinity level – both in river water and soils. It is evident, diurnal tidal water goes more inland than before. Regularly recharged affluent soil supports reliable crop production and helps economic growth. Growing human pressure induces increasing cropping intensity for meeting needs in market which functions through economic competition, and contributes in the production value chain. This obviously influences on the land use changes which is related to demographic condition and feed backs on the economic system as a chain of drivers. The overall system works under the changing nature of climate extremities (Tawsif et al. 2022), which interrupts the regular system and force stakeholders toward vulnerability.

Land use change in the entire Bangladesh coast may be divided into four parts. Land use change in the eastern part is settled and attached to the Tertiary Hill is under high economic competition. Central-eastern part is very fragile and heavily influenced by geomorphic processes. The central-western part is geomorphologically settled with matured soil

R. Ahmed et al. (eds.), *Transforming Bangladesh*, World Regional Geography Book Series,
https://doi.org/10.1007/978-3-031-45093-8_3

and under heavy agricultural pressure. The western coast is under inter-tidal active delta-growing process and economic activity is tide-related. The coastal transformation system may be observed through two examples from the Bangladesh coast – change in agriculture, which largely linked to local economic value chain; the second one is the shrimp farming, which reflects economic competition between local and international demand linked to global value chain. Both the processes are linked to the transformation in land use and socio-demographic change under the influence of physical systems.

The present discussion is focused to see the (i) pattern of the change of coastal agriculture and land use to indicate economic transformation process (ii) land use changes due to the expansion of shrimp farming, and (iii) the pattern of influence of physical dynamics those tend to impact on the economic transformation process.

3.2 Study Area

Bangladesh, the largest delta of the world, has 47, 201 km^2 of coastal area, which constitutes about 32% of the country's total land area. The coast is about 750 km long and may be divided into four parts: (i) the western mangrove forest, (ii) the west-central coast region, (iii) the central estuarine region with islands, and (iv) the eastern coast. All these regions have distinct physical and economic features. For the present investigation the west-central agricultural land and lands attaching to the north of the mangrove forest have been considered. The west-central coastal region, covering Jhaokathi, Pirozpur, Barguna, Patuakhali, and Barisal, is predominantly an agricultural region with rice as major crop, followed by jute and value-added crops. This region is influenced by regular coastal hazards with limited tidal impact. The lands attaching to Sundarbans' western part is dessicated by thick river networks and under regular inter-tidal effect which is good for captivating salt water in lands – necessary for shrimp farming. Entire coast has an elevation not more than 3 m from MSL.

The prevailing rapid change in land use resulting in high pressure on land turns land to a critical level of its productivity, forces social instability, and sometimes out-migration due to loss of owning land. The coastal region is featured predominantly with agricultural practices of different types and time-phased crop-combinations, extensive shrimp farming and urbanization. Other than these there exists a gradual natural reclamation of lands from sea due to fluvial sedimentation process (Ali et al. 2007). The other land uses are natural and social forest cover, shrimp farming, and settlements. There are still some more land uses such as water bodies (including marshy land), fallow land, and land under industrial use, but these play insignificant contributory role in the general trend.

Land under agriculture (about 50%) and shrimp farming (about 10%) are the main contributions to economy, to environment, and to socio-demographic changes. These are the two activities linked to external inputs, and also sources of human out-migration. Agriculture dynamics include crop combination, cropping intensity, cropping change, changes toward value-added crops, and introduction of modern inputs and cultural habits. Whereas shrimp farming dynamics include higher economic competitiveness, added value in production chain, external financial inputs, and policy decisions. Both the uses run under a competitive process in context to time and increase in intensity of land uses – i.e., more economic return from same unit of space. This has three clear impacts, first is an increasing impact on environment and the second is tendency to displace human population due to lesser need of labor input. The third dimension is the changes in land, soil, and climate. Overall salinity increase, fluctuations in climate phenomena, and decreasing soil fertility are significant. Ever-rising dependence on ground-water irrigation for increasing demand of crop, expansion of shrimp farming induced salinization of lands and lesser discharge of fresh water from up-streams are the key issues behind soil salinity increase. Regular visit of cyclone including water surge (saline sea-water generally) exerts significant damages in crop lands. Untimely rain makes crop production vulnerable. Over-rain inundates cropped areas. Drought decreases productivity of both crop and shrimp.

3.3 General Land Use Transformation

General land use pattern in the coast west of the estuarine region is strongly dominated by agricultural lands followed by natural mangrove forest and shrimp farms. Other than these are settlements (including urban built-up areas), afforested lands, accreted lands, inland water bodies, and open/uncovered/fallow land (Fig. 3.1).

Mangrove forest and surrounding shrimp farm dominates the western part where an extended northern edge of the forest being experienced with saline tidal water gives scope for shrimp cultivation (economically more competitive) to take over croplands. The most significant change during the last about 50 years is visible in the conversion of crop lands to shrimp farming has contributed to national economy. On the other hand, there is a general decrease in agricultural land. There is also a phenomenal increase in population grown with an increase in GDP. GDP per capita in 1972 was 94.38 US$, in 2018 was US$ 1754 and total GDP was US$ 9457 billion as per The World Bank report. Growth of GDP has a direct impact on people's consumption pattern. The current per capita GDP in 2020 was US$ 1969 (World Bank Report 2020). Also, consumption pattern is strongly related to globalization of cultural innovations and globalization of trade

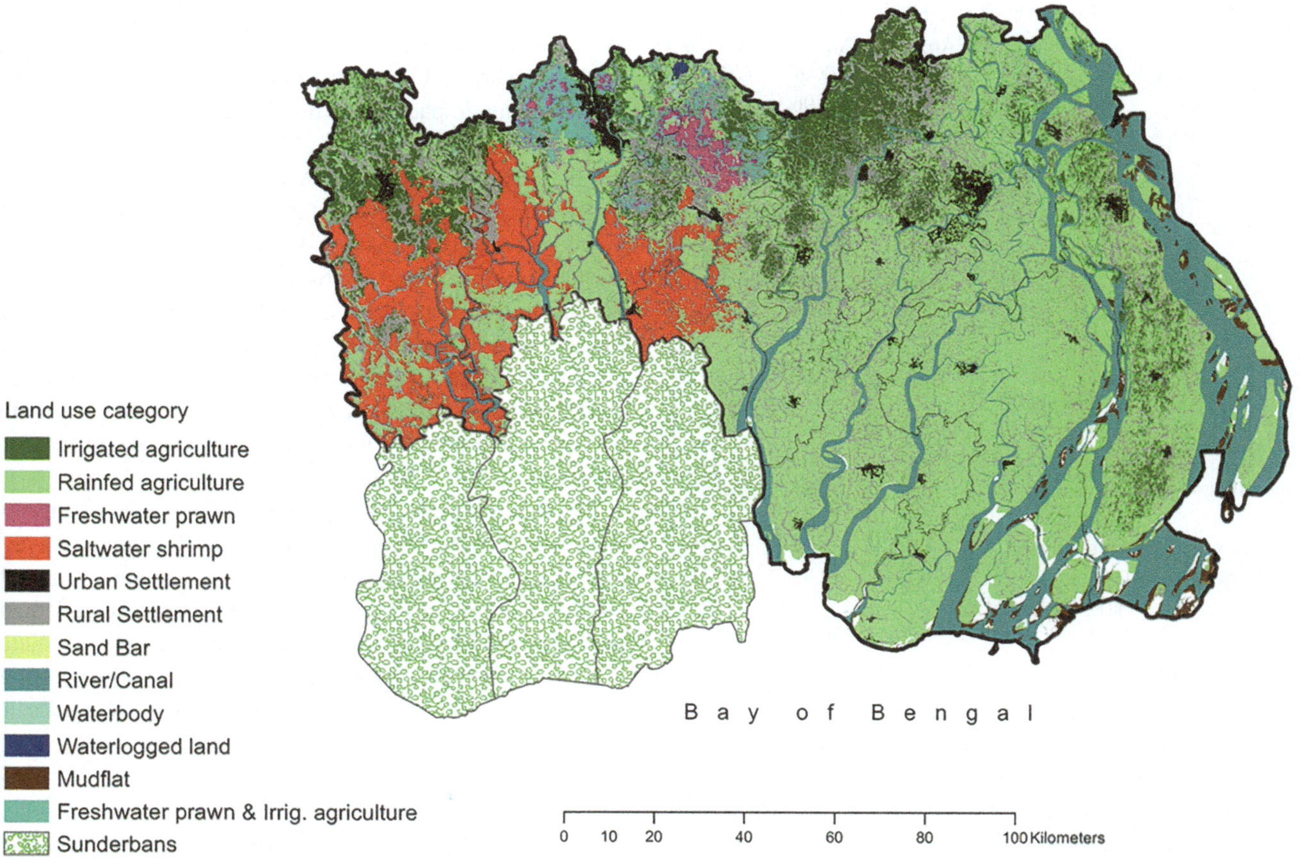

Fig. 3.1 Coastal land use. (Image source: Adams et al. 2016)

under market-oriented economic system. So, the change is inducted not only by its people's economic need, also by collaboration of international business. This is reflected on the economic activities, particularly on agriculture which still employs about 40% of its population but contributes about 15.96% to GDP in 2017. On the other hand, service and industry sectors are rising. Coastal land use changes have the capability of modifying the ecosystems which provide services including biodiversity and other resources such as food, fiber, and water for human society. Therefore, unsustainable change to land use may directly affect the coastal ecosystems and thus humans living in those areas (Chase et al. 2000). Though both natural and anthropogenic factors play important roles in coastal land use change, anthropogenic factors have gradually become prominent over the last half-century and coastal land uses of Bangladesh have gone through major changes during this period.

Land conversion from agriculture to industrial purposes is more dominant in the eastern part of the coast, whereas in western part of the coastal lands it is toward shrimp farming – a product of international demand with strong up-scale value change. Changes in the west-central part are still within itself and are passing through a selection by competitive markets resulting in more economically viable shrimp farming to take over agricultural lands by using advantage of availability of sea water from inter-tidal flow. The land use change in the estuarine islands is strongly influenced by regular erosion from the head-ward side of river current and accretion generally on the other side. Through this erosion and accretion, the estuarine islands have a general shifting from northern to south-western direction. When calculated, it is found that gain is more – about 1368 km^2 during 1973–2016 (Hassan 2017). This increase of land is actually supported by conceptual delta-building geomorphologic process – through which Bangladesh grew since the late *Cretaceous Period* (during about 66 million years and to be more precise since *Pliocene Epoch* of about 2.4 million years). This is still continuing to create debate among scientists whether the delta's stability is under the threat of possible sea-level rise.

Table 3.1 shows the general trend of land use change in Bangladesh between 1976 and 2010. It was prepared by the Soil Resources Development Institute of Bangladesh using aerial photographs and field survey. The table indicates, during 34 years, land under crop decreased by nearly 7%. During that time population increased about double and additional lands for settlement purposes conceptually, which can only come traditionally from agricultural lands

Table 3.1 Land use (in percentage) of Bangladesh

Land cover/use		Area in % (1976)	Area in % (2000)	Area in % (2010)
Agricultural land	Cropland	67.38	64.96	60.04
	Forest	12.11	9.02	9.84
	Mangrove	3.12	3.35	3.03
	River	6.29	6.11	6.44
	Lake	0.35	0.40	0.35
	Beel-Haor	1.66	1.73	1.72
	Aquaculture	0.004	0.99	1.21
	Tea	0.83	0.95	0.66
	Salt pan	0.08	0.17	0.25
	Total agricultural land	91.83	87.69	83.53
Non-agricultural land	Rural settlement	6.11	10.03	12.12
	Urban and industrial	0.18	0.33	0.60
	Accreted land	1.87	1.95	3.75
	Total non-agricultural land	8.17	12.31	16.47
	Total land area	100.0	100.0	100.0

Source: SRDI (2013)

(Al-Maruf et al. 2022). But this did not happen. A massive migration saved agricultural lands to some extent but exploded through urbanization (from about 7 million in 1976 to 52 million urban population in 2010) (Haque et al. 2021; Rahman 2013). Increasing urban population density is a significant phenomenon in Bangladesh. A significant amount of migration sources from the coastal lands reflect extensive transformation in land use change. A study conducted on coastal land use change taking 4 small geographical units (Hatiya, Koyra, Kalapara, and Rampal) in 2018 shows similarity with the general land use pattern in Bangladesh (Bernzen et al. 2016). Figures 3.2 and 3.3 show a general change and Table 3.2 indicates the change pattern.

The information clearly indicates an increasing intensity of land use and increasing pressure on environment. One important fact is that Bangladesh has recently attained self-sufficiency in food despite decreasing amount of land available for cultivation. This is due to modernization of agriculture. But soil is getting exhausted making production system more vulnerable. Indeed, new settlements (including urban built-ups at regional scale) take up certain number of fertile lands. Generally fallow or unusable lands are insignificant.

In the coastal region any land generated from accretion becomes usable in just 3–5 years by repeated siltation process. Right from a short delay the land is ready for selected crops. This is more visible in the estuarine islands where islands experience a shift in their shape and size which indicate the growth of lands in and around the estuary (Ghosh et al. 2015).

This belongs to the general geomorphologic process through which the delta and the floodplain evolved during the Quaternary and Pleistocene Period. The estuarine islands suffer strong impact from regular coastal cyclone, water surge, erosion, and accretion. All these increase vulnerabilities in agriculture and living. A significant part of the islands (particularly newly accreted lands) are taken up by governments and make afforestation to increase forest cover. In the western side, the mangrove forest is protected to some extent but inside the forest there exists forest change. Occasional illegal logging is restored by government initiative but it cannot bring diverse nature of species. The most dominant species is Sundari (*Heritiera littoralis*) from which the name appeared, and also the most threatened species in terms of logging.

By far the most controversial land use change in Bangladesh coastal land has been taking place in the western part around the *Sundarban* mangrove forest since the early 1970s as shrimp farms grew through the new market demand in European and western countries (Table 3.3). It is estimated that about 2000 km^2 land is under saltwater shrimp farming currently. Shrimp contributes to big part of Bangladesh's GDP (Shamsuzzaman et al. 2020).

3.4 Changes in Agriculture

Nearly about three-fourth of the coastal lands are still under agriculture, and it is under transformation. The transformations are of three types: (i) decreases in the agricultural lands, (ii) changes in cropping pattern, and (iii) conversion of land uses due to physical impact. The total population of Bangladesh is 166.5 million, of which the urban population is 36.6% and rural population is 63.37% as of 2018. Annual growth rate of population is 1.37%. Bangladesh is one of the most densely populated countries in the world and its population density is 1062.5 per km^2. It was estimated using Bangladesh Bureau of Statistics data that there has been a general decrease of land under agricultural of about 2.1% (it is yet to confirm) during 1988 to 2000 (Jaim and Begum 2003). The most dominant crop is rice, which occupies about three-fourths of the available land, and also linked to economy and country's food security and political system with strong influence. In the southern part of Bangladesh cropped land is rain-fed and in the northern part it is irrigation-dependent. Situation is changing through crop combination and time pattern. Bangladesh has attained self-sufficiency in food production by introducing modern technology, genetically modified breeds and farm mechanization despite decreasing land available for cultivation (Al-Maruf et al. 2021a, b). At the same time, it is bringing lands from single crop to double crop to triple crop at many places and in combination of value-added crops where possible. The result is

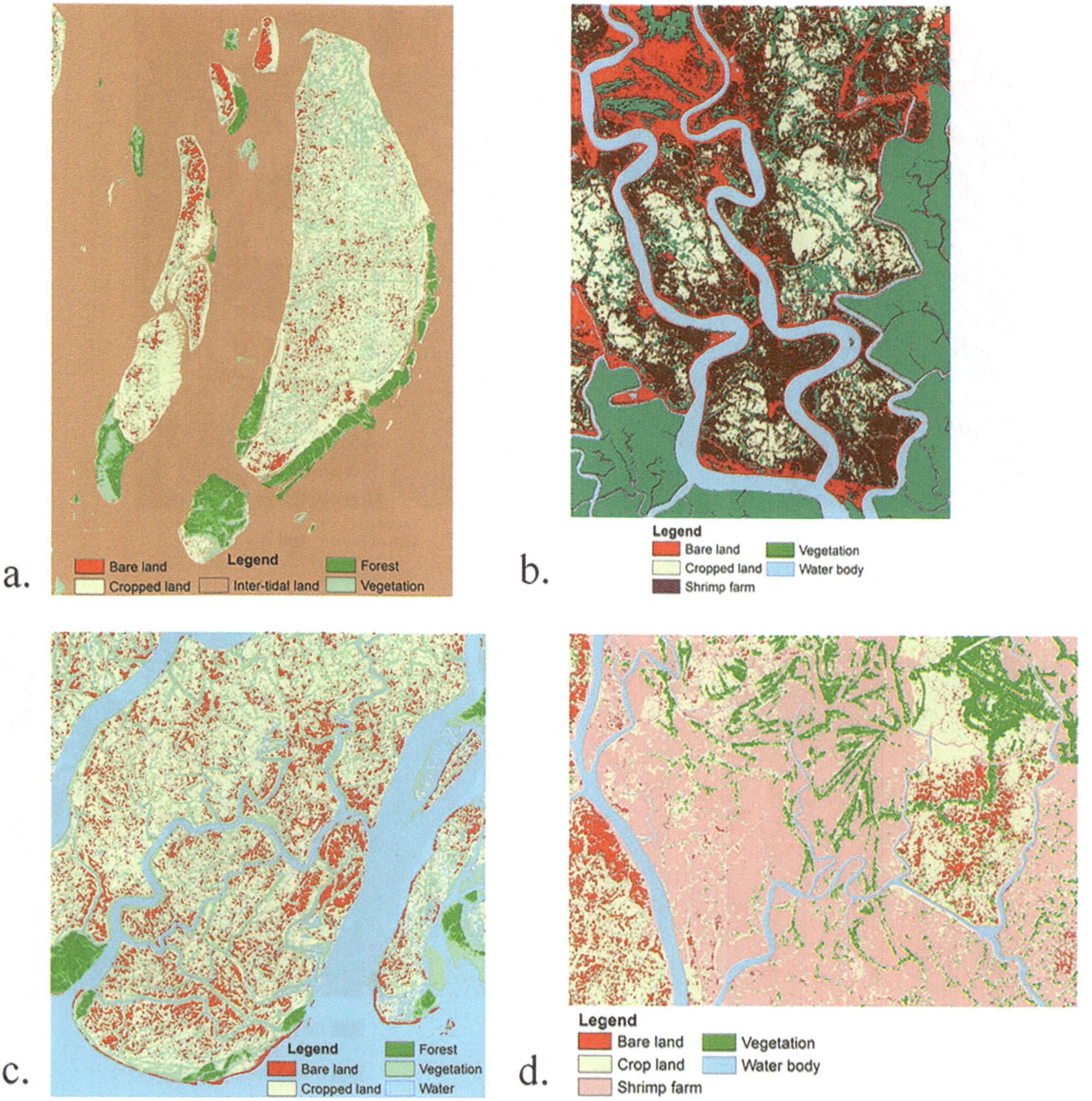

Fig. 3.2 Land use map of 1989; (**a**) Hatiya, (**b**) Koyra, (**c**) Kalapara and (**d**) Rampal. (Source: Researchers' own investigation)

increased crop production at the cost of land quality which is getting under intense pressure and approaching to a critical level (Pyka et al. 2020). Due to marketization of economic system bigger investors increasingly dominate by replacing small landholders who subsequently either change occupation or migrate to other places. A recent study noted that a significant occupational shift from agricultural to non-agricultural sectors is taking place more visibly (Al-Maruf 2017). Increasing value-added crops are a new phenomenon which is more a commercial product such as potato, spices, vegetables, and fruits. Economic competition decides but rice as basic food might face challenge in future. Ever increasing intensity of cropping is getting more dependent on irrigation in rain-fed areas bringing changes in availability of surface and also ground water. Growing demand on fresh irrigable water induces saline water to penetrate further north appears one of the causes of increasing salinity in water and land. So, the changes are becoming visible not only on land uses but also on quality of lands (Sanzidur 2010; Rahman 2013).

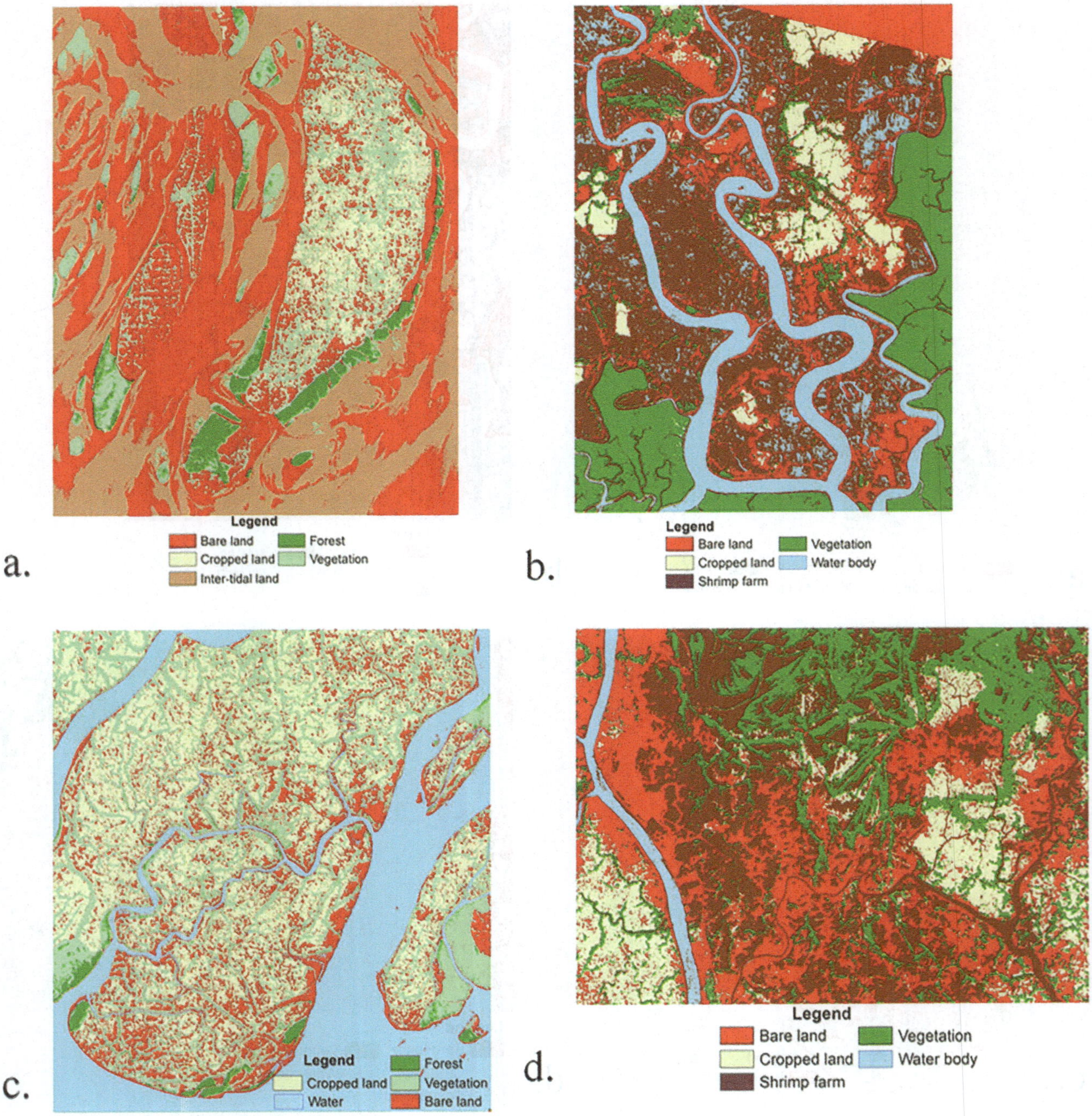

Fig. 3.3 Land use map of 2018; (**a**) Hatiya, (**b**) Koyra, (**c**) Kalapara and (**d**) Rampal. (Source: Researchers' own investigation)

Table 3.2 Area under different land use types (in km^2) in 1989 and 2018

	Hatiya		Koyra		Kalapara		Rampal	
Area (in km^2)	1989	2018	1989	2018	1989	2018	1989	2018
Cropland	306	141	80	33	432	366	354	188
Vegetation	135	146	156	106	214	288	126	110
Bare land	49	168	81	86	137	187	33	68
Waterbody	–	–	63	80	439	333	43	48
Inter-tidal land	73	91	–	–	–	–	–	–
Shrimp farm	–	–	129	175	–	–	123	269

Table 3.3 Export of shrimp from Bangladesh

Year	Wild spp. (mt.)	Farmed giant tiger shrimp (mt.)	Farmed giant river prawn (mt.)	Farmed brown shrimp (mt.)	Farmed Indian white prawn (mt.)	Others farmed spp. (mt.)	Total exports (mt.)
1982	3542	2347	294	89	64	569	6903
1983	3421	3993	561	165	99	1041	9312
1986	3345	6134	1886	733	241	1528	13,613
1987	3542	7287	2458	975	216	1985	16,275
1990	3876	7988	2468	1263	532	1937	17,505
1991	3679	8393	2238	1698	621	1916	17,985
1995	4568	12,816	4919	1288	719	1678	26,277
1996	5023	14,221	4026	528	521	936	25,255
2000	4876	17,678	3800	550	498	1112	28,514
2001	4656	20,413	3715	324	315	290	29,713
2005	5295	24,980	6282	2528	822	2500	42,943
2006	5654	25,182	7158	2178	982	5379	46,533
2009	5805	28,658	9282	1928	678	4017	50,368
2010	5365	30,172	11,322	1623	876	2241	51,599
2011	6217	31,371	11,980	1820	996	2507	54,891
2012	5666	28,556	10,228	1529	835	1193	48,007
2013	5977	30,124	10,615	1126	927	1564	50,333
2014	5929	32,188	11,615	1218	955	2588	54,500
% Total	11%	59%	21%	2%	2%	5%	–

3.5 Expansion of Shrimp Farming on Cropped Lands

This is a major transformation of land uses, particularly in the western part of the coast around the northern borders of the mangrove forest. Low-lying and flat topographic nature of land and international demand, traditional cropped lands are transformed into shrimp farms by trapping saline water in the fields from the tidal rivers. The shrimp production grew at a faster speed (Portley 2016) due to its global demand and an upper hand in local competitive price index. The land under shrimp farming was just 39 km^2 in 1980 to more than 200 km^2 currently. Shrimp farming is less labor-intensive than conventional agricultural practices. Hence, it induces two processes, changes in occupation of the local people from agriculture to shrimp farming or migrate to other places. Also, the livelihood pattern is changed due to overall system's change. The shrimp farm expansion is shown of an area called Rampal in the coast (Figs. 3.2d and 3.3d). Visibly, the consequence is retention of salt water for a long time on fields turns soil steadily saline, and in time shrimp productivity reduces. Often the lands are returned to former cropping practice, but it takes at least 3–5 years to bring it back for growing of crops, results in unsustainable land conversion – brings debate on environment and economic benefit. The possible solution may be intensive and integrated shrimp farming where less amount of land is used.

3.6 Increase in Soil Salinity

According to a recent study conducted by the Soil Resource Development Institute, Bangladesh shows that the salinity-affected area has increased from 8330 km^2 in 1973–10,560 km^2 in 2009 (SRDI 2010). Also, the rate of salinity intrusion in coastal Bangladesh is faster than it was predicted a decade ago. On the basis of the observed symptoms, it is therefore logical to assume that more agricultural lands in the coastal area will be affected by salinity in the days to come. As a result, quality of productivity of soil will be degraded. This will eventually contribute to the loss of agricultural land and thus pose threat to food security (Agarwala et al. 2003). Modern agriculture is largely dependent on irrigation, but irrigable surface water is being contaminated often due to tidal inundation and salinity intrusion. Besides constraining agricultural production, salinity limits freshwater availability for drinking purpose and industrial production in the coastal areas. Even though salinity intrusion is a slow process, the impact is long-lasting and devastating as well. In coastal regions of Bangladesh, factors involved in the salinization include degree of flooding, precipitation, the effect of the river system and its discharges, the nature of the soil, depth of groundwater table and salt deposits, land relief, tidal action, slope of the ground, and proximity to drainage channels (Huq and Shoaib 2013). In the coastal areas of Bangladesh, saline soils occur in the river deltas in a strip of land a few km to 180 km wide along the seacoast. Drastic reduction of upstream flow through the Ganges-Brahmaputra-Meghna river systems has contributed to an increase in salinity level in the tidal rivers, decreased surface water availability in the rivers and canals, lowering of groundwater table, and reduced soil moisture content. Though salinization of soil has restricted crop cultivation on coastal lands, the increase of water salinity of these areas has created a suitable habitat for shrimp cultivation (SRDI 2010).

3.7 Physical Dynamics and Impact on Economic System

The most important physical dynamics that influence on the economic and environmental processes in the Bangladesh coastal region are the (i) delta growing process, (ii) increase in salinity in river water and soil, and (iii) coastal cyclone and water surge (Al-Maruf et al. 2021a, b). Except eastern mountainous and hilly regions Bangladesh is broadly a land formed through delta process under the influence of hundreds of distributaries and tributaries. The stream courses are the Ganges, the Brahmaputra, and the Meghna in terms of carrying down sediments. The coast at the estuarine region is highly vulnerable by erosion and accretion (source: researcher's own investigation). Temporal satellite images suggest a considerable amount of land is eroded away and at the same time about 70–100 km^2 land is accreted in about 10 years' time. This land gain is the basis of the theory of delta growing process and at the same time contributes to changes in land use causing out-migration. Stream orientation and islands shape tend to follow north to south-west direction in a gentle curve. A general land use change in the decreasing of croplands and increase in bare land reflects the above fact. Also, the new lands usually take just 3–5 years to be used for copping.

In most of the shoreline regions newly accreted lands in attached to the islands have been taken under afforestation program by government which shows an overall increase in the coastal lands (Fig. 3.4). The western part of the coast, until about 180 km from the sea shore is under the influence of daily tidal flow. This part includes the mangrove forest in the south including a belt of area, which is known as active delta region. Due to flat and low land-level character of the lands around the mangrove forest, it is easy to grow shrimp farm by capturing salt water from tidal rivers. At one time this makes the land saline and becomes difficult to bring back for reuse in agricultural purposes. Finally, it takes more than 3–5 years in general environmental condition to bring the soil back to its original status. Extensive expansion of shrimp farming is a cause of salinity increase in the coastal lands. It is evident from Figs. 3.2 and 3.3 that there is a significant conversion of croplands into shrimp farm. Gradual decrease in water flow through the desiccating rivers due to extensive withdrawal at the upper stream causes saline sea water to come further inside to exert its influence. Overall result is the salinity increase both in ground (Fig. 3.5) and surface water (Table 3.4).

This is further deteriorated by growing dependency on irrigation that makes environment vulnerable. SRDI reports that salinity affected areas increased from about 8330 km^2 in 1973 to 10,560 km^2 in 2009 (SRDI 2010). Also, the rate of growth of this affected area is faster than thought before (Agrawala et al. 2003).

Coastal cyclone and water surge has double effect on coastal economy – on land and on livelihood. It devastates economy and makes the land flooded and saline resulting in environmental consequence. Historically, Bangladesh

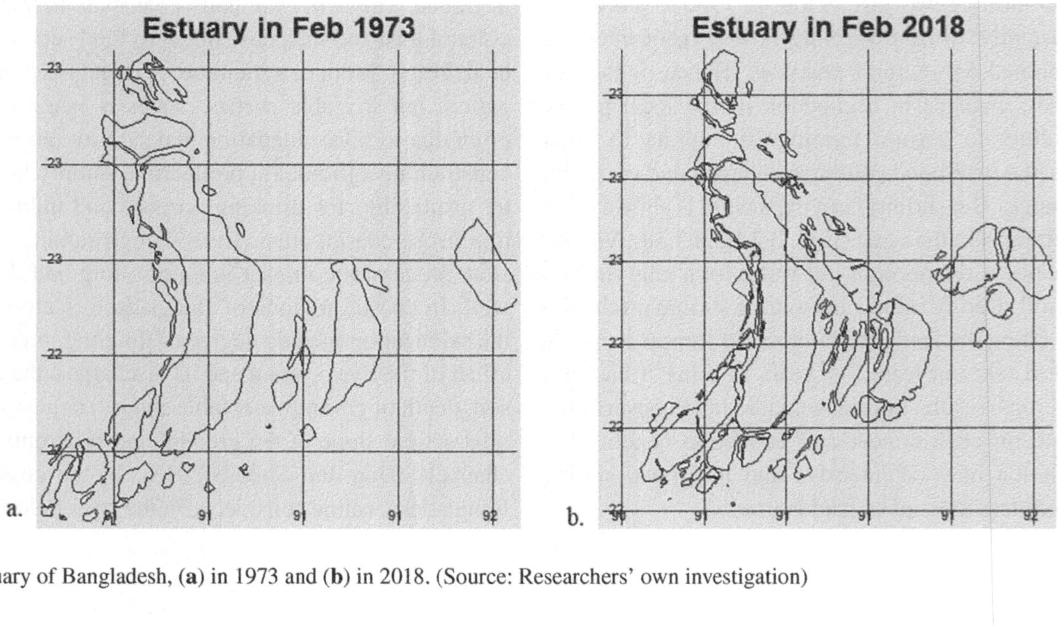

Fig. 3.4 Estuary of Bangladesh, (**a**) in 1973 and (**b**) in 2018. (Source: Researchers' own investigation)

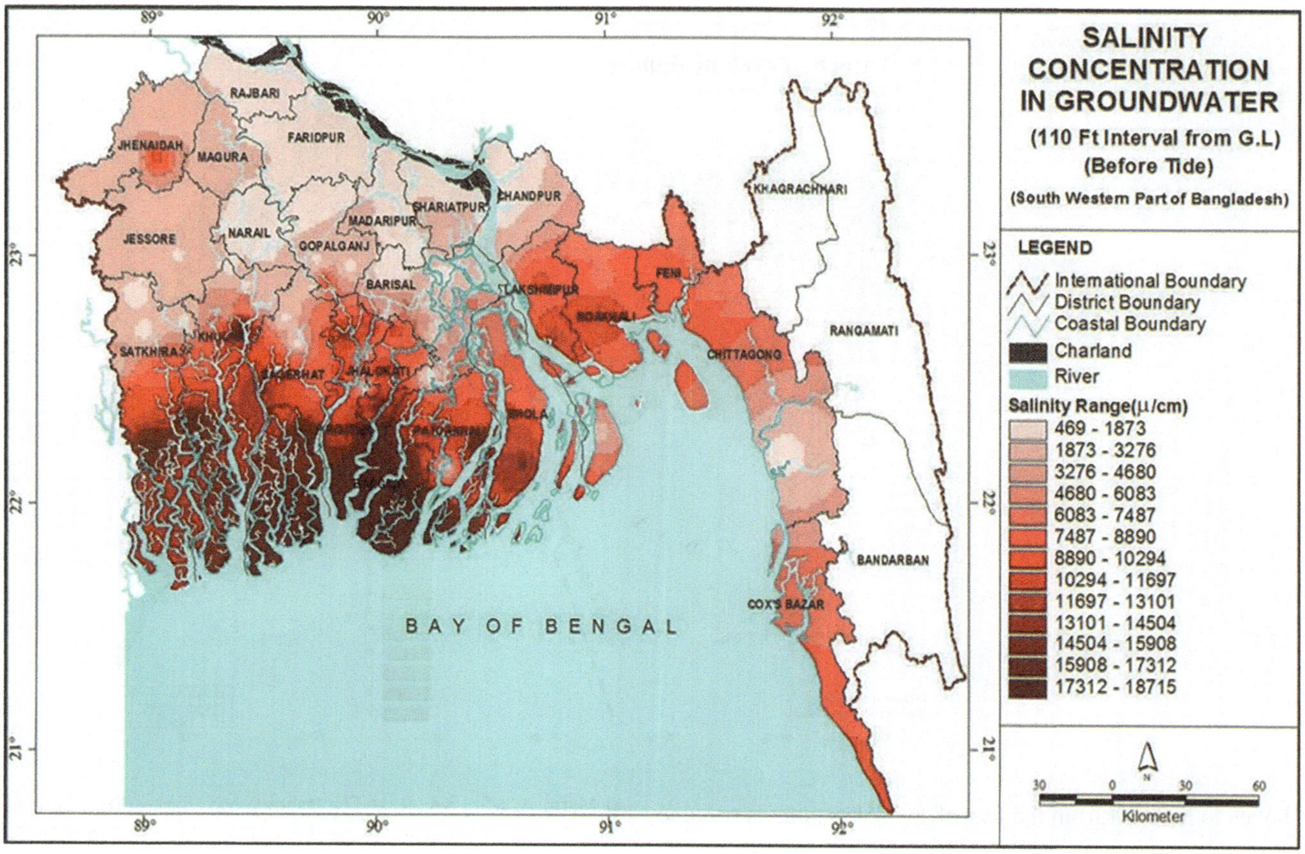

Fig. 3.5 Salinity concentration in groundwater. (Source: Researchers' own investigation)

Table 3.4 Water salinity at Daratona station of Alaipur Khal in Bagerhat district

	Salinity level (in EC) during years				
Months	2006	2007	2009	2014	2015
Jan	1700	3600	4050	4900	5750
Feb	3850	3950	6500	7500	9800
Mar	8200	8100	10,650	11,000	16,000
Apr	9150	10,700	12,500	14,000	17,200
May	9700	14,600	15,600	16,050	14,800
Jun	4000	12,000	13,600	14,800	10,200

During monsoon due to heavy rainfall and huge water discharge through the rivers, salinity becomes low. An increasing trend of salinity is observed with reduction of river water flow and delineation of rainfall. Considering salinity measurements of these 6 months seems enough to elicit these variations

coast has been hit by 2–4 coastal cyclones along with water surge every year. Their general direction is from south-west to north-east makes the estuarine islands most vulnerable. These two phenomena used to claim lives of hundreds of thousands every year, but now loss of lives is significantly reduced, although their magnitude is found closely similar. This is due to development of awareness among people, forecasting technology, post cyclone/water surge evacuation and restoration, and improvement in disaster management system. But their impact on land remains closely similar as land use and land utilization is getting intensified due to growing population and economic demand. A spatial pattern of a cyclone's possible impact is shown in the above-simulated map (Fig. 3.6) using a range of factors.

The index runs from 0 as lowest and 10 being the highest. Estuarine islands have less diversified crops where lands away from the major cyclone path and inside the main lands have more diversified crops. Also crops depend more on rain in the lands close to coast than inland.

3.8 Conclusion

Bangladesh's poverty rate fell from 82% in 1972, to 18.5% in 2010, to 12.9% in 2016, and 5.6% in 2019 as measured by the percentage of people living on the equivalent of US$1.90 or less per day in purchasing power parity terms. Based on the current rate of poverty reduction, Bangladesh is projected to eliminate extreme poverty by 2030, as well by becoming upper-middle income economy, according to the World Bank. But the growing salinity, pressure on land, continuous impact of coastal hazards, and population growth are significant

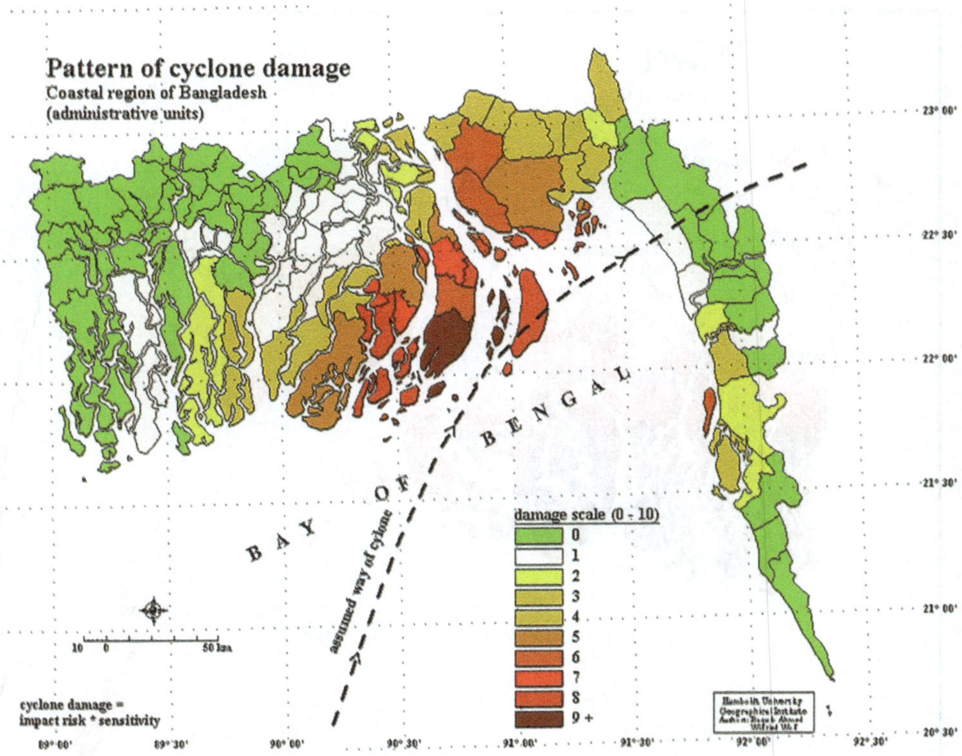

Fig. 3.6 Tropical cyclone damage pattern in the coastal region of Bangladesh. (Source: Researchers' own investigation)

threats to it. Apart from the fear of possible sea-level rise due to global warming this looks apparent that weather extremities and increase in the frequencies due to climate change are making coastal lands more vulnerable both for living as well as for agriculture despite there is positive changes in the adaptation mechanism. Decrease of dependency on agriculture is a global phenomenon which is also visible in Bangladesh. But it has to continue still for a long time for its sustainable economy with a fact of decreasing availability of cultivable lands. Interaction between human need and capacity of nature is nearing to a critical level of productivity very fast. Possible consequences might be future out migration from the region potential to create pressure on cities and shift poverty from regional level to city level.

References

Adams H, Adger WN, Ahmed S, Ahmed A, Begum D, Lazar AN, Mathews Z, Rahman MM, Streatfiled PK (2016) Spatial and temporal dynamics of multidimensional well-being, livelihoods and ecosystem services in coastal Bangladesh. Sci Data Nat Res J 3:160094

Agrawala S, Ota T, Ahmed AU, Smith J, Aalst MV (2003) Development and climate change in bangladesh: focus on coastal flooding and the Sunderbans. Organization for Economic Co-Operation and Development (OECD)

Al-Maruf A (2017) Enhancing disaster resilience through human capital: prospects for adaptation to cyclones in coastal Bangladesh. Doctoral dissertation, Universität zu Köln

Al-Maruf A, Jenkins JC, Bernzen A, Braun B (2021a) Measuring household resilience to cyclone disasters in coastal Bangladesh. Climate 9(6):97

Al-Maruf A, Mira SA, Rida TN, Rahman MS, Sarker PK, Jenkins JC (2021b) Piloting a weather-index-based crop insurance system in Bangladesh: understanding the challenges of financial instruments for tackling climate risks. Sustainability 13(15):8616

Al-Maruf A, Pervez AK, Sarker PK, Rahman MS, Ruiz-Menjivar J (2022) Exploring the factors of farmers' rural–urban migration decisions in Bangladesh. Agriculture 12(5):722

Ayub A, Mynett AE, Azam MH (2007) Sediment dynamics in the Meghna estuary, Bangladesh: a model study. J Waterw Port Coast Ocean Eng 133(4):255–263

Bernzen A, Al-Maruf A, Lin A, Ahmed R (2016) Landnutzungswandel im Küstenraum von Bangladesh: Management von Land auf Gemeinde-und Haushaltsebene. Geogr Rundsch 68(7–8):16–22

Chase TN, Pielke RA, Kittel TGF, Nemani RR, Running SW (2000) Simulated impacts of historical land cover changes on global climate in northern winter. Clim Dyn 16:93–105

Ghosh M, Kumar L, Roy C (2015) Monitoring the coastline change of Hatiya Island in Bangladesh using remote sensing techniques. ISPRS J Photogramm Remote Sens 101:137–144

Haque AKM, Al-Maruf A, Rida TN (2021) Governance for sustainable development to combat the impact of climate change in a medium-sized city in Bangladesh. In: Bangladesh II: climate change impacts, mitigation and adaptation in developing countries. Springer, Cham, pp 357–373

Hassan SMT (2017) Estimating erosion and accretion in the coast of Ganges-Brahmaputra-Meghna delta in Bangladesh. In: 6th International conference on water & flood management, pp 115–124

Huq SMI, Shoaib JUM (2013) The soils of Bangladesh, World soils book series, vol 1. https://doi.org/10.1007/978-94-007-1128-0_7

Jaim WMH, Begum R (2003) Changes in land use pattern in Bangladesh over the last two decades. Bangladesh J Agric Econ XXVI(1–2):87–96

Mostafa Shamsuzzaman M, Mozumder MMH, Mitu SJ, Ahamad AF, Bhyuian MS (2020) The economic contribution of fish and fish trade in Bangladesh. Aquac Fish 5(4):174–181

Portley N (2016) Report on the shrimp sector Asian shrimp trade and sustainability. Asian Shrimp Trade and Sustainability, p 42

Pyka L, Al-Maruf A, Shamsuzzoha M, Jenkins C, Braun B (2020) Floating gardening in coastal Bangladesh: evidence of sustainable farming for food security under climate change. J Agric Food Environ (JAFE) 1(4): 161–168. ISSN (Online Version): 2708–5694

Rahman HZ (2013) Urbanization in Bangladesh: challenges and priorities Hossain Zillur Rahman. Bangladesh Economists' Forum, p 3

Sanzidur R (2010) Six decades of agricultural land use change in Bangladesh: effects on crop diversity, productivity, food availability and the environment, 1948–2006. Singap J Trop Geogr 31:254–269

Soil Resources Development Institute (SRDI) (2010) Saline soils of Bangladesh. SRDI/Ministry of Agriculture, Dhaka

SRDI (2013) Agricultural land availability. Ministry of Agriculture, GOB

Tawsif S, Alam MS, Al-Maruf A (2022) How households adapt to heat wave for livable habitat? A case of medium-sized city in Bangladesh. Curr Res Environ Sustain 4:100159

World Bank Report (2020). https://www.macrotrends.net/countries/BGD/bangladesh/gdp-gross-domestic-product. Bangladesh GDP 1962–2021. www.macrotrends.net. Retrieved 2021-10-01

4 Bangladesh Delta Plan: Anatomy of a Long-Term Development Strategy

Rejaul Karim Bakshi and Shajeda Aktar

Abstract

The Bangladesh Delta Plan 2100 is formulated with a target of achieving a safer, secured, resilient and prosperous country by the end of the twenty-first century. The primary goals of the delta plan are to attain food and water security in the long run in this delta along with environmental sustainability and continuous economic growth. The country has been considered as one of the next 11 emerging economies of the world. However, the country is also one of the most climatic vulnerable countries across the globe. Ensuring food security for all remains a key policy challenge for Bangladesh for over a long period. The country also set 2030 as the year to eradicate extreme poverty and to attain the Sustainable Development Goals. In this perspective, the Bangladesh Delta Plan 2100 is unique in its nature that sets goals and identifies strategies for a period up to the year 2100. However, the delta plan has missed certain issues that are critical for the sustainable development of the delta. This paper, therefore, offers recommendations for consideration as there are scopes to explore these issues being the delta plan is adaptive in nature. Our recommendation ranges from education and research; from exploring the potential of the blue economy to trans-boundary cooperation for water and energy security; from ensuring food security to food-quality security, water and air quality security and public health; from ensuring credibility of the government institutions to providing a national database with comprehensive household information suitable to enable effective policy intervention during shocks, disasters and pandemic like Covid-19 situation.

Keywords

Delta plan · Climate change · Sustainable development · Economic growth · Middle-income country

4.1 Introduction

The Government of Bangladesh has devised a long-term plan in 2017 titled 'The Bangladesh Delta Plan 2100' what it dubbed as the plan for a safer, secured, resilient and prosperous country by the end of the century. It considers the approach of adaptive delta management (ADM) at the core of this comprehensive plan. The delta plan is a long-term development framework that takes adaptive management technique due to the uncertainties involved in the climatic phenomenon and taking climate change issues into consideration in adapting policy interventions. The primary goals of the delta plan are to attain food and water security in the long run in this delta along with environmental sustainability and persistent economic growth.

Bangladesh is a developing country with ample opportunities as well as certain challenges facing for long-term development. With a population exceeding 168 million (BBS 2022), it has been considered as one of the next 11 emerging economies of the world (Goldman Sachs 2019). Governments in the past four decades put emphasis on improving standard of living, achieving self-sufficiency in food production, reducing illiteracy, child mortality and improving per-capita income and food security (BBS 2022; Mainali et al. 2018). The country has seen GDP growth rate of above 7% consistently in the past few years. Per capita income has grown up from USD 200 in 1980 to USD 2575; poverty has declined from about 83% 1972 to 18.5% in 2021; life expectancy has increased from 55 years in 1974 to 72 years in 2021; child

R. K. Bakshi
Department of Economics, University of Rajshahi, Rajshahi, Bangladesh
e-mail: rkbakshi@ru.ac.bd

S. Aktar (✉)
Department of Public Administration, University of Rajshahi, Rajshahi, Bangladesh
e-mail: aktar96@ru.ac.bd

R. Ahmed et al. (eds.), *Transforming Bangladesh*, World Regional Geography Book Series,
https://doi.org/10.1007/978-3-031-45093-8_4

mortality has dropped from 240 per thousand in 1974 to 21 in 2021; fertility rate has also gone down from 7.3 in 1974 to 2.00 in 2021; and adult literacy rate has seen a rise from a mere 22% to 72.3% between 1974 and 2021 (BBS 2022). With modernization and technological improvement, the economy is transforming from agriculture dominant one to manufacturing and services sectors dominated one. Bangladesh has met the lower middle-income country requirements of the World Bank in 2015 and aspires to reach the upper middle-income country status by 2030. Yet, the country with a land mass of 1,47,570 square km and 168 million population, Bangladesh is the country with highest density in the world with about 1270 people per square km, excluding the island nations and the city states.

In terms of natural hazards, the country is also ranked as the fifth most vulnerable of the world (IPCC 2013). Cyclone, tidal surge, flood, salinity, river erosion and droughts take heavy toll every year in terms of lives and property damages. In addition, there is concern of rise of sea level to engulf the coastal belt of the country displacing millions of people of the south-western districts. The region's huge population is severely asset poor to combat against climate change (Johnson and Hutton 2018; GED 2017b). The coastal belt of the country is already suffering from cyclone, tidal surge, river erosion, flood, water logging and salinity—and the sea-level rise would deteriorate the situation in the coming days. Saving the coastal belt from future climate change remains a big challenge for the country. The rising sea level along with deceleration of freshwater flows of rivers in the coastal belt is damaging agriculture and public health in the area. The country in large is facing rising temperature in the recent years causing increase in monsoon rains triggering overflow of rivers and floods. On the other hand, in the North-Western region of Barind Tract, low rains and lower level of groundwater are causing water shortage and droughts damaging agriculture and livelihood of the people badly. Supplying water in the urban areas also appears a challenge as water layers have fallen low in many areas specially in Dhaka city. Ensuring adequate freshwater for growing number of households and rapidly expanding industries is a big concern from the urban management perspective.

In addition to these climatic challenges, ensuring food security for all remains a key policy challenge for Bangladesh for over a long period. This is particularly important as the country has set 2030 as the year to eradicate extreme poverty and to reach the upper-middle income level. Furthermore, the country is thriving to attain the Sustainable Development Goals (SDGs) by 2030 and many of the SDG goals are interlinked (GED 2017c; Hutton et al. 2018). For example, the SDG goal #2 aims at ending hunger, goal # 3 attempts to ensure quality health and healthcare, goal #4 aims at expanding education; goal #6 plans to ensure availability of water and sanitation, and goal #13 targets to take action against climate change—and all these goals are dependent on achieving goal #8 which envisages to attain sustainable economic growth.

In this perspective, the Bangladesh Delta Plan 2100 is unique in its nature that comes up with a long-term vision instead of shot-cuts to shape courses and paths of the nation in every sphere for a period up to the year 2100 (GED 2017a). Linking the plan objectives with well orchestral action strategies, combined with visionary leadership and productive manpower is essential in order to harness the benefit from the delta plan and achieve the desired level of living and being for the people of this dynamic delta of the world by 2100.

4.2 Challenges of the Bangladesh Delta

The Bangladesh delta is facing certain challenges that need to be addressed for sustainable development. The BDP 2100 identified these challenges in six specific categories as described below.

4.2.1 Climate Change

Global warming becomes a growing concern for the entire world and the country is not to this exception. The International Panel on Climate Change (IPCC) projected that the average temperature of Bangladesh may go up by about 1.4–1.9 °C due to climate change if business as usual prevails and can go up by 2 °C for extreme cases by the year 2050. The temperature will continue to increase in the rest of the century. The rainfall will be erratic following the climate change. While the monsoon rain is expected to increase due to climate change, the rainfall is expected to drop in the rest of the year—causing excess flood during monsoon and drought scenario in other seasons. On the contrary, the southern coast of the country and the eastern hill tracts might see drop in rainfall by 2050. The effect of climate change will be fatal on the economy and the livelihood of the people of the delta. According to the Global Climate Risk Index (GCRI) 2010, about 8241 people in Bangladesh lost their lives due to climate change, while economy suffers a damage by USD 1.2 billion (1.81% of the GDP) every year (MoEF 2012; GCRI 2010). Another study conducted by the Asian Development Bank reported that Bangladesh is likely to suffer a loss of 2% of its GDP every year by 2050 and the loss can rise up to 9% of GDP by the year 2100 (Ahmed and Suphacholasai 2014). Study also predicts that due to climate change shock the incidence of poverty in Bangladesh might go up by 15% by 2030 (IPCC 2014).

4.2.2 Floods

The delta consists of floodplains of some big rivers like the Ganges, the Brahmaputra and the Meghna. About 70% of the total land mass of the country lies not more than 1 m above the sea level. In addition, about 10% of the total land mass consists of rivers and water bodies. About 60% area of the country gets flooded due to overflow of the three big rivers. The mega floods of 1987, 1988, and 1998 were devastating in their size and impact in the recent history. The 1988 flood submerged 61% area of the country resulting in over 45 million people homeless and damaging the economy by about USD $1.2 billion (MoEF 2009). The 1998 flood was devastating which took 1100 lives, inundated over 100,000 square kilometres areas, damaged half a million homes, caused 30 million people homeless, shattered major roads, hundreds of bridges and ruined the economy by about USD 2.8 billion (Ahmed and Suphachalasai 2014; MoEF 2009). Empirical evidence suggests that the size and frequency of big floods are increasing in the recent time due to climate change (CEGIS 2013). The study further predicts that the extent of flood will continue to rise for all areas of the country by 2050.

4.2.3 Drought

Despite the delta is carrying about 700 rivers, the north-western Barind tract of the country is considered as the drought-prone one. The concept of drought is defined as the period when soil wetness is insufficient and retards crop production during a crop-growing season. The Barind tract remains wet for 4–5 months of the year mainly during monsoon, with an average annual rainfall of 1250–1750 mm (BDP 2100, 2017). The spell of dry days and drought is acute during April–May having soil with low moisture and high temperature exceeding 40 °C. The central region of Madhupur tract is also considered a drought-prone area. Lengthy spell of dryness and droughts are a big threat in these areas' agricultural production. It is predicted that if climate change continues the rice production, specially the Aus rice, would decrease by at least 27% in these areas while wheat production would drop by 39%. In case of a severe climate change situation even the Boro rice production may go down by 62%. Drop of soil moisture might force farmers to decrease Boro cultivation significantly.

4.2.4 River Bank Erosion

River bank erosion is another challenge in this delta. Among the three big rivers the Jamuna River has erosion of about 1770 hectares per year; for Padma it is about 1298 hectares and for Meghna it stands at 2900 hectares (BDP 2100; CEGIS 2016). The Jamuna is widening every year at an average of 130 m by eroding its banks. Just in the last 40 years, Jamuna erosion corresponded to loss of land by about 120,000 hectares.

4.2.5 Sea-Level Rise and Salinity Intrusion

Sea level rise (SLR) and resulting salinity intrusion appear a big concern for the south-western coastal area of the country. The current trend of sea-level rise in the coastal area is 6–20 mm per year. IPCC (2013) predicts the sea level to rise by up to 1 m in extreme cases (and 0.2 m for minimum emission) by the year 2100. Sea-level rise reduces availability of freshwater in the coastal zone significantly, causing flood and intrusion of salinity. With rising sea level, the area under 1 ppt[1] salinity will increase from current 10% to 17.5% by 2050, and the area under 5 ppt salinity will see a rise from current 16% to 24% over the same period of time (CGEIS 2013). The situation will be deteriorated further if the flow of the freshwater declines from upstream which is a very likely possibility due to climate change and dying of river flows.

4.2.6 Cyclones and Storm Surges

The deadly cyclones and resulting tidal surge put Bangladesh as the fifth most climatic vulnerable country of the world (IPCC 2017). Rising sea level and increase in the sea surface temperature would increase cyclone and tidal surges further in the coming years. It is observed that a 10 m high tidal surge resulting from deadly tropical cyclone hits Bangladesh in almost every 10 years, and that of a 5 m high tidal surge in every 5 years. The IPCC 2017 predicts that due to sea-level rise and increase in sea surface temperature would result in both the high frequency and intensity of the storm and tidal surge in the coming years.

4.3 Vision, Mission and Goals of the Delta Plan

The Bangladesh Delta Plan 2100 envisages transforming the delta with a long-term vision. The vision of the plan emerges from the challenges it is facing and the strategic goals it wants to achieve including sustained increase of per capita income, eliminating extreme poverty, enhancing human development and reducing climate change vulnerability.

[1] *ppt* stands for '*parts per thousand*'. 1 ppt means 1 g of salt per 1000 g of water. Higher the ppt, the higher is salinity.

4.3.1 Vision

Achieving Safe, Climate Resilient and Prosperous Delta.

4.3.2 Mission

Ensure long-term water and food security, economic growth and environmental sustainability while effectively reducing vulnerability to natural disasters and building resilience to climate change and other delta challenges through robust, adaptive and integrated strategies, and equitable water governance.

The vision and mission are to be achieved through specific medium-term and long-term goals set by the plan. The delta plan sets 3 policy goals for the national level that can be considered as medium-term goals; and 6 delta specific long-term goals in addition to these national-level goals. The delta plan envisages that the vision of the plan would be materialized through these sets of goals.

4.3.3 Goals at National Level

Goal 1: Eliminate extreme poverty by 2030
The poverty rate in Bangladesh currently stands at 18.5% with extreme poor of 7.9% (BBS 2022). The rate of extreme poverty has declined consistently over the past 20 years from 34.3% in 2000. The livelihood of the extreme poor depends on farming, fisheries, forestry and informal jobs. They are the most vulnerable to climate shocks including flood, cyclone, drought and salinity. The delta plan aims at eliminating extreme poverty from the country by the year 2030. This goal is also in line with the Sustainable Development Goals (SDGs) goal #1.

Goal 2: Achieve Upper Middle-Income Country status by 2030
Bangladesh has reached to the lower middle-income country threshold in 2015, and has fulfilled all the three eligible criteria for graduation from the least developing country to the lower-middle income country (World Bank 2018). The delta plan envisages to consolidate the growth rate of GDP further and to attain the threshold of the upper-middle income country by 2030. The plan aims at reaching GDP growth rate of 8% by FY2020 and 10% by the FY2024 to achieve this objective; however, the goal has been shattered by external shocks including Covid-19 and Ukraine war.

Goal 3: *Being* a prosperous country by 2041
The country sets its goal to be a high-income country by 2041. According to World Bank estimates in FY2041 the threshold income level for a high-income country would stand at USD 19,793. The delta plan aims at achieving this level of per capita income by 2041 by achieving sustained economic growth, transforming age-old agriculture, modernizing manufacturing and service sectors, ensuring energy, infrastructure, food and water security.

In addition to these national-level goals, the delta plan also sets up specific goals to achieve by the year 2100. These delta-specific goals are outlined as:

4.3.4 Delta Specific Goals (BDP 2100)

Goal 1: Ensure safety from floods and Climate Change-related disasters
The delta plan sets this goal to manage risks of floods during the rainy season as well as other climatic disasters including cyclone and tidal surge in the coastal zone. The aim is to plan, mitigate, prepare and recover disaster risk at a minimum and acceptable level such that the loss of lives and properties can be kept at a minimum. The goal outlines to facilitate: (i) an environment safe for everyone's living across rural and urban settings; (ii) sustainable water supply through reliable water system for economic development in the long run; and (iii) improvement of functioning of the key societal and economic mechanisms.

Goal 2: Ensure water security and efficiency of water usages
This goal aims at efficient management of water resources during the summer and dry spells with shortage of water. The goal targets providing sufficient and reliable availability of freshwater for livelihood, environmental sustainability and economic activities. The specific targets are: (i) provision of safe and sufficient supply of water for drinking, farming, fisheries, environment, transport and industry; and (ii) enhancing water quality through effective control of pollution considering environment, industry and health need; and ensuring access to standard sanitation for all.

Goal 3: Ensure sustainable and integrated river systems and estuaries management
The delta plan aims at fostering sustainable river system in consideration with the morphological dynamics for cost-effective recovery and improvement of land. The specific goals are: (i) ensuring long-term solutions for water ways; (ii) supply of sediment to floodplains in a planned way; (iii) reclamation of land for economic and social development; and (iv) fostering connection between major and minor rivers for economic and environmental development.

Goal 4: Conserve and preserve wetlands and ecosystems and promote their wise use
This goal attempts to provide safeguard in preserving the wetlands, to protect sensitive ecosystem areas including mangrove forest, water bodies and rivers, coastal green-

belt, seasonal and perennial wetlands, tidal freshwater frontier and all water bodies including swamps, lakes, haors and canals.

Goal 5: Develop effective institutions and equitable governance for in-country and trans-boundary water resource management

The delta plan aims at improving the governance system to combat the exogenous as well as internal challenges in the management of water resources. The aim includes development of integrated approach in water resource management in local, regional and national level by enhancing information and knowledge base through consultation, dialogue, information sharing and modelling, trans-boundary communication, information sharing and negotiations for water resource management; and building multi-level cross-border platforms across the basin for mutual cooperation.

Goal 6: Achieve optimal use of land and water resources

This goal attempts in optimizing efficiency using scarce land and water resources of the delta. The use of land, water and water transport should be decided and designed with integrated approach to maximize the potential of these resources. Encouraging and fostering interaction between various modes of resources are given priority in this goal.

4.4 Natural Hazards Hotspots and Strategies Under BDP 2100

The delta plan refers hotspot to "a broad grouping of districts and areas facing similar risks evolved by Hydrology, climate change and natural hazards". It defined hotspot as "a place of significant activity or danger. Hotspots are prototypical areas where similar hydrological and climate change vulnerability characteristics and problems converge also influenced by natural hazards".

The Bangladesh Delta Plan (BDP 2100) identifies 6 such hotspots in this delta, described below:

(i) Coastal Zone (27,738 square km).
(ii) Barind and Drought Prone Areas(22,848 square km).
(iii) Haor and Flash Flood Areas (16,574 square km).
(iv) Chattogram Hill Tracts (13,295 square km).
(v) River Systems and Estuaries (35,204 square km).
(vi) Urban Areas (19,823 square km) (Fig. 4.1).

The remaining area of the delta is termed as "Cross-cutting" areas featured with various hazards and challenges including floods, drought, storm, river bank erosion, groundwater depletion, pollution and sanitation. The hotspots of the Bangladesh delta, the unique challenges these hotspots face and the potential Delta Action Plans outlined in the BDP 2100 are described below:

(i) *Coastal Zone Hotspot*

The south-west coastal zone faces diverse challenges including cyclone, tidal surge, salinity, and flood. The rising sea level due to climate change further aggravates the problems affecting livelihood of the people negatively. Most of the south-western districts are exposed to cyclone, storm surges and salinity. There are also problems of river bank erosion and damage of polder embankments. Some rivers are suffering from navigation problems due to siltation and some areas witness prolonged water-logging because of encroachment of river banks and tidal channels.

It is projected that tidal fluctuation will affect over 50% of coastal zone by 2050 aggravating the salinity problem and destroying the polder embankments. The CEGIS 2016 predicted that by 2100 the sea level in the Bay of Bengal may rise by 1 m devouring 17–21% of land area of the country. This will also cause severe shortages of water supply household purpose and irrigation; loss of land and livelihood; damage the mangrove forest; and fisheries sector. The intrusion of salinity due to sea-level rise could affect 7000–8400 square km in the districts of Barishal, Bagerhat, Bhola and Barguna distressing 2.9–5.2 million people. Increase in the frequency and intensity of cyclones could see storm up to 250 km per hour leaving catastrophic effects on infrastructure, crops, livestock, trees, homes and livelihoods.

Delta Action Plan The delta plan aims at combating storm surge and salinity intrusion through effective management of existing polders, reclaiming new land in the coastal zone and conservation of the Sundarbans (6017sq km). Specific measures in this regard include:

- Revival of the regional rivers and improvement of drainage in local rivers and canals.
- Acceleration of tidal river management (TRM) activities in the coastal polders and in other areas with appropriate feasibility study.
- Building Naoakhali-Urir Char cross dam to reclaim new lands from the Bay of Bengal.
- Fostering land accretion steps in the Meghna estuary.
- Appropriate planning, development and mapping of the newly accreted land.
- Continuous dredging of all channels of Sundarbans including the Ghasiakhali channel.
- Maintaining and improvement of mangrove species in Sundarbans, plan to improve the islands and green belt development.
- Use of mangroves to accelerate afforestation process in the newly accreted lands.

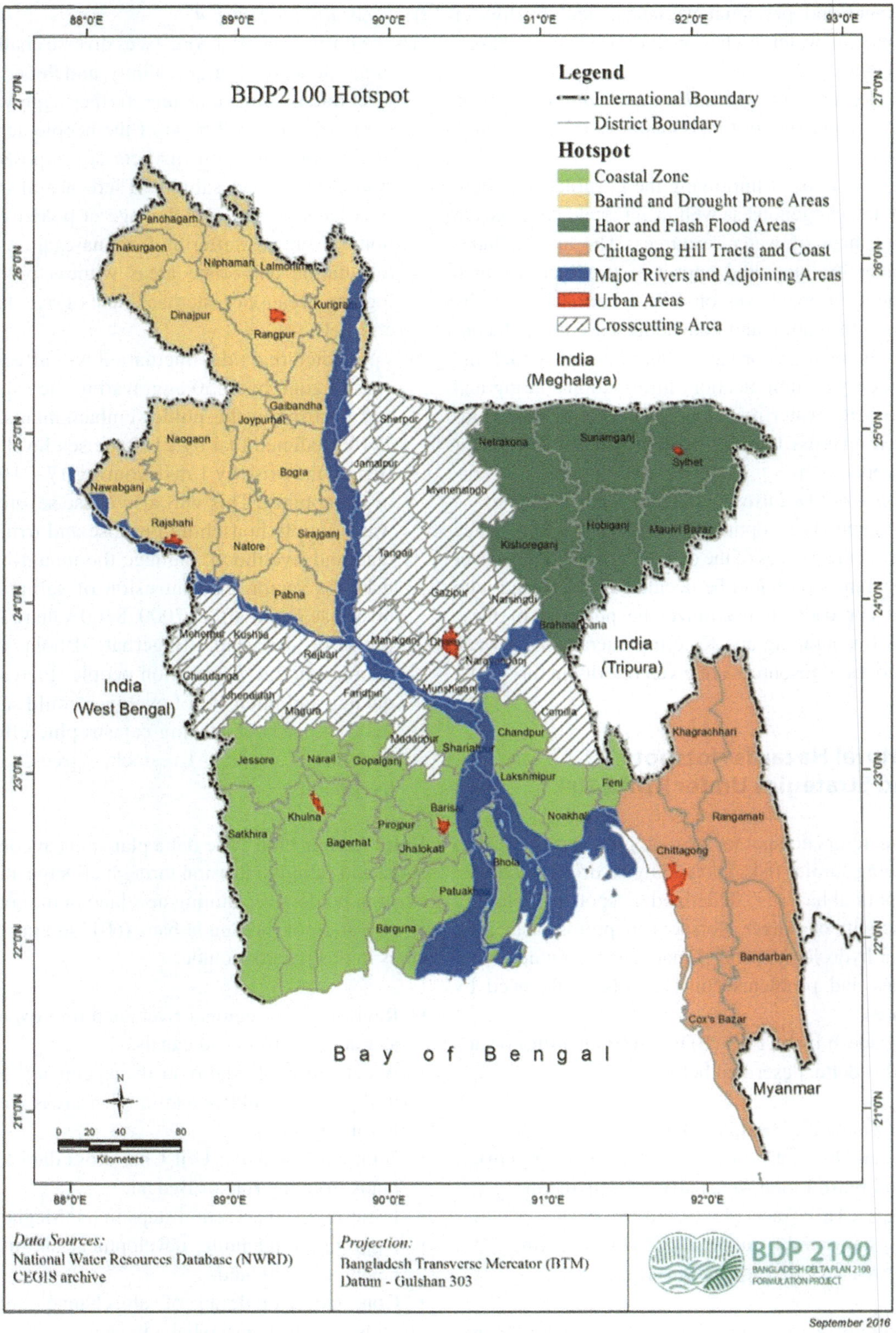

Fig. 4.1 Major hotspots of the Bangladesh Delta. [Source: Bangladesh Delta Plan 2100]

(ii) *Barind and Drought Prone Hotspot.*

The drought-prone Barind Tract is in the north-western Rajshahi division of the country. The area has few small river flows consisting of a fragile ecosystem with very low vegetation cover. The region practices intensive agriculture especially Boro rice which requires high amount of irrigation extracted from groundwater source. This causes depletion of groundwater levels further. In addition, blocking of upstream water flows by India through Teesta Barrage (at Gazaldoba) and in Dalia barrage in Bangladesh severely reduced the water flow in the dry season. The climate change will intensify the water crisis further in this region as the dry season will be longer which would affect 39 million people in this Hotspot. The agriculture will be the hardest hit. The CEGIS 2016 predicts that the temperatures of this region would go up to 2 °C and would reduce production of crop and livestock, cause water scarcity, increase salinity and badly affect livelihood of the people.

Delta Action Plan The delta plan envisages at promoting sustainable growth in the drought-prone Barind tract by ensuring freshwater supply, reducing losses due to flood and improving sanitation. The specific strategy outlined in the delta plan includes:

- Use of surface water from Padma and other regional rivers for irrigation.
- Designing strict regulations in using and extracting groundwater.
- Planning for efficient basin management for Barind tract.
- Creation of manmade reservoirs for retaining surplus water during wet season (including ponds, rubber dams, and rainwater harvesting).
- Excavation of Chalan Beel and Baral river.
- Recovering river banks, canals and ponds from encroachment to facilitate freshwater flow.
- Improvement of the flood management and control system for this hotspot.
- Establishment of a full functioning centre for the management of flood and drought in this area.
- Set up of sewerage treatment plants in all cities and municipalities, industrial waste treatment, supply of drinking water.

(iii) *Haor and* Flash *Flood Areas Hotspot.*

Haors and flash flood areas are found in the Sylhet division of the country's north-eastern part of Bangladesh. During rainy season the area witnesses flood up to 6 m high but during dry season water goes down significantly. Climate change will severely affect the surface water level of this area of the country including the Haor hotspots. This area is posed by number of threats due to climate change including shrinking wetlands, declining water from river flows, declining groundwater level, and flash floods as water holding reservoirs goes down significantly. The livelihood of the people will be in great danger as Haors are mostly occupied by the poor and destitute people. The CEGIS 2016 further predicted that earthquake can also go up in this active seismic area further deteriorating the ecosystem and waterways badly.

Delta Action Plan The delta plan targets at protecting agriculture, industry and livelihood of this hotspot from floods, erratic rainfall and climate change. The measures outlined in the delta plan include:

- Facilitating irrigation using surface water from regional rivers.
- Building submersible embankments and their proper maintenance to protect from flash floods.
- Use dredged soils to elevate village bases and platforms in the Haor and flash flood areas.
- Introduce farm mechanization and adoption of climate-friendly technologies in agriculture.
- Ensure proper development and implementation of the Flood Management, Drainage and Irrigation (FMDI) structures.
- Enforce the wetland preservation laws and regulations without compromise.
- Formulate and apply proper plan for the development of fisheries and preserving extinct species.
- Popularizing fisheries, aquaculture and cash crops instead of rice-only production focus.

(iv) River *Systems and Estuaries Hotspot.*

Bangladesh is a land of rivers including three big rivers: the Ganges, the Brahmaputra and the Meghna. These rivers not only carry huge amount of water every year but also about 1.1 billion tons of sediment. About 90% of the land mass of the country is constructed by these deltaic sediments. Climate change posits serious challenge to the flow of these rivers and associated livelihood. Erosion of river bank is causing about one million people homeless every year. Floods are damaging lives, property and infrastructure. Deposited sediments due to erosion of the river bank and from mountains are damaging navigation through waterways.

Erosion of Jamuna river is categorically dangerous which is devouring hundreds of homes and thousand hectares of farming land every year in Sirajganj and Kurigram. The river bank erosion increases the number of people under poverty as households lose cultivable land, homesteads and employment spaces. River bank erosion is a big challenge for this hotspot. The CEGIS 2016 further predicts that river bank erosion would increase in this hotspot by 2050 as there is a chance of 13% increase in precipitation. This hotspot is facing further challenges from unplanned industrialization in the river banks and encroachment by the influential quarters reducing water flow, loss of water quality and damage of ecosystem.

Delta Action Plan The delta plan aims at reviving rivers in danger and recovering encroached water bodies to smooth flow of water, reduce flood risk and proper sediment management. It also outlines strategies for effective management of the newly accredited lands. The specific measures include:

- Necessary capital dredging in major rivers including the Padma, Meghna, Jamuna, Brahmaputra, Teesta, Dharla, etc.
- Strategies to control salinity of the rivers flowing in the south-western region specially during dry season.
- Flood management plan according to flow of the river and its erosion and accretion pattern.
- River bank protection for stability of the river flow.
- Ensuring natural flow of rivers for efficient sediment management.
- Action to increase the water flow of the river specially during dry season including strategic dredging and sediment management.
- Proper initiatives for the development and maintenance of the newly accreted lands.
- Development of efficient urban management including river and water body management.

(v) *Chattogram Hill Tracts Hotspot.*

The Hill Tracts (CHT) area of the country consists of three districts: Rangamati, Khagrachhari and Bandarban. This hotspot faces different kind of challenges from the rest of the country. Landslides are the biggest threat in this hotspot due to intense rain and loss of topsoil and vegetation. Certain human activities are also responsible for the landslide including cutting of hills for housing, unplanned farming, and deforestation. Landslide puts about half a million people's lives in danger. On the other hand, irregular rainfall and decline in water bodies due to climate change cause the search for drinking water difficult for its people. The hotspot is at great risk of losing forest, vegetation, farmland, crops and livelihood due to land degradation, deforestation and flash floods.

Delta Action Plan The delta plan envisages to ensure ecological balance, improvement of the ecosystem, biodiversity, water security and sustainable development in this hotspot. Specific measures include:

- Promoting indigenous agriculture for sustainable development and reforestation with indigenous species.
- Conservation of hills, waterfalls and canals and ensuring biodiversity.
- Building multipurpose dykes, dams, and canals to control and divert water flows in the erosion-prone areas.
- Construction of embankments, dam, etc., to protect locality from flash floods from the rivers of Karnafuli, Sangu, Chengi, Bakkhali and Matamhuri.
- Provision of piped water from reservoirs for household purposes.
- Introducing and maintaining sewage and waste treatment.
- Preservation of Kaptai lake, its fisheries and discharge of additional silt.
- Preserving forest resources of the hills, control deforestation and increase afforestation.

(vi) *Urban Areas Hotspot.*

Bangladesh is a 'rapidly urbanizing country'. The country's urban population has gone up from 9% in 1974 to 39.4% in 2020 (Worldometers 2020). The urban area is already crowded with over 60 million people with more than 1800 people per square km. With rapidly increasing population and industrial growth, the urban areas are already under pressure from groundwater extraction, deteriorating water quality, difficulty in wastewater management, and treatment of sewage for making urban life safe and secure. The rivers of big cities including Dhaka, Chattogram, Narayanganj and Khulna are polluting at alarming rates by households and industries. The environment is challenged with rising air pollution, water pollution, noise pollution and so on from polluting industries, smokes from vehicles and brick kilns and poor management of solid waste and sewerage. The urban area hotspot is facing further challenges due to climate change including rising sea level, water logging, flash floods, erratic rainfall, depleting groundwater level, surface water contamination, etc. Wetlands are declining rapidly due to urbanization and encroachment by land grabbers which reduces water absorption causing fresh floods in the inner city centres.

Delta Action Plan The delta plan outlines certain strategies for sustainable urban development, efficient land and water resources, improved urban utility services including water supply, sanitation and waste management, conserve and preserve urban wetlands and ecosystems and promote their wise use. The specific objectives outlined in the delta plan include:

- Designing effective plan to control water logging in the urban areas.
- Develop sustainable management of solid waste. Separate treatment of hospital waste, household waste and e-waste.
- Development of drainage system and water management.
- Preservation of natural water bodies and reservoirs.
- Ensure quality of water supplied for households and industrial consumption.

- Enhancement of greenery, vegetation, lakes and water bodies.
- Effective governance for financial sustainability and green city approach.
- Encourage private sector initiative in waste management, green space and water body reservation.

4.5 Financial Arrangement for Combating Climate Change, Water Resource Management and Environmental Sustainability

The present investment of Bangladesh on climate change, environmental management and water resources is very low, roughly about 0.6–0.8% of GDP. Creation of the Delta Fund is proposed in the Delta Plan with budget allocation of at least 2.5% of GDP by the year FY2031. It is suggested that among this budget about 2% of GDP will be required for new investment projects while 0.5% of the GDP for operation and maintenance. The delta plan also emphasizes on public-private partnership in this endeavour. Accordingly, about 80% of the total expenditure (or 2% of GDP) of the delta fund should come from government allocation but the rest 20% should come from private sector (which is 0.5% of GDP). In addition, the development partners should also come forward to finance the delta plan for the sustainable development of the delta. Some policy mixes are also suggested for the funding strategy. Accordingly, the delta fund can be mobilized using both the tax and non-tax revenue sources. The recovery of cost is suggested by applying *Beneficiary Pay Principle* as well as the *Polluters Pay Principle*. It is projected that the country's budget will go up from current 10% of GDP to 25% of GDP by FY2031 as there will be increasing pressure from other thrust sectors including education, health, infrastructure, energy and social security. The funding of the proposed allocation for climate change, environmental sustainability and water resource management will, therefore, remain challenging. Table 4.1 outlines the proposed budget allocation in the initial stage of the delta plan:

The Delta Plan 2100 also offered projection for infrastructure development and knowledge base institutional capacity development projects necessary for the long-term development goals of the delta. The table below presents such investment plan for the delta in the first phase of the plan (Table 4.2).

Among the 80 projects the delta plan proposes, about 65 projects are listed as infrastructural development and 15 projects for knowledge and capacity building necessary for the execution of the mission and vision of the delta plan for a long-term, sustainable and prosperous delta.

4.6 Areas Unexplored in the Delta Plan: Scopes for Improvement

The Bangladesh Delta Plan has set certain goals to achieve in the short, medium and long term up to the year 2100. Many of these goals are consistent with the Sustainable Development Goals (SDGs). For example, the goal to eradicate poverty, ensuring food security, addressing the climate change issues and environmental sustainability, fostering economic growth and employment, social development, ensuring biodiversity and preservation of forestry and agriculture. The delta plan also identified six hotspots on the basis of climatic characteristics and challenges. However, the delta plan has missed certain issues that are critical for the sustainable development of the delta. These are summarized below:

(i) *Research and Education Plan*: The Bangladesh delta plan is a long-term plan based on cutting-edge science and technological fervour. However, the plan has not outlined any idea how the delta can accelerate in research and education in the twenty-first century. The overhaul and restructuring of education and research is not only desirable for a knowledge-based society but also essential for survival in this rapidly changing world. The recent pandemic of Covid-19 and the mismanagement in handling the diverse challenges from the pandemic has shown how ill-prepared the nation was in facing a critical time like this. Only appropriate research institutions and cutting-edge scientific research culture led by able researchers could correctly guide a society in a critical situation like this. But what we have seen in Bangladesh is that despite the outbreak of Covid-19 in December in China, and the first reported case of corona infection on March 8, 2020 in the country, it is still lagging in terms of supplying necessary equipment to the doctors, nurses and the patients. The country is the second largest exporter of RMG products in the world including PPE (personal protective equipment) yet is struggling to provide PPE to the hospitals in due course and with appropriate standard (The Daily Star 2020). The similar is true for the Covid-19 testing kit. The country exports medicines including viral medicines to at least 45 countries of the world, yet is dependent on import for testing kit for Covid-19 (The Daily Star 2020).

In another dimension, lack of quality education is hindering the prospect of country's growth significantly. The country has about 150 public and private universities producing nearly half a million graduates every year. While in one hand the majority of these graduates remained unemployed for years, on the other hand the growing business sector of the country is increasingly

Table 4.1 Investment requirement for water resource management, climate change impact and environmental sustainability (% of GDP)

Sources of Finance/Sector for Expenditure	FY2016	FY2020	FY2025	FY2031
Total Delta investment required	0.8	1.8	2.5	2.5
Total public investment required	0.8	1.3	1.5	1.5
Baseline public investment	0.8	0.8	0.8	0.8
Additional public investment	–	0.5	0.7	0.7
Maintenance cost (recurrent budget)	0.0	0.3	0.5	0.5
Private investment required	0.0	0.2	0.5	0.5

Source: Bangladesh Delta Plan 2100

Table 4.2 Hotspot wise estimated cost for the Investment Plan (2017–2030)

Hotspot	No of projects	Billion Taka	Billion USD
Coastal zones hotspot	23	884.36	11.14
Barind and draught-prone hotspot	9	163.15	2.06
Haor and flash flood hotspot	6	27.98	0.35
Chattogram hill tracts hotspot	8	59.87	0.75
River system & estuaries hotspot	7	482.61	6.08
Urban areas hotspot	12	671.52	8.46
Cross-cutting areas	15	688.79	8.68
Total	**80**	**2978.27**	**37.53**

Source: Bangladesh Delta Plan 2100

occupied by Indian, Sri Lankan, Chinese, Pakistani and Vietnamese skilled manpower. It is estimated that around 1 million Indian workers are working in the RMG, power and infrastructure sectors of the country remitting out an astonishing figure of about USD 4 billion every year (Dhaka Tribune 2018). This clearly shows that the graduates produced by the country are not matching with the job market requirements and hence are kicked out by foreign workers (The New Age 2019). Together with the Covid-19 experience, these unfortunate scenarios offer a bitter lesson that a country without cutting-edge education, research and innovation cannot cope with the changing environment of the world. A country like Bangladesh may grow with a business-as-usual approach for a short period of time but in order to sustainably grow over a longer period of time it must have effective plan for boosting quality education, cutting-edge research and innovation.

(ii) *Tapping the Potential of the Blue Economy*: Bangladesh has gained its maximum claim of maritime boundary settling its long-standing disputes with Myanmar and India in the international court in 2012 and 2014 respectively (The Hindu 2014; Al Jazeera 2014). After the settlement by the International Court, the country has a total of 118,813 square km of maritime area. In addition, the country has 200 nm exclusive economic belt and 354 nm along the coastal line of Chattogram. All the mineral and non-mineral resources and all sorts of economic activities in this maritime area are sovereign authority of the country according to these settlements. This has offered the country a huge potential for the development of the Bangladesh delta.

However, in order to fully materialize the potential of the maritime boundary, clear vision and a blue economy approach are urgent. Several sectors can be considered for exploiting the potential of the blue economy including: marine fisheries and aquaculture; oil, gas and mineral exploration; marine biotechnology; exploration of renewable energy; production of sea salt; development of marine shipping, trade and transport; marine tourism; marine education and research, etc. With well-planed approach under the blue economy framework, appropriate planning, investment and management could help tapping billions of dollars for the country and provide thousands of jobs. It is important to note that the Bay of Bengal is literally the backbone of the country's economy. Proper planning for the blue economy can also help mitigate climate change issues facing by the country including salinity, tidal surge, loss of lives due to cyclone, river bank erosion and so on.

(iii) *Trans-National Water Sharing*: Bangladesh is a land of rivers. About 700 rivers are flowing over this small land and most of the big rivers are originated from Himalayas. Among the 57 trans-boundary rivers, Bangladesh shares 54 rivers with India and three with Myanmar. The average water flows in Bangladesh are projected at 40,000 cubic metres/second. The three big rivers Ganges, Brahmaputra and Meghna contributes to about 31,000 cubic metre/second. These river flows are critical for the country for irrigation, industrial demand and waterway navigation. However, the rivers are seen to get dry during summer and overflowing water during wet seasons. The upstream countries namely China and India have built dams in rivers of Brahmaputra and Ganges respectively to divert water flows during the dry season which causes serious consequences for the downstream country Bangladesh. The country, therefore, needs strong negotiation with these countries at multilateral approach for the due share of water in the dry season. The country can also think of international arbitration like it sought settling disputes over maritime boundary against Myanmar and India.

(iv) *Food, Water and Air Quality Security*: The Bangladesh Delta Plan has talked about ending hunger and ensuring food security. But it didn't focus on ensuring *food quality security* which is very important for public health, labour productivity and household vulnerability. Bangladesh is currently self-sufficient in the production of chicken, eggs, fish, beef and mutton. However, the study finds that the poultry industry of the country is heavily dependent on processed feed containing high antibiotic, chromium and hormone at a dangerous level. Researchers found that broiler chicken in Bangladesh to have chromium of 20 to 40 times higher than the permissible level. Chromium is a major cause of cancer the scientists often dubbed (BD-Protidin 2018). The eggs produced in Bangladeshi farms also contain 4–8 times higher chromium than the maximum limit. Similar situation prevails for the fishes currently produced by the commercial fisheries of the country. Because of excess presence of antibiotic in the fish, meat and eggs, people are becoming antibiotic-resistant over time which raises serious health concern (Kaler Kantho 2017).

Likewise, the water supplied by the WASA in Dhaka and other major cities are undrinkable. This causes spread of water-borne diseases including cholera, diarrhoea, hepatitis, etc., causing hundreds of death every year in the country and incurring huge financial costs for health care. The air quality in Dhaka is the second worst of the world with the Air Quality Index (AQI) score of 178 (UNB 2020). Dust and smokes from vehicles, brick kilns, factories, and poor urban management lead to such a situation. The dusty poor air quality is causing number of airborne diseases including respiratory malfunctions, lung infections, asthma, nose bleeding, and damages to sense of smell among others. The newborn babies and the senior citizens are the worst hit of such contamination.

(v) *Energy Security*: About 75% of population of Bangladesh has access to electricity with a meagre per capita consumption of only 320kWh compared to 705kWh in South Asia. The country currently has a capacity of 24GW electricity but actual production is about 14GW. By the year 2041, the country aspires to achieve the status of a Developed Country which will need to produce 52GW electricity to meet the growing household, business and industry demand for its tentative 200 million people at that time (Energy for Growth 2019). For this, the country needs to invest by around USD 35 billion by the year 2041 for which private-public partnership and foreign investment are critical. The country is already suffering economically from power shortage and electricity outage. The country is heavily dependent on domestic gas for electricity production—which has already started depleting. Currently, 61% of its electricity comes from this source which indicates that the country will need to find alternative sources including import of LNG, coal-fired power, nuclear power and renewable energy sources. In addition, the transmission loss is estimated at 14% right now and the major grids are already old enough disrupting the power supply. There is huge requirement for an overhaul refurbishment of the national and regional transmission grids. The country can also seriously and actively move forward to build a regional power partnership comprising Bangladesh, India, Nepal, Bhutan and China for developing hydro-power to be produced from the strong current of the Ganges, Brahmaputra and other trans-boundary rivers.

(vi) *Ensuring Credible Government Institutions*: The delta plan has only talked about governance and institutions for water resource management but didn't mention anything about the challenges of overall governance and institutional capacity building. The country aspires to be a developed nation by 2041 but in order for this, building credible institutions, their efficacy and acceptability to the mass people are the prerequisites. Without acceptability of the government machineries no policies work in the long run. It is possible that in the short run one can run a society without proper institutional mechanism but to lead the society to a path of sustainable development we must need sustainable institutions too. Country's recent struggle against Covid-19 raised this question further. The poor institutional capacity was first observed from under-preparedness of the Ministry of Health in combating against the Corona virus despite the outbreak of the virus months earlier in China. The poor governance and poor institutional efficacy were also noticed with the rampant corruption of the relief materials across every corner of the country during the Covid-19 emergency response initiatives. The local government and political institutions appeared so fragile that it forced the prime minister to appoint 64 Secretaries to coordinate relief distribution in 64 districts—a rare backtrack of an elected government first time in country's history since independence (Dhaka Tribune 2020). Mismanagement and sudden closure of many government and private hospitals in providing regular health care services during corona pandemic and resulting sufferings of the masses also pointed out that country might have been able to increase its per capita income three-fold in the past 12 years since 2008 (CEIC 2019), yet the delta is seriously fragile in terms of delivering quality and credible governance for the mass people.

(vii) *National Database*: The country seriously lacks a national database for implementing major national policies and programmes. Currently, the government is trying to register all citizens under national ID card system. This may help counting heads or providing demographic statistics but surely is far away for policy implication. A detail database of all citizens and all households including household-level information are important. Such database can help the government and non-government organizations to identify the vulnerable groups in a given point of time such as during corona pandemic, cyclone or flood; helps in determining the beneficiaries and their requirements; identify genuine farmers for agricultural inputs and subsidies; potential beneficiaries for social security net, etc. The country can learn from its neighbour West Bengal government's *Aadhar Card*, for example, how it is helping effectively screening households in a shortest possible time for any policy intervention based on household occupation, income, vulnerability and adaptability.

4.7 Conclusion

The Bangladesh Delta Plan 2100 aims at achieving a safer, secured, resilient and prosperous country by the end of the ongoing century. The primary goals of the delta plan are to attain food and water security in the long run in this delta along with environmental sustainability and persistent economic growth. The country has been considered as one of the next 11 emerging economies of the world. The economy is transforming from agriculture-based one to manufacturing and services sectors dominated one. It has achieved the lower middle-income country requirements in 2015 and aims at reaching the upper middle-income country status by 2030. However, the country is also ranked as the fifth most vulnerable country of the world. Cyclone, tidal surge, flood, salinity, river erosion and droughts take heavy toll every year in terms of lives and property damages. Saving the coastal belt from future climate change remains a big challenge for the country. Low rains and lower level of groundwater are causing water shortage and droughts damaging agriculture and livelihood of the people badly in the North-Western region of Barind Tract. Ensuring food security for all remains a key policy challenge for Bangladesh for over a long period. This is particularly important as the country sets 2030 as the year to eradicate extreme poverty from the country and to attain the Sustainable Development Goals (SDGs). In this perspective, the Bangladesh Delta Plan 2100 is unique in its nature that comes up with a long-term vision instead of shot-cut ones to shape courses and paths of the nation in every sphere for a period up to the year 2100. However, the delta plan has missed certain issues that are critical for the sustainable development of the delta. For example, the plan has not outlined any idea how the delta can accelerate in research and education in the twenty-first century. The overhaul plan and restructuring of education and research is not only desirable for a knowledge base society but also essential for mere survival in this rapidly changing world. The delta plan can also benefit from focusing on materializing the potential of the maritime boundary by exploiting the potential of the blue economy. The plan can also benefit from proposing a trans-boundary multilateral approach to ensure due share of water in the major rivers flowing over the country in order to ensure water security in the long term. Trans-boundary cooperation is also advocated in this paper to be included in the delta plan, specially in producing hydro-electricity for energy security of the country. The paper further emphasizes on ensuring food quality security, and water and air quality security for public health which is not addressed in the current draft of the delta plan. Ensuring good governance and achieving credibility of the government institutions is emphasized as the sustainable and long-term development strategy of the country. Finally, a national database with detailed household information is advocated to be included in the delta plan as a prerequisite to design and implement appropriate national policies including policy interventions during shocks, natural disasters and pandemic like the Covid-19 outbreak.

References

Ahmed M, Suphachalasai S (2014) Assessing the costs of climate change and adaptation in South Asia. ADB and UK Aid, Manila

Al Jazeera (2014) Bangladesh wins sea border dispute with India. Al Jazeera, 9 July 2014. https://www.aljazeera.com/news/asia/2014/07/bangladesh-wins-sea-border-dispute-with-india-20147881221238354.html

BBS (2022) Bangladesh economic review. Bangladesh Bureau of Statistics, Dhaka

BDP (2017) Bangladesh delta plan 2100, General Economics Division (GED), Bangladesh Planning Commission, Government of the People's Republic of Bangladesh. https://www.bangladeshdelta-plan2100.org/

BD-Protidin (2018) Mach-Murgite Bishakto Pushti. Bangladesh Protidin, 28 March 2018, Dhaka. https://www.bd-pratidin.com/last-page/2018/03/28/317625

CEGIS (2013) Bank erosion prediction of different mighty rivers of Bangladesh, Dhaka

CEGIS (2016) Technical notes on assessment of BDP2100 indicators using GIS data layers. Climate Change and Disaster Management Division, Center for Environmental and Geographic Information Services (CEGIS), Dhaka

CEIC (2019) Bangladesh GDP per capita: 1960:2019. https://www.ceicdata.com/en/indicator/bangladesh/gdp-per-capita

Dhaka Tribune (2018) Illegal foreign workers and tax evasion hurting the Bangladesh economy. Dhaka Tribune, 4 April 2018, Dhaka. https://www.dhakatribune.com/opinion/special/2018/04/04/illegal-foreign-workers-tax-evasion-hurting-bangladesh-economy/

Dhaka Tribune (2020) PM: 64 secretaries to monitor relief operations in 64 districts, 20 April 2020. https://www.dhakatribune.com/bangladesh/government-affairs/2020/04/20/pm-64-secretaries-to-monitor-relief-operations-in-64-districts

Energy For Growth (2019) The future of Bangladesh's power sector, May 8, 2019. https://www.energyforgrowth.org/memo/the-future-of-bangladeshs-power-sector/
GED (2017a) Bangladesh state of the economy by august 2017. General Economics Division (GED), Government of the People's Republic of Bangladesh, Dhaka
GED (2017b) Data gap analysis of Sustainable Development Goals (SDGs): Bangladesh perspective. SDGs Publication #2. General Economics Division (GED), Planning Commission, Government of the People's Republic of Bangladesh, Dhaka, Bangladesh. http://plancomm.gov.bd/wp-content/uploads/2017/01/SDG-DATA-Gap-Final-Draft.pdf
GED (2017c) The Bangladesh delta plan 2100. Draft circulated for expert comments. General Economics Division (GED), Planning Commission, Government of the People's Republic of Bangladesh. http://www.bangladeshdeltaplan2100.org/bangladesh-delta-plan-2100-draft/
Global Climate Risk Index (GCRI) (2010) Who is most vulnerable? Weather-related loss events since 1990 and how copenhagen needs to respond. http://www.germanwatch.org/cri
Goldman Sachs (2019) Beyond the BRICS: Look at the next 11. www.golmansachs.com
Hutton CW, Nicholls RJ, Lázár AN, Chapman A, Schaafsma M, Salehin M (2018) Potential trade-offs between the sustainable development goals in coastal Bangladesh. Sustain Sci 10(4):1008. https://doi.org/10.3390/su10041108
IPCC (2013) Climate change 2013: the physical science basis. In: Working Group I to the fifth assessment report of the intergovernmental panel on climate change. Cambridge
IPCC (2014) Climate change: impacts, adaptation, and vulnerability. Part A: global and sectoral aspects. In Working Group II to the fifth assessment report of the intergovernmental panel on climate change. Cambridge University Press, Cambridge/New York
IPCC (2017) Climate change 2017: impacts, adaptation. Intergovernmental Panel on Climate Change, Cambridge
Johnson FA, Hutton CW (2018) A geospatial analysis of the social, economic and environmental dimensions and drivers of poverty in south-west coastal Bangladesh. In: Nicholls RJ, Hutton CW, Adger WN, Hanson SE, Rahman MM, Salehin M (eds) Ecosystem services for well-being in Deltas: integrated assessment for policy analysis. Springer, Cham, pp 383–403. https://doi.org/10.1007/978-3-319-71093-8_21
Kaler Kantho (2017) Broiler chicken can cause cancer, reduce immune system significantly. Kaler Kantho, September 13, 2017, Dhaka. https://www.kalerkantho.com/online/prescription/2017/09/13/542181
Mainali B, Luukkanen J, Silveira S, Kaivo-oja J (2018) Evaluating synergies and trade-offs among sustainable development goals (SDGs): explorative analyses of development paths in South Asia and sub-Saharan Africa. Sustainability 10:3. https://doi.org/10.3390/su10030815
MoEF (2009) National forest and tree resources assessment (2005–2007). Dhaka
MoEF (2012) Second National Communication of Bangladesh to the UNFCCC. Ministry of Environment and Forest, Government of the People's Republic of Bangladesh
The Daily Star (2020) Govt now scrambles for testing kits, PPE: Experts slam 'failure to act timely'. The Daily Star, March 19, 2020, Dhaka. https://www.thedailystar.net/frontpage/news/govt-now-scrambles-testing-kits-ppe-1882633
The Hindu (2014) Bangladesh wins maritime dispute with India. The Hindu, July 9, 2014. https://www.thehindu.com/news/national/bangladesh-wins-maritime-dispute-with-india/article6191797.ece
The New Age (2019) Univ graduates in Bangladesh struggle in job market for lack of skills. The New Age, 20 October 2019, Dhaka. http://www.newagebd.net/article/88133/university-graduates-struggle-in-job-market-for-lack-of-skills
UNB (2020) Daka's air 2nd worst in the world. United News of Bangladesh, April 14, 2020, Dhaka. http://unb.com.bd/category/Bangladesh/dhakas-air-2nd-worst-in-the-world/49530
World Bank (2018) Bangladesh overview, challenges and strategies. https://www.worldbank.org/en/country/bangladesh/overview
Worldometers (2020) Bangladesh population. Worldometers. https://www.worldometers.info/world-population/bangladesh-population/

Urbanisation in Bangladesh: Geographically Limited and Socially Divided

5

Shahadat Hossain

Abstract

While urbanisation leads to economic growth and development, the benefits are not necessarily distributed evenly. This unevenness is strongly pronounced in Bangladesh where rapid urbanisation accompanies growing and deepening of urban poverty. This paper explains this contextual character of the urbanisation in Bangladesh presenting its geographical limitation and social reach. Absence of nation-wide urban development strategies, neglected rural development, administrative limitations and weakness, anti-poor attitude, and the non-recognised rights and needs of the urban poor are some of the issues discussed here to explain the particular relations of urbanisation and urban poverty in Bangladesh.

Keywords

Urbanisation · Urban poverty · Regional disparity · Development · Growth · Social division

5.1 Introduction

The World Bank's *Bangladesh Poverty Assessment 2019* reads:

> As the country is rapidly urbanizing, its rural and urban areas did not experience the same level of poverty reduction. The rural areas reduced poverty *impressively*, [...] in urban areas, progress has been *slower* and extreme poverty has *not* decreased. [...] About 1 in 4 Bangladeshi still live in poverty [...] there are now more people living in extreme poverty in urban Bangladesh than in 2010. [..] More than half of the population can be considered vulnerable to poverty, as their levels of consumption are close to the poverty threshold [...] The national slowdown in poverty reduction has occurred largely due to an inability of urban Bangladesh to sustain progress (World Bank 2019: v, 11–13 & 26, emphasis added).

The report analyses the poverty situation and changes in the period between 2000 and 2016. This period represents a structural transformation and rapid urbanisation in Bangladesh marked by the fallen share of agriculture in GDP (24–15%), the transfer of a quarter of the agriculture labour to industries and services, and an average 3.9% growth rate of urban population (ibid, p. 48). Despite the structural transformation and rapid urbanisation, poverty is reduced predominately in the *rural* areas (90% of the total reduction) and only slowly and marginally in the urban areas (p. 26). Also, the urban poverty reduction is not only the worst performance but also deeply socially divided – no poverty reduction in the service sector where about 44% of the urban poor are engaged (p. 16&18), increase among self-employed in service-sector (predominately poor households) (p. 26), and slower among female labour force (p. 16). The rural domination in the poverty reduction may be explained by what Banks et al. (2011) termed the perceived 'image and identity' and the obstructive 'ideologies of a rural Bangladesh' that privilege the rural poor as more deserving of support and investment. In reality, urban poverty is far more severe and vulnerable than what usually reported – the poverty line fails to acknowledge the many non-food expenditures of urban living and doing so misrepresents and underestimates the severity of urban poverty in research, policy and actions (ibid). This picture of urban poverty contrasts the century-long belief in the role of cities in enhancing economic growth and its translation into policy formation promoting urbanisation. What worked in a part of the world may not function in other parts and there are local specificity and particularities that are important determinants. What are these specificities and particularities that may help understand urbanisation and its relations with the persisting poverty in Bangladesh? This chapter gives a little reflection to it.

S. Hossain (✉)
University of Duhok, Duhok, Iraq
e-mail: shahadat.hossain@tu-dortmund.de

R. Ahmed et al. (eds.), *Transforming Bangladesh*, World Regional Geography Book Series,
https://doi.org/10.1007/978-3-031-45093-8_5

There are some difficulties that need to be shared from the outset. Urban primacy of Dhaka – the capital and the prime city of Bangladesh – extends to urban research and studies. Limited information (if not an absent) from other urban areas makes it difficult to properly reflect on the diversity and dynamics that one may find in medium and small urban areas. Bangladesh is very slow in publishing census reports and, also that there are changing definition of urban and adjustment in urban areas between censuses (Jones et al. 2016, X) – this means research works/publications need to be based on old or projected data. Needless to say, data production (e.g. project, aggregation, adjustment) is also not free from the interests of the involved parties and the dependency in-between. Neither a current situation nor an actual condition can be delivered based on old and projected data – the picture is rather indicative and an approximation.

5.2 Urban Population Growth in Bangladesh

Bangladesh is historically known as a rural-agrarian country and its level of urbanisation was slow at least until the middle of the twentieth century. Until the partition from India in 1947, Bangladesh experienced the highest urban population at 3.66% of the national population (or about 1.54 million) recorded in 1941 (BBS 2014, xii). After the partition, the immigration of vast Muslim people from India and the emergence and expansion of new centres of trade, commerce, industry and the government offices led to a steady positive growth of urban population (Panday 2020). During the Pakistan period, the East Bengal received very little attention in terms of development, as the resources generated in this part were transferred to finance large-scale industrial development in the West Pakistan (CUS 1976). While the 1961 census recorded 5.19% urban population (about 2.64 million), the two censuses following the independency of Bangladesh reported a much higher number at 8.78% (about 6.27 million) and 15.54% (about 13.54 million) in 1974 and 1981, respectively (BBS 1988). With these counts, the 1960s and 1970s marked the highest rate of urbanisation in the history of Bangladesh.

Substantial increase in the number of urban areas since 1961 (Laskar 1996, 210), some industrial development in 1960s (Rouf and Jahan 2007), and the migration to cities after the liberation of the country (census delayed by 3 years for 1971 war) were the main reasons for the urban population increase in the 1974 census. The drastic increase of the urban population in the later census in 1981 cannot be explained completely by the post-liberation economic penury, the 1974 feminine and the massive flooding along the Brahmaputra (Jamuna) river unless the extended definition of 'urban' introduced in the 1981 census is brought in the scene. Removing the earlier urban definition of an agglomeration of at least 5000 people, presence of public urban utilities and a population of mostly non-agricultural involvements, the 1981 census included areas like sub-districts (Thana/Upazila headquarter), development centres, railway stations, tourist centres and (rural) marketplaces with distinct urban characteristics as urban areas and thus the number of the urban population increased dramatically (BBS 1988). The urban definition was extended (and consistently applied till 2001) for the advantages of the politico-administrative decision-making and happened without any extension or improvement of urban facilities and utilities in those newly included urban areas (Panday 2020, 45) (Table 5.1).

After 1981 Bangladesh experienced a steady average annual rate of urbanisation in the range between 3.04 and 5.19%. Despite this increase in urban population, Bangladesh presents a low level of urbanisation, only slightly more than India: "it was not until 1985 that Bangladesh reached the level of urbanisation reached by South and South Asia in 1950" (Jones et al. 2016, 15). Experiencing an increase of about 35% in the period between 2001 and 2011, the urban population of Bangladesh in 2011 (41.94 million[1]) stands at an adjusted 28% of the total nation population (BBS 2014, 9). Thus, the urban population of 2011 is about 6.7 times increase in 37 years since 1974. On an average about *one million* people added to the urban population *annually* between 2001 and 2011 (BBS 2014, 9). A projection of this number on the rate of urban population growth results into an estimated one-third of the total population of Bangladesh living in urban areas now. A 2016 estimation claims that at least some 52 million population will be added in the urban areas in the next 40 years (Jones et al. 2016, 17).

The rural-urban migration, high natural increase of native urban population, and the territorial expansion and redefinition of urban areas are some major reasons for the growth of urban population (Islam 2001; Islam et al. 2006). The rural-urban migration is dominated by people affected by natural disasters (climate refugee), wage farmers and women. Migration to urban areas is the only option for the millions of families who become landless and unemployed for natural disasters like cyclone in the coastal southern districts and river erosion and flooding especially in the northern districts. In rural Bangladesh, land is mostly hold by families who are not directly involved in agriculture. The inflow of remittance to the rural areas increases the number of such families and thus adds challenges to holding land by those who need it for agriculture (Hossain 2013c). The government policies regarding redistribution of land among landless were not

[1] Jones et al. (2016) discuss the complications in the analysis of the trends in urbanisation and the characteristics of urban and rural population of Bangladesh for reasons of changes in definitions (pp. 11–14, pp. 17–22, etc.). Disputes exists on the percentage of urban population, as, for example, United Nations and the World Bank see 28% level as the underestimation of the actual percentage of population living in urban areas (Jones et al. 2016, 14).

Table 5.1 National and urban population growth, 1901–2011

Year	Total national population (million)	Annual growth rate (exponential)	Total urban population	Percentage urban (%)	Inter-census variation	Inter-census variation (%)	Annual growth rate (%)
1901	28.93	1.59	702,035	2.43			
1911	31.56	0.87	807,024	2.55	104,989	14.95	1.4
1921	33.25	0.52	878,480	2.64	71,456	8.85	0.85
1931	35.60	0.68	1,073,489	3.66	195,009	22.20	2.02
1941	42.00	1.65	1,537,244	3.66	463,755	43.20	3.66
1951	42.06	0.02	1,819,773	4.33	282,529	18.38	1.70
1961	50.84	1.90	2,640,726	5.19	820,953	45.11	3.79
1974	71.48	3.41	6,273,602	8.78	3,632,876	137.57	6.88
1981	87.12	1.98	13,535,963	15.54	7,262,361	115.76	11.61
1991	106.31	1.99	22,455,174	20.15	8,919,211	65.89	5.19
2001	124.33	1.57	31,077,952	23.81	8,622,778	38.40	3.30
2011	144.04	1.47	41,943,532	28.0	10,865,580	34.96	3.04

Sources: Total national population & annual growth rate (exponential) from BBS (2015, 85); Total urban population & percentage urban from BBS (2014, xii & 9); others are calculated by the author

successful in making any changes in the landholding and in resolving rural poverty (ibid, 371). With expansion of the families and consequently the inability of the land to meet family-needs despite modernisation of the agriculture, increasing number of families see no alternatives but to search for livelihoods opportunities in the cities. Also in the recent decades, agriculture does not promise any stable and adequate income. Huge inflow of imported agricultural products, especially from the neighbouring India, has added extra competition for the local farmers and made some agricultural production non-profitable (or a net loss). Farmers in the remote districts of the northern Bangladesh are additionally challenged by adverse climate and the fact that the price of agricultural products often fall too low that harvesting does not make any more sense. Farmers of the districts in the periphery of the Megacity Dhaka are challenged with the difficulties to keep landholding in the face of the increasing trend of land leasing and forced land selling at whatever amount the powerful lessees/buyers pay (Feldman and Geisler 2012; Haque 2012; Arefin 2018).

Landholding is particularly difficult for the women for the wide and unquestioned application of discriminatory inheritance laws and social and customary practices.[2] The practice that such rights are 'not claimed by the women but offered to them' creates a situation of mercy, uncertainty, and delays. A large number of garment factories located mostly in and around Dhaka have created employment opportunities for many women alongside other options like informal domestic service, street-food vending, prostitution, and begging (Ullah 2003; Ward et al. 2004; Haque 2015). Informal sectors like transportation (primarily rickshaws), construction workers, street vending, shop attendants, security guards, private cars driving, begging, etc. are primary involvements of majority of the male migrants who lack necessary skills, training, network/connections and support-money (bribes) to get jobs in the formal sectors (Siddiqui et al. 2010; Begum and Sen 2005). Contrasting the severe and widespread poverty in the major cities, as described later, the rural-urban migration is fastened by the 'myths' that the returnees describe exaggerating a portray of the 'city as a place of opportunities', and the assistance that a new migrant receives from the earlier migrants in securing a foothold in the city (Siddiqui et al. 2010; Kuhn 2003). Overly presenting Dhaka City and its cultural life in the films, *natok* and entertainment industry has also created additional attraction of urban life to the rural folks. The recent years also mark a very drastic increase of temporary migrants, especially rickshaw drivers, who come to city for very short period, stay in rickshaw garage organised overnight stays and leave the city after a certain amount of cash income is earned. These cash earnings are spent mainly to pay children's private tuition fees and private healthcare-related expenditure.[3] The number of such temporary migrants is increased in the time of the year when there is no cultivation or harvesting-related works.

5.3 Unbalanced Urbanisation: Geographical Restriction

The level of urbanisation varies widely between districts, from as low as 8.85% in Gaibandha district (Ranjpur division) to as high as 77.36% in Dhaka district (BBS 2014,

[2]As practiced in Bangladesh, a daughter gets half of the father's property than what a son receives, a woman receives an eighth of the husband's property and a sixteenth of the son's property in the case the son predeceases the mother.

[3]Private tuition and private medical consultancy have become a common practice in Bangladesh. Despite teachers and doctors are blamed for not being sincere in the regular service, it may be misleading to grossly make them responsible for this practice. Parents and patients seem to have higher level of satisfaction when services are offered privately (despite additional costs).

19–20). In 2011, in 44 of the 64 districts less than 20% of the total district population lived in urban areas against the national average of 28% (ibid). Dhaka Megacity alone contains one-third of the national urban population.[4] The very high level of urbanisation in the district of Dhaka can be explained by its long presence in the political and economic life, historically. That Dhaka and Chittagong together contain more than two-fifths of the total national urban population indicates a two-city primacy in the urbanisation in Bangladesh. Contrasting the 35% increase of the national urban population between 2001 and 2011, the third largest city, Khulna, reported a 10.8% decline of its urban population.

Now, the country is clearly divided into two distinct regions separated by the Jamuna-Padma-Meghna River – the eastern and western regions.[5] The eastern region is far more urbanised compared to the western region, in both percentage and absolute number. In 2011, the percentage of people living in urban areas of the eastern region is simply double compared to that in the western region[6]; more than two-thirds of the total urban population of the country live in the eastern region (BBS 2014, 21). Jones et al. (2016) summarise the eastern and western differences:

> The sharp difference in growth between the two largest (and Eastern) cities – Dhaka and Chittagong – and decline in the next two largest (and Western) cities – Khulna and Rajshahi – serves to underline the apparent gulf between urbanization trends in Eastern and Western wings of Bangladesh (p. 21).

In the eastern region, most urban growth in the last decades happened in the Dhaka Megacity, in Chittagong SMA and along the road and rail line that connect the two cities. While Dhaka and Chittagong were connected by rail since 1937, the travel time between them reduced significantly on roadway especially after the opening of the Meghna bridge in 1991. The transportation (rail and road) links facilitated development of the Dhaka-Chittagong growth corridor creating incentives to locate industries along this corridor (e.g., Comilla and Feni). This is the longest corridor served much better with infrastructures necessary for industrial development than other parts of the country. Seven of the sixteen districts reported a growth rate of at least 15% and all the ten most densely districts in Bangladesh are located along this corridor (Jones et al. 2016, 72). Two other districts have shown rapid urbanisation outside of the corridor: the large amount of remittance speeded up urban growth in Sylhet district, despite low level of industrial development; the massive touristic development geared up rapid urban growth of the beach city and the most attractive touristic destination of Bangladesh, Cox's Bazaar district. Of the total seven divisions of Bangladesh, about half (46.44%) and a fifth (20.57%) of the urban population live in Dhaka division and Chittagong division, respectively (BBS 2014, 21). While the Export Processing Zone (EPZ) is usually seen as fastening the urban growth, this is not the case for all regions. Despite having an EPZ, Bagerhat district in Khulna Division reported losing urban population in the period 2001–2011.

The western region shows very slow growth of urban population other than Bogra and Barisal districts where, however, the increase of urban population is a result of a significant increase of urban area in the 2011 census[7] (BBS 2014). Despite Rajshahi and Rangpur divisions are better connection with Dhaka Division since the opening of the Bangabandhu Bridge in 1998, urban growth in these regions is stagnant for reasons like no industrial development, limited coverage of gas supply, unstable power supply, poor road and rail infrastructure, no functioning airport, remoteness to seaports, and limited educational and health facilities. Despite Barisal division is better connected with the capital city compared to Khulna division thanks to the river-based transport, low industrial development accompanied by frequent environmental-based disasters hinder the level of urbanisation in this coastal division. Khulna division have had a better road-based connection with the capital very recently after the opening of the Padma River in mid-2022. Other factors for the decline of urban population of Khulna city (SMA) include the closure of jute processing industry, which was the major employer in the division, unstable power supply, salinity in water, lack of critical infrastructure like an airport and poor educational and healthcare facilities[8] (Jones et al. 2016, 21–22). The divisions of the eastern region remain a major supply of agricultural and fishery products and labourers for the Dhaka Megacity and other parts of the country.

Six of the seventeen cities that reported a population above 200 thousand in 2011 are in the Dhaka mega-urban region and three of them reached to this population level

[4]In 2011, the total national urban population was 41,943,532, of which 14,171,567 in Dhaka Megacity (the largest metropolitan and mega city), 3,724,433 in Chittagong (second largest metropolitan city), and 1,046,341 in Khulna (third largest metropolitan (BBS 2014, 10–11).

[5]Eastern region: Dhaka, Chittagong, Sylhet and now Mymensingh divisions; Western region: Khulna, Barisal, Rajshahi and Rangpur divisions.

[6]In 2011, 34% population of the eastern region live in urban areas compared to 17% in the eastern region. A similar difference in the level of urbanisation was reported in 2001–29.8% for the eastern region compared to 15.8 percentage for the wester region (Jones et al. 2016, 17). The calculation needs to be read with the foot note 7 description on page 24.

[7]Between 2001 and 2011, the increases of urban population in Bogra district and Barisal district are 72.31% and 31.54% against the increase of urban areas from 80.28 sq. km. to 185.14 sq. km, and from 134.88 sq. km. to 172.10 sq. km, respectively (BBS 2014, 14–15, 19–20).

[8]Before the partition in 1947, Khulna was closely connected with the nearby city, Kolkata, and the regional transport infrastructures were developed accordingly. Only after the partition it needed to establish a connection with Dhaka.

doubling the population reported in 2001 (Saver's population increased 2.32 times). In 2011, the 14.17 million population of Dhaka Megacity comprised more than one-third of the total urban population of the country and three times of the combined population of the next three largest metropolitan cities (Chittagong, Khulna and Rajshahi) (BBS 2014, 10–11). Following the policy concerning the relocation of garment factories and leather tanneries from the core city of Dhaka, the areas outside of the DCC have already started receiving population at much higher rate. With the population of about 20.3 million (WPR 2020) and the average population density of 44,500 people per sq. km (Wright 2020), Dhaka megacity is now the seventh largest, the fastest growing and the most densely populated city in the world. Adding an estimated 0.7 million people every year until 2030 (United Nations 2019, 79), Dhaka is expected to accommodate more than 28 million population – to be the fourth largest city in the world in 2030 (United Nations 2018, 4; 2019, 77). The second largest city, Chittagong, has a much smaller population size (3.724 million in 2011) compared to Dhaka Megacity (BBS 2014, 11). For reasons of boundary change in the 2011 census, it may be misleading to bring Chittagong's urban population in regional and national comparison.

The imbalance and the regional disparity in the level of urbanisation is, however, linked with the absence of nationwide urban development strategies and an integrated urban and regional development policy. In 1965, the Urban Development Directorate (UDD) was formed to address a nationwide urban and regional planning. While its earlier master plans, mostly for areas outside Dhaka, lost context after the independence, its proposed draft Urban and Regional Planning Act is now awaiting legislation following its approval by the cabinet, after years of delay, in the beginning of 2017 (as the Urban and Regional Planning Act 2017). Such a legal planning document provides a framework and a common vision, and its absence may be blamed for the haphazard urban developments with individual social and economic developments not oriented to a common urban future. The absence of the necessary resources, capacities, interdepartmental coordination, a decentralised authority, and a functional local government system may jeopardise the implementation and thus the success of the Act.

Other than a few large cities, most urban areas are limited in their planning capacity in terms of expertise and resources necessary for planning and plan implementation. Additionally, the local government of these urban areas are challenged by poor governance, limited political authority, power contestation between local and national government representatives, poor tax base and accordingly financial dependency on national government, ambiguity in the flow of resources, imposition of priorities and needs from the top, systematically downplaying the local knowledge, etc. (Aminuzzaman 2010). In cooperation with the local government department, several international organisations are offering training to urban authorities on issues related to governance and democracy, climate change, infrastructure development, etc. How far these external driven efforts change the skills and capacity of the local authorities, given the centralised political and decision-making system, needs independent investigation. The capacity of the local urban government is also limited by its subordination to the central administration and its power and responsibilities defined in general and vague terms (Rahman 2002; Rahman and Robinson 2006). Decentralisation of authorities has not been surfaced, despite the need has been repeatedly highlighted in the 5 Years Plans and the Poverty Reduction Strategy Paper. There are master plans for Dhaka, Chittagong, Khulna and Rajshahi city developed by the responsible development authorities, but their implementation is challenged by spatial and institutional disintegration, the steering of decisions by powerful interests including the real estates, and the violative practices even by the public authorities (Hossain and Hackenbroch 2019). Also here, absence of an integrated nation-wise urban and regional planning guideline means that piecemeal planning interventions do not bring intended result, but rather intensify the problem.

Roughly speaking, it is not the urban and regional planning, but the infrastructural development, especially the road-network to Dhaka that shapes the pattern of an unbalanced urbanisation in Bangladesh. Urbanisation as such increases the urban population but without creating sufficient economic opportunities and ensuring the infrastructures and amenities necessary for a decent urban life. Uddin (2016, 214) gives an estimation of the very poor status of major urban infrastructures and facilities in the urban areas of Bangladesh. The situation looks much worse when the heterogeneity in access between groups is added on the picture. The absence of a national and regional planning and the practice of fragmented thinking (i.e., absence of looking the whole and the interdependency between parts) lead to the fact that benefits of both the combined and multiplier effects and the potentiality of the regional advantages remain unexplored. The regional disparity in the level of urbanisation is a reason of the fact that the increase of gross domestic product (GDP) is the central consideration in the decision about investment, e.g., in industrial growth, infrastructural improvement and service provisions. That these decisions rarely consult non-economic factors means a severe damage to the ecology, environment, and the society. The scope of this paper does not allow any elaboration on all different non-economic dimensions; therefore, the discussion in this paper is limited to the particular relationships between the present form of urbanisation and some social issues.

5.4 Urbanisation: Deepening Social Division

Let us now return to deepening urban poverty reported in the World Bank's document. Bangladesh's slow performance in the urban poverty reduction and the high presence of extreme poverty in its urban areas necessitate a closer analysis of the relationship between urbanisation and poverty in Bangladesh. Given that Dhaka with its political, institutional and administrative domination and privilege in research and practice dictates the urban development practice in the country, this analysis along with the challenges for other urban areas as already presented may suffice to portray a picture of the relation between urbanisation and poverty[9] in the country. I acknowledge the fact that scale not only multiplies the problem, but also complicates it in all dimensions.

In Bangladesh (then East Pakistan), the need for a guided urban development was recognised already in the 1950s by the establishment of the development authorities for Dhaka, Chittagong, Khulna and Rajshahi and then the several master plans[10] prepared to guide their development. Implementation of the plans was however challenged by violative practices (also by the public authorities), enforcement difficulties, corrupt bureaucratic practices, and the powerful real estate interests (Hossain and Hackenbroch 2019). There are mushrooming of speculative land development (Haque 2012) and housing projects (by both public authority and private real estate companies) in the urban periphery implemented violating the structure plan and environmental regulations and thus causing huge environmental challenges (Hossain and Hackenbroch 2019; Hoque 2004; Alam 2016). Hossain (2013c) summarises the planning practices in Dhaka:

> The urban renewal and redevelopment affecting Dhaka like many other cities favours a very small portion of urban population. The city in short span of time has transformed, the landscape now dominated by new developments and an array of real-estate advertisements offering lucrative land and housing deals dot the city. The majority of these development projects are of high-end real-estate for the powerful and affluent classes; with the exception of a few housing projects where access was determined on having connections with political parties. (p. 374)

Already in 2005, more than 35% of the total urban population of the major cities of Bangladesh lived in informal settlements representing the massive unplanned development in major cities (Islam et al. 2006). It is difficult to give a realistic count of the poor living in urban areas of Bangladesh. The application of different definitions of slum leads to very different number of people living in urban slums of Bangladesh, ranging from the lower end of 2.23 million people reported by the Government of Bangladesh to 16.73 million people calculated based on the definition of the Sustainable Development Goal (SDG). Following the criteria used by the Millennium Development Goal, between 45% and 70% households living in the city corporation of the seven divisions are deprived of at least one of the five criteria namely, water, sanitation, living space, building structure and tenure security (ibid, 95). The percentage increases to between 71 and 99% when the definition of the SDG is applied. Patel et al. explain the persisting urban poverty with the underestimation by the government: "Such underestimation by country governments could mean a significantly lower allocation of budgetary resources for slums and adjacent communities [...] Such inadequate targeting is one of the main reasons that slum policies fail to make their expected impacts" (ibid, p. 82).

Official planning in Bangladesh does not recognise the needs and priorities of the urban poor. The implementation of the official plans only adds sufferings and hardships resulting in the process of making place for the 'desired' development (e.g., through eviction of information settlements). The preparation of the master plans is guided by the ideology of order and beautification which necessitates getting rid of everything that does not fit in the ideological image. Hossain (2013c) describes how the ideology of beautification redefines the urban space in Dhaka evicting the poor from the urban centre and 'dumping' them in the urban periphery. That the recent policy intervention about the relocation of the production and garment factories in the periphery of Dhaka keeps complete blind eyes about the necessities to create housing and utilities provision in the peripheral locations for the factory works (the urban poor) is an exemplary case. The peripheral settlements are developed quickly by private interests in an unplanned and haphazard way responding to the growing demands for housing and thus multiplying the persisting severe environmental problems. Banks et al. (2011) explain the growing urban poverty with the poverty reduction policies that are informed by the perception that equates urban poor with issue of crime and squalor, the belief that investment in the urban poor results in further migration and the narratives that identify the poor with the rural areas.

In the major cities of Bangladesh, the citizenship of the urban poor is not guaranteed officially but negotiated within an informal governance created locally by the local supporters of the ruling political party, lower level officials of the public departments, supporters of the national and local

[9] Poverty is a very broad topic and comprises various dimensions. There is no intention to elaborate all many dimensions and perspectives; the scope of this paper also does not allow it.

[10] It may be noted that two earlier master plans for Dhaka – one by the Civil Surgeon of Dhaka, Henry Charles Cutcliffe, in 1868 and the other by town planner Patrick Geddes in 1917 — were not implemented arguably for reasons of funding shortage in the case of the first one and for not taking the other one seriously. Geddes (1917) and Polin and Alam (2020) for further details.

political, and influential local residents (Hossain 2013b; Banks et al. 2011; Suykens 2015). This informal governance operated within clientelist relations now negotiates access to land (Suykens 2015), public space (Hackenbroch and Hossain 2012), water supply (Hossain 2012, 2013a; Hackenbroch and Hossain 2012), electricity (Hossain 2012), street food vending (Etzold 2013), etc. for the urban poor who do not have the means to claim citizenship officially. Since 1994 when the urban poor were granted voting rights, the informal settlements become an important place for the politicians to mobilise votes. Given the large size of urban electorates – up to 100,000 voters per ward in Dhaka (Banks et al. 2011) and much more for the local members of the parliament (MP) – and the challenges for the local elected representatives as already discussed, the political leaders have no choice but to outsource the mobilisation activities to local supporters in return for privileges and protections. Hossain (2013a, b), Suykens (2015) and Banks (2012) give a detailed account of the externally connected local leaders, the hierarchical network structure and the dependent structure that govern the urban life in the informal settlements in urban Bangladesh. While the informal governance allows options to access urban utilities, such options come at higher financial costs, long-term dependency and reinforcement of political and economic inequality. For the poor inhabitants, high cost for housing and utilities leads to constant struggle to earn enough to pay for healthcare, education, and food expenditures. Varies between uses and connection types, water price in the informal settlement, for example, is up to 30 times higher than the official price (Hossain 2013a, b, c).

Despite hugely diversifying household labours, at least two-thirds of the urban poor do not have enough income to pay for the routine expenditure (Banks 2012). Following the adaptation of the neoliberal policy since the late 1980s and 1990s (Nuruzzaman 2004), the partial withdrawal of the state and subsequently the privatisation of the major part in the healthcare sector and the mushrooming of private educational institutions have worsened the situation even further (Rahman 2019). The shockingly poor education and public health in the informal settlements of the metropolitan cities of Bangladesh (Quattri and Watkins 2019; Adams et al. 2015; Cameron 2010) is a result of the diversification of household labour (children works instead of going to school) and the shortage of money for needs other than housing and utilities. There is a high incidence of working children in the informal settlements and the garment factories are the major employers of female child labour (Quattri and Watkins 2019). The 2.6 times more under-five mortality in urban slums compared to the urban non-slums (NIPORT et al. 2008, 339) presents the poor healthcare services offered to the slum inhabitants primarily by pharmacies, non-formal/traditional 'doctors' and the limited service of a few NGOs (Adams et al. 2015). The poor people's dependency on this informal governance and the absence of any alternative makes it difficult for them to counter the system (Hossain 2013a), nor is it possible for them to use the system for any long-term improvement (Banks 2012).

The Global Liveability Index 2019 has placed Dhaka – the city that alone accommodates one-third of the total national urban population – as one of the least liveable cities after Damascus and Lagos indicating the city's very poor combined performance in stability, healthcare, culture & environment, education, and infrastructure (EIU 2019). The city authorities do not have the capacity to guide development and manage the diverse urban needs of the growing population. The challenges to coordinate the activities of some 19 ministries and 40 government organisations[11] that are directly involved in the development and administration of the city (Islam et al. 2000) is now joined by additional burden about the coordination of activities between the two newly created city corporations (i.e., Dhaka North City Corporation, DNCC and Dhaka South City Corporation, DSCC). The recurring outcry for a metropolitan government with necessary power for decision-making and coordination does not receive any political attention. The power of the mayor and the elected local government representatives (ward commissioners) are limited by, among others, the overlapping of institutional responsibilities, independent and separate operation of the departments responsible for urban development (RAJUK) and utilities (DWASA, DESCO, TITAS), the domination of the central government ministries, and the irregularity and delay in holding local government election supports.

Since the start of the COVID-19 pandemic in the country in the early 2020, the urban areas, especially Dhaka, are going to be different. While the government has promised seven stimulation packages, other than to the garment sector, these remained largely on paper. Even the government's stimulation package of $13 billion to the RMG sector (Antara and Syed 2020, 2) could not change the situation that more than a million garment workers have already been fired or furloughed due to cancellations of order and the buyers' failure to pay for the cancellations (Anner 2020). About three in every four furloughed workers were sent home without paying their severance (ibid) and many are still fighting to get the unpaid loan (Perera 2020). In terms of economy, the hardest hit is the self-employed service sector, informal economic sectors and those working outside any legal protection – 21.7% *new* poor have been added in the society (Rahman 2020). About 16% of the poor and low-income people have left the capital losing employment or for uncertain income and unrelenting expenditure burdens (ibid). That

[11]The number of government department directly involve in the urban affairs of Dhaka may be increased by now – Uddin (2016) reported 51 such government organisations.

the government's stimulation packages are GDP focused and targeted to support those having power to steer the government decision means that such packages will not be available to benefit the poor. What will be future of the poor and low-income group and how COVID-19 is going to change the pattern of urbanisation are yet to be seen.

5.5 Conclusion

Despite the few urban settlements whose origin can be traced as early as in the ancient time, Bangladesh was predominantly a rural-agrarian country at least until the participation of India and Pakistan in 1947. The increase of urban population happened primarily after the independency of the country in 1971 and without necessary institutional development and policy formation. The increase is a combination of the natural growth and the rural-urban migration of the surplus agriculture labour and the climate refugees. While the urban growth can be explained by the absence of major rural development, modernisation of agriculture (surplus agriculture labour) and difficulties for the farmers to access agricultural, it is also a result of the territorial expansion of urban areas and the redefinition of 'urban' in census reports. For no nation-wide urban and regional policy till date, the growth of urban population did not happen evenly in the urban areas of the country, but concentrated in the eastern districts that are well connected to Dhaka or Chittagong, or located in the Dhaka-Chittagong growth corridor. The city authorities do not have the capacity to guide growth of the city and to offer a minimum level of municipal services to the increasing population. The urban expansion is haphazard and unregulated and happens severely damaging the environment. While the non-poor migrants start urban life further densifying the inner city, the poor migrants (the majority) find their place in the informal settlements in the city and increasingly in the peripheral locations. Prevailing anti-urban-poor policy attitudes limit investment and efforts to improving the living conditions of the urban poor. The urban poor continue depending on local leaders and political connections (informal governance) to access land, housing, and utilities unofficially, however accepting high material costs, dependency, deepening inequality and sacrificing any possibilities for long-term improvement. Improvement in policy and institutions is necessary for a guided urbanisation and thus to realise the benefits of urbanisation. Such urbanisation however may not reduce the social division unless a change in the prevailing anti-poor ideology guides policy implementation. The recent COVID-19 pandemic is going to write a different chapter in the urbanisation of Bangladesh, as evident by increasing unemployment of a segment of people and a massive out-migration from the capital. The coming years will give us more details about the changes.

References

Adams AM, Islam R, Ahmen T (2015) Who serves the urban poor? A geospatial and descriptive analysis of health services in slums in Dhaka, Bangladesh. Health Policy Plan 30:i32–i45

Alam MJ (2016) Peripheral land conversion for land and housing development projects and its impacts upon the affected people: a study in greater Dhaka. Int J Sustain Soc 8(1):73–89

Aminuzzaman SM (2010) Local government and development in Bangladesh. Lessons learned and challenges for improving service delivery of Union Parishad (UP). Local Government Support Project – Learning and Innovation Component (LGSP-LIC), Local Government Division, Government of Bangladesh

Anner M (2020) Abandoned? The impact of Covid-19 on workers and businesses at the Botton of global garment supply chains. Research report. PennState Centre for Global Workers' Rights. https://www.workersrights.org/wp-content/uploads/2020/03/Abandoned-Penn-State-WRC-Report-March-27-2020.pdf. Accessed 9 Oct 2020

Antara IJ, Syed SK (2020) COVID 19 crisis impact on RMG wages and employment and role of trade unions media tracking report. Series: Gender 10, 24 August 2020. BRAC Institute of Governance & Development, BRAC University, Dhaka

Arefin S (2018) The real estate and capital accumulation: an inquiry into Harvey's model in Dhaka. Peoples' Preface J Res Initiatives 2(2):153–165

Banks N (2012) Urban poverty in Bangladesh: causes, consequences and coping strategies. BWPI working paper no. 178. University of Manchester, Manchester

Banks N, Roy M, Hulme D (2011) Neglecting the urban poor in Bangladesh: research, policy and action in the context of climate change. Environ Urban 23(2):487–502

BBS (1988) Bangladesh Population Census 1981, National Series, Report on Urban Area. Bangladesh Bureau of Statistics, Dhaka

BBS (2014) Bangladesh population and housing census 2011. National report, vol. 3, Urban area report. Bangladesh Bureau of Statistics, Dhaka

BBS (2015) Population distribution and internal migration in Bangladesh, Population monographs: volume 6. Bangladesh Bureau of Statistics, Dhaka

Begum S, Sen B (2005) Pulling rickshaws in the city of Dhaka: a way out of poverty? Environ Urban 17(2):11–25

Cameron S (2010) Access to and exclusion from primary education in slums of Dhaka, Bangladesh. Research monography no. 45, Centre for International Education, University of Sussex

Centre for Urban Studies (CUS) (1976) Squatters in Bangladesh cities: a survey of urban squatters in Dhaka, Chittagong and Khulna – 1974. University of Dhaka, Dhaka

Economist Intelligence Unit (EIU) (2019) The global liveability index 2019. The Economist Intelligence Unit, London

Etzold B (2013) The politics of street food. Contested governance and vulnerabilities in Dhaka's field of street vending. Franz Steiner, Stuttgart

Feldman S, Geisler C (2012) Land expropriation and displacement in Bangladesh. J Peasant Stud 39(3–4):971–993

Geddes P (1917) Report on town planning, Dacca. Bengal Secretariat Book Depot, Calcutta

Hackenbroch K, Hossain S (2012) "The organised encroachment of the powerful" – everyday practices of public space and water supply in Dhaka, Bangladesh. Plan Theory Pract 13(3):397–420

Haque KNH (2012) The political economy of urban space: land and real estate in Dhaka city. IGS working paper series no. 03, July. Institute of Government Studies, BRAC University, Dhaka

Haque R (2015) Voices from the edge: justice, agency and the plight of floating sex workers in Dhaka, Bangladesh. LIT, Wien

Hoque MJ (2004) Impact of private land development on the environment of the eastern fringe area of Dhaka. PhD thesis submitted at the Department of Urban and Regional Planning, Bangladesh University of Engineering & Technology, Dhaka

Hossain S (2012) The production of space in the negotiation of water and electricity supply in a bosti in Dhaka. Habitat Int 36(1):68–77

Hossain S (2013a) Contested water supply – claim making and the politics of regulation in Dhaka, Bangladesh. Steiner, Stuttgart

Hossain S (2013b) The informal practice of appropriation and social control: experience from a Bosti in Dhaka. Environ Urban 25(1):209–224

Hossain S (2013c) Migration, urbanization ad poverty in Dhaka, Bangladesh. J Asiatic Soc Bangladesh (Hum) 58(2):369–382

Hossain S, Hackenbroch K (2019) Whose interest finally count? The statutory production of urban space at the fringes of Dhaka, Bangladesh. Plan Theory 18(2):167–190

Islam N (ed) (2001) Urbanization, urban planning and development and urban governance: a reader for students. Centre for Urban Studies, Dhaka

Islam N, Khan MM, Nazem NI, Rahman MH (2000) Reforming governance in Dhaka, Bangladesh. In: Islam N (ed) Urban governance in Asia: sub-regional and city perspectives. CUS and Pathak Shamabesh, Dhaka, pp 135–162

Islam N, Mahbub AQM, Nazem NI, Angeles G, Lance P (2006) Slums of urban Bangladesh: mapping and census, 2005. Centre for Urban Studies, Measure Evaluation, National Institute of Population Research and Training

Jones G, Mahbub AQM, Haq I (2016) Urbanization and migration in Bangladesh. United Nations Population Fund, Dhaka

Kuhn R (2003) Identities in motion: social exchange networks and rural-urban migration in Bangladesh. Contrib Indian Sociol 37(1–2):311–337

Laskar SI (1996) Urbanization in Bangladesh: some contemporary observation. Bangladesh Dev Stud XXIV(1–2):207–216

National Institute of Population Research and Training (NIPORT), MEASURE Evaluation, International Centre for Diarrhoeal Disease Research, Bangladesh (ICDDR,B), and Associates for Community and Population Research (ACPR) (2008) 2006 Bangladesh urban health survey. NIPORT, MEASURE Evaluation, ICDDR, B, and ACPR, Dhaka/Chapel Hill

Nuruzzaman M (2004) Neoliberal economic reforms, the rich and the poor in Bangladesh. J Contemp Asia 34(1):33–54

Panday PK (2020) The face of urbanization and urban poverty in Bangladesh. Palgrave Macmillan, Singapore

Perera W (2020) Garment workers protest in Bangladesh as COVID-19 spreads. World Socialist Wesite, 15 April 2020. https://www.wsws.org/en/articles/2020/04/15/bang-a15.html. Accessed 9 Oct 2020

Polin F, Alam D (2020) The first town plan of Dhaka (Part 1). ICE Business Times, 10 March 2020. https://ibtbd.net/the-first-town-plan-of-dhaka-part-1/#:~:text=Henry%20Charles%20Cutcliffe%2C%20who%20was,region%20and%20organic%20and%20historic. Accessed 25 Sept 2020

Quattri M, Watkins K (2019) Child labour and education – a survey of slum settlements in Dhaka (Bangladesh). World Dev Perspect 13:50–66

Rahman HZ (2002) Re-thinking local governance: towards a livelihoods focus. In: Toufique KA, Turton C (eds) Hands not land: how livelihoods are changing in rural Bangladesh. Bangladesh Institute of Development Studies, Dhaka, pp 113–119

Rahman R (2019) The privatisation of healthcare system in Bangladesh. Int J Health Care Qual Assur 32(2):97–107

Rahman HZ (2020) 6 months of Covid in Bangladesh: Looking back and forward. The Business Standard, 07 September 2020. https://tbsnews.net/analysis/6-months-covid-bangladesh-looking-back-and-forward-129769. Accessed 10 Oct 2020

Rahman HZ, Robinson M (2006) Governance and state effectiveness in Asia. IDS Bull 37(3):130–149

Rouf MA, Jahan S (2007) Spatial and temporal patters of urbanization in Bangladesh. In: Jahan S, Maniruzzaman KM (eds) Urbanization in Bangladesh: patterns, issues and approaches to planning. Bangladesh Institute of Planners, Dhaka, pp 1–24

Siddiqui K, Ahmed J, Siddique K, Huq S, Hossain A, Nazimud-Doula S, Rezawana N (2010) Social formation in Dhaka, 1985–2005. A longitudinal study of society in a third world megacity. Routledge, London

Suykens B (2015) The land that disappeared: forceful occupation, disputes and the negotiation of landlord power in a Bangladeshi Bastee. Dev Chang 46(3):486–507

Uddin AKMR (2016) Urbanization. In: Riaz A, Rahman MS (eds) Routledge handbook of contemporary Bangladesh. Routledge, New York, pp 207–219

Ullah AKMA (2003) Prostitution and AIDS in Bangladesh: is the ignorance endangering us? Asian J Women's Stud 9(2):95–114

United Nations (2018) The world's cities in 2018 – data booklet (ST/ESA/ SER.A/417). Department of Economic and Social Affairs, Population Division, United Nations

United Nations (2019) World urbanization prospects. The 2018 revision (ST/ESA/SER.A/420). Department of Economic and Social Affairs, Population Division, United Nations

Ward K, Rahman F, Islam AKM, Akhter R, Kamal N (2004) The effects of global economic restructuring on urban women's work and income-generating strategies in Dhaka, Bangladesh. Crit Sociol 30(1):63–102

World Bank (2019) Bangladesh poverty assessment: facing old and new frontiers in poverty reduction. World Bank, Washington, DC

World Population Review (WPR) (2020) World city population 2020. https://worldpopulationreview.com/world-cities. Accessed 26 Sept 2020

Wright S (2020) The world's most densely populated cities. World Atlas, published on 31 August 2020. https://www.worldatlas.com/articles/the-world-s-most-densely-populated-cities.html. Accessed on 26 Sept 2020

Shrimp Farming and the Question of Land and Social Environment

6

Craig A. Meisner and Saleh Ahmed

Abstract

Shrimp cultivation in Bangladesh, while having a historical perspective as a farmer-driven experiment, has become much more of a capital-intensive cultivation system in Bangladesh during its 60-year history. Its value chain is now driven by the export markets that the processors have tapped worldwide. Currently, there are approximately 150,000 farms producing shrimp and prawns in Bangladesh, employing over 600,000 workers. Most of the shrimp and crab farming ponds present in the country are located in the south-west of the country, in the districts of Satkhira (75,867 ha), Bagerhat (72,684 ha), Khulna (58,367 ha) and Jessore (11,234 ha) due to polderization of these districts in the 1970s. Of course, before polderization, the silt carried upstream by the tide was deposited on the plains. However, preventing the inflow of silt to the plains eventually led to deposition in the river beds, causing them to rise rapidly. As a result, saline water began to flow over the high embankments, submerging the low-lying agricultural lands within the polders, blocking the sluice gates with silt, and creating areas of saline water-logged land. Thus, growers experimented with shrimp due to their salinity. While worldwide transformations of forestlands into shrimp farms are considered one of the primary reasons for declining mangrove forests, in Bangladesh, the salinity intrusion into rice fields drove the industry. This chapter highlights the history of the shrimp industry and its impacts on the local environment and social systems. However, this chapter wishes also to dwell on the sustainability and future of shrimp cultivation, marketing, and its inclusive value chains rather than its somewhat painful past.

Keywords

Prawn farming · Embankments · Water-logged · Saline water

6.1 Introduction

Bangladesh is one of the world's largest producers of shrimp, with exports of more than 80,000 metric tons in 2016, with a value of more than $459 M[1] – of which 90% are tiger brackish water shrimp (*P. monodon*). The number of hectares of shrimp farms in Bangladesh has grown steadily from 135,000 in 2008 to 265,000 in 2019 that produced 240,000 tons of shrimp that year. The total value of Bangladeshi shrimp exports in 2014 was almost US $450 million but declined to $365 million by 2019.

There are approximately 150,000 farms producing shrimp and prawns in Bangladesh, employing over 600,000 workers. The Tiger shrimp productivity, at an average of 450 kg/ha/year in Bangladesh, remains very low in contrast to other countries (Department of Fisheries, GOB 2015). However, one cannot compare to other countries as the shrimp species is quite different (whiteleg), and the other countries use the intensive cultivation compared to Bangladesh which uses the extensive cultivation. There are approximately 130 shrimp processing plants in Bangladesh, 65 of which are operational. These plants run at about 40% capacity due to supply and quality constraints. Most plants clean, preserve, and package shrimp. Some plants are also

C. A. Meisner (✉)
Retired Agricultural Researcher from IFPRI, CIMMYT, FAO, Chapel Hill, NC, USA

S. Ahmed
School of Public Service, Boise State University, Boise, ID, USA
e-mail: salehahmed@boisestate.edu

[1]Media New Age Limited, June 14, 2017 http://www.newagebd.net/print/article/17740

R. Ahmed et al. (eds.), *Transforming Bangladesh*, World Regional Geography Book Series,
https://doi.org/10.1007/978-3-031-45093-8_6

beginning to cook or semi-cook the shrimp to increase value-added as well as get around the rejection of shrimp due to salmonella contamination, food fraud, and adulteration of seafood.

In addition to shrimp and crawfish/prawn, the country exports nearly US$50 million of crab meat, lobster, and hard shell (oysters and/or clams) to Thailand and India. Crab (both hard and soft-shelled) exports rose to 8520 tons in 2013–2014, earning $22.91 million.[2] This industry is expanding as a result of public and private investment in crab hatcheries, aimed at replacing the unsustainable crablet extraction from the Sundarbans World Heritage Site. Crab fattening is a gender-friendly sector, with women mostly involved.

The two main species cultivated for export are: (1) Black Tiger shrimp or Giant Tiger Prawn (*Penaeus monodon*) or *Bagda* in the Bengali language and (2) Giant River Prawn (*Macrobrachium rosenbergii*), also known as *Galda*.[3] The other important shrimp species exported are wild: the Indian White shrimp (*Penaeus indicus*) is third by value and volume in export value, and the Speckled shrimp (*Metapenaeus monoceros*) is fourth by volume. Most of the country's shrimp and crab farming ponds are located in the southwest of the country, in the districts of Satkhira (75,867 ha), Bagerhat (72,684 ha), Khulna (58,367 ha), and Jessore (11,234 ha). These five districts cover 95.3% of the country's total shrimp farming ponds. In the east of the country in the Coxs Bazar District there are another 44,465 ha. However, one cannot compare the hectares of the others with Coxs, since the effective pond area is only about 1/10 of these hectares, the rest is fallow land in Coxs. Cox's Bazar district is also where most of the shrimp hatcheries that produce seeds for the farms are located (Map 6.1).

An interesting finding of a study by Winrock was the increasing leasing (here called leasing-in) that is occurring. Shrimp farmers in Satkhira, Bagerhat, and Khulna typically have their own ponds/ghers, but also have leaseholds (48% of farmers in Khulna have leaseholds; 47% in Bagerhat and 63% in Satkhira). In Cox's Bazar, ownership is less common and 100% of farmers surveyed are tenant farmers. Among shrimp farmers, most of the ponds are both owned and leased (23% of shrimp farmers in Khulna are leased; 63% in Bagerhat, 57% in Jessore, and 78% in Satkhira). Most leases are individual, not joint, and have terms of 2–5 years (SAFETI Baseline Report 2017).

Shrimp cultivation, while having an historical perspective as a farmer-driven experiment, has become much more of a sustainable cultivation system in Bangladesh during its 60-year history. Its well-developed value chain is now driven by the export markets that the processors have tapped throughout the world. But how sustainable can this system be for the future? This chapter wishes to address the current status of shrimp cultivation, its potential future, and its effects on the socio-economic status of the rural population whose lives and futures are centered on its ultimate sustainability.

6.2 History of Shrimp Farming

Prawn and shrimp farming in Bangladesh first started in the 1970s. The coastal aquaculture sector began expanding rapidly in the 1980s (Muir 2003). Over the last three decades, prawn and shrimp culture have undergone revolutionary development (Ahmed et al. 2011). A number of historical events are associated with the emergence of commercial shrimp farming in Bangladesh. It was first tested by local farmers in the 1960s, when a government project created permanent water logging with saltwater in coastal cultivable lands.

In the 1950–1960s, existing river embankments deteriorated such that saline intrusion and tidal surges caused routine crop damage. Crop failures caused by saline intrusion or monsoon flooding were reported in most areas once every 3 years (Nishat 1988). In the 1960s, the government looked to increase domestic rice production to accommodate the growing population using Green Revolution technologies with the development of High Yielding Varieties (HYV) of rice. To increase rice production, the government recognized the need to convert the southwestern coastal tidal plains into permanent freshwater areas by 1961. To dramatically increase rice production, the government implemented a project called "the Coastal Embankment Project (CEP)" in southwest Bangladesh, with funding from the World Bank (Islam 2004). Borrowing from the Dutch, the term "polder" was used to designate areas surrounded by dykes or embankments, separating them from the river, offering protection against tidal floods, saline intrusion, and sedimentation.[4]

Ninety-two polders were constructed with 4022 km of embankments and 780 drainage sluices during the first phase. By June 1971, the gross polder area was 1.01 million ha (Talukder 1991). Crops were saved from salinity and flooding, with some rice yields increased by 200–300% (Nishat 1988). The dominant land use during this period was paddy cultivation of primarily traditional local varieties – which were changed by introducing modern paddy varieties and technological packages. Still salt production, mangrove forest, and traditional shrimp farming remained the same.

[2] Bangladesh Export Promotion Bureau.

[3] Department of Fisheries, GOB.

[4] Polder means to create a large embankment around delta islands to keep out river/ocean water from the enclosed land).

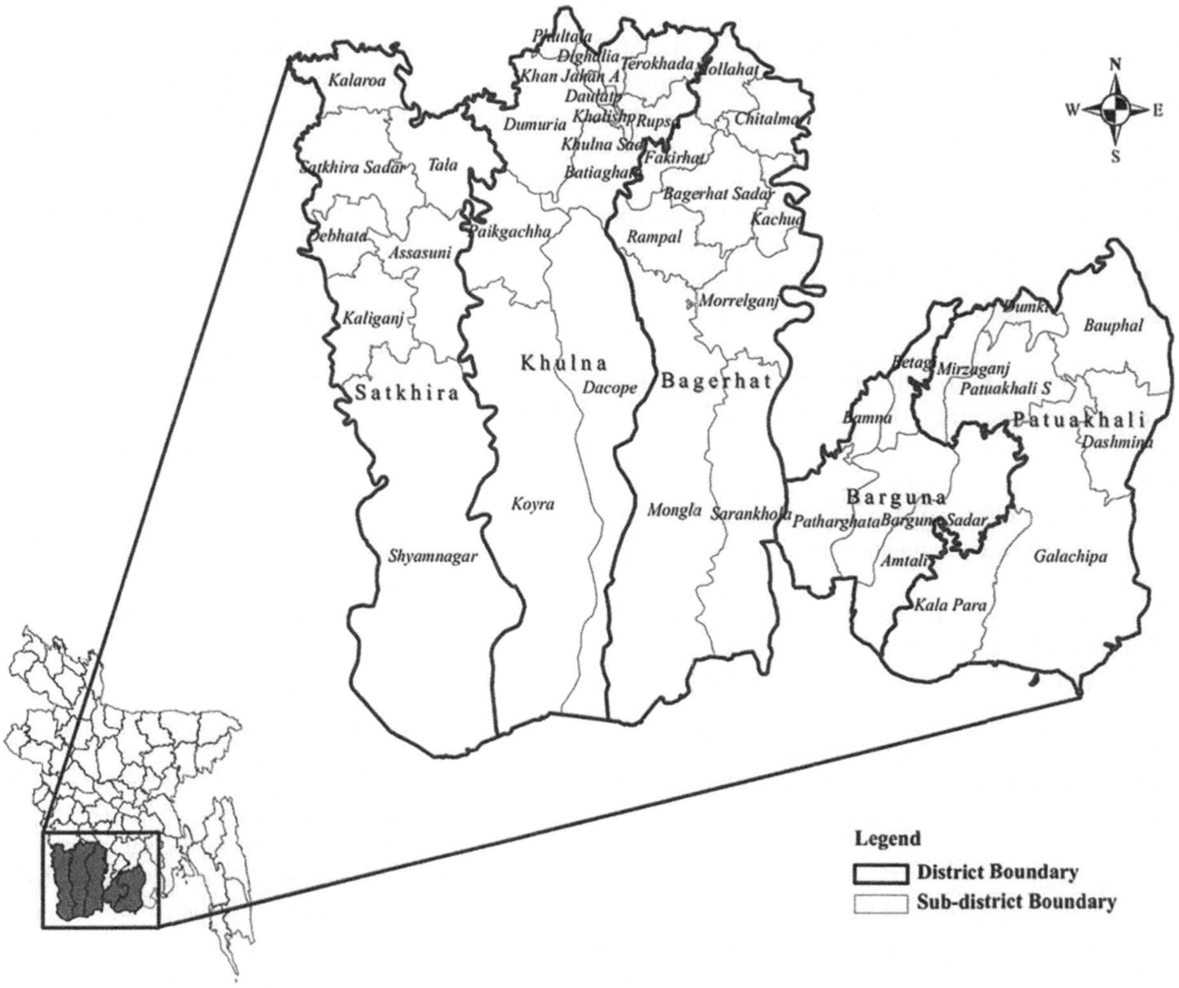

Map 6.1 Cultivated shrimp production areas. (From Ahmed 2018)

The World Bank and other donors continued supporting large-scale polderization of coastal Bangladesh in the 1970–1980s. Finally, a total of 123 coastal polders were implemented by the end of 1980s (PDO-ICZMP 2004). The government soon realized that internal water management had to be established within these polders to further increase production. Changes in land use occurred due to the intensification of paddy cultivation with an expansion of modern crop varieties, and conversion of agricultural land to non-agricultural use (Sereno 1981). During this time, coastal afforestation efforts were initiated to help protect the land from cyclones and foreshore erosion.

Southwestern coastal Bangladesh experienced drainage congestion inside and heavy siltation outside the polders in the 1990s because of extensive polderization of the hydrodynamically active delta. As a result, the achievements from polderization gradually evaporated. Those polders in the region were designed to eliminate all-natural tidal inflow onto the plains from the rivers or canals. This process aimed to maintain lower level of salinity for farm activities. Before large-scale polderization, the silt carried by the rivers from upstream regions was deposited onto the downstream plains. In fact, this is how these islands in the south were mostly emerged. Preventing the inflow of silt on to the plains led to silt deposition in the river beds, causing them to rise rapidly. As a result, salinity began to increase in the area by submerging a large swath of low-lying coastal farmlands within the polders. This had also blocked the sluice gates with silt, and created region-wide water logging, which ultimately contributed to further salinity in land and groundwater. Many smallholder farmers in the regions forced to switch from rice cultivation to shrimp cultivation in their saltwater-logged small plots. After their initial success, shrimp cultivation attracted larger attention, and began to spread in the region. However, this had contributed to further increase of salinity in the surrounding areas, and large areas of farmlands became unproductive (Swapan and Gavin 2011).

There have been many papers written on the history and impacts of shrimp cultivation (Deb 1998; Karim et al. 2014; Ishtiaque et al. 2017; Parvin et al. 2017). Each has brought varying perspectives to light. This chapter will be highlighting the current aspects of the shrimp industry as it moves to the future.

Gher[5] shrimp/fish farming in Bangladesh can be categorized into three categories based on the level of intensity (descriptions of these categories from World Bank publication 1999):

- Extensive culture: shrimp that depend entirely on naturally produced organisms in the ponds for their food and growth.
- Improved extensive: shrimp that depend on natural food and the application of fertilizer in the pond water, sometimes combined with supplemental feeds to enhance growth.
- Semi-intensive culture: shrimp that depend entirely on supplemental feeds using intensive management practices (i.e., aeration).

Seventy percent (70%) of the shrimp farms in Bangladesh use traditional or extensive, 25% semi-intensive, and 5% intensive culture techniques (Hossain et al. 2013).

(a) *Extensive Culture*
  - Pond sizes average from 0.5 to 2 ha.
  - Growers open sluice gates to their fields when the tides contain high densities of young shrimp (crabs and tilapia also enter).
  - Growers can stock wild juveniles they have caught or purchased.
  - No additional feed is provided; thus, dependent on natural fee.
  - Stocking rates rarely exceed 10,000 per ha.
  - Harvesting is commonly from continuous or semi-continuous, ranging from 50 to 200 kg/ha (head-on weight).
  - Water exchange is provided by tidal exchanges.
  - Operating costs are low.
  - Harvest is done by cast-nets or bamboo traps.
  - During the monsoon, rice production is possible.
  - Some use salt production (dry) and shrimp production (wet).

Growers adopted more modern techniques in the extensive cultivation by stocking with post larvae (PL) sold in the market. Additionally, Specific Pathogen Free PLs (free of major pathogens) are available in the market, and growers are willing to pay a premium for these. Some growers have a continuous stocking practice whereby there are multiple harvests and multiple stocking during the season.

(b) *Semi-intensive*
  - Usually at the upper end of tidal range or above the high tide line.
  - Use of well laid-out pond of 0.1–5 ha.
  - Use of shrimp feed.
  - Water is pumped using diesel or electrical pumps.
  - Stocking rates from 25,000 to 200,000 juveniles per hectare.
  - Construction costs vary from $10,000 to $25,000 per hectare.
  - Some growers use grow-out ponds.
  - Harvest is done by either draining the pond through a net or using a harvest pump.
  - Better practice is to restock the ponds once a year.
  - No rice production in this system.

(c) *Intensive Cultivation*
  Intensive shrimp farming usually involves small ponds (0.1–5 hectares) or large commercial farms (100 + ha), high stocking densities (more than 200,000 juveniles per hectare), around-the-clock management, intensive feeding, waste removal, and mechanical aeration. Mechanical aeration – the addition of oxygen to the water – permits much higher stocking densities and feeding levels. Construction costs range from $10,000 to $25,000/ha for simple pond systems.

Harvest of shrimp is usually throughout the season, based on the multiple stocking of the shrimp throughout the season. Table 6.1 indicates the length of the seasons for the various shrimp cultivated including fish culture with the shrimp in the same ponds. For the shrimp industry, the stocking of Tiger shrimp occurs from March through November with multiple harvests from July through December. Thus, food processors that deal exclusively with tiger shrimp have only a 6-month period to capitalize on shrimp processing. Once frozen, usually in water blocks, they can be stored to wait for a buyer or more favorable price. The Department of Fisheries (DoF) conducts laboratory tests on each lot. The labs are ISO certified, but still do not have the standards required to keep up with industry needs (DoF 2010).

Karim et al. (2014) research showed that compared to the "traditional"' shrimp production, two modified (MMT1 and MMT) and one closed system (CST) technologies increased yields and profitability. The CST was double the profit of just MMT. Feed use, stocking density, gher size and white spot syndrome virus incidence were key factors associated with the economic returns of CST farms, while various supplementary feed inputs made a significant positive contribution

[5]Gher farming represents a traditional agricultural system found in Bangladesh, where ponds are dug into small fields (usually of rice) to allow for shrimp/fish farming. The scooped soil is then used as a wide embankment (dikes) for vegetable production around the pond.

Table 6.1 Season of shrimp/ freshwater and brackish water fish

http://www.marinefreshbd.com/

towards increased return for the MTT farms. Shrimp yields show a slow increase over the years as the shrimp growers adopt improved technologies. USAID and GIZ have expansive projects in the shrimp industry currently, with intensive farmers' training a key component of their projects.

6.3 Economic Trend in Shrimp Farming

The annual net profits of the surveyed shrimp farmers averaged $735 per ha at Khulna, $792 per ha at Bagerhat, $980 per ha at Satkhira, and $167 per ha at Coxs Bazar. The shrimp farmers surveyed reported annual net gains of $2664 per ha at Khulna, $2667 per ha at Bagerhat, $1941 per ha at Jessore, and $717 per ha at Satkhira, again including all pond farms, shrimp, prawns, fish and vegetables (Winrock SAFETI Baseline Report 2017).

Shrimp Exports: The total value of Bangladeshi shrimp exports in 2014 was almost US$ 450 m with production of 220,000 tons. The European Union (EU) has traditionally been the most important market, consistently accounting for 75–80% of the total export volume and value. Bangladeshi shrimp exports are shown to be going to Belgium or the Netherlands but are mostly re-exported from there to other EU trade hubs and European countries. A recent visit by EU shrimp certification agencies (2016) to Bangladesh to examine the shrimp supply chain first hand, claimed Bangladesh will no longer require third-party inspections or testing for shrimp export to the EU. Currently, Department of Fisheries (DoF) does this testing in their Khulna lab. This is in contrast to India and Myanmar which still retain testing and third-party inspections of all their shrimp exports to the EU. This indicates how far the industry has come in Bangladesh since the 1997 debacle of prawn shrimp rejections by the EU and the USA.

The Bangladesh shrimp supply chain faces strong competition from other regional shrimp-producing countries including India ($3.1 billion industry), Vietnam ($1.6B), Indonesia ($1.2B), China ($970 M), Thailand ($620 M), Myanmar ($420 M), and Malaysia ($200 M).[6] India is the top shrimp supplier to the United States, the world's biggest buyer of shrimp. China is the top producer of farmed shrimp in Asia, growing almost 1.8 million tons in 2015, up from 1.01 million tons in 2014 and 910,000 tons in 2013. But, there are signs that China's dominance in the marketplace will not continue. Reasons include the Government of China enforcement of greater environmental restrictions, diseases, indiscriminate use of antibiotics (leading to rejection in some markets), growing domestic markets, and poor brood stocks. In fact, China now imports shrimp from Vietnam. China's shrimp exports fell 13.6% in 2014, while its shrimp imports rose by 10%.[7]

The Government of Bangladesh (GOB) regulates livestock, fish, and shrimp from hatcheries to feed, with 16 laws and acts in place.[8] The GOB also maintains an export incentive of ~10% for shrimp exporters, compared to India, where subsidized inputs led the US to put a 5% tariff on imports, and reduced to 2.2% in 2017.

There are strong government regulations in terms of international trade. Importing is permitted only through irrevocable letters of credit, and certain documents are needed for doing business under this regulatory structure. Export regulations have been published for 2015–2018 but are in Bangla. These are summarized in English on a resource website from a law firm in Bangladesh.[9]

The GOB passed a Code of Conduct (first in S. Asia) on 10 selected sectors of the shrimp supply chain in July 2015.[10]

[6] 2015 data except Myanmar, 2010 http://www.worldstopexports.com/big-export-sales-for-frozen-shrimps

[7] https://www.seafoodsource.com/news/aquaculture/china-s-shrimp-industry-still-leads-but-problems-loom

[8] http://www.mofl.gov.bd/site/view/law/Law

[9] https://resource.ogrlegal.com/export-policy-2015-18-published/

[10] Bangladesh Shrimp Industry Code of Conduct Overview.

Each sector is supposedly following this code. A quick perusal of the document shows each sector basically does its own monitoring, but no one entity is "in charge" of overall monitoring in each sector.

The Bangladesh Department of Fisheries (DoF) is reticent to allow the importation of *vanemai* or whiteleg shrimp as India and Vietnam did. These species greatly increase shrimp productivity by good responsiveness to semi- and intensive cultivation, but also have the greatest negative environmental impacts. In fact, some in Bangladesh consider the whiteleg shrimp an invasive species. In the fall 2018, DoF allowed the Bangladesh Fisheries Research Institute to raise whiteleg shrimp, but only confined to protected ponds for research and observation of this species under Bangladesh conditions. DoF has allowed the import of Specific Pathogen-Free (SPF) Tiger brood shrimp from Hawaii. It has established an SPF hatchery (MKA Hatcheries) through WorldFish[11]/USAID who markets their Post-Larvae (PL) through an NGO in Khulna and Satkhira. That said, the number of SPF or Polymerase Chain Reaction (PCR) tested (i.e., tested negative for viruses) PLs is only ~10% of the market. MKA hatcheries have been able to domesticate the brood shrimp from those SPF PLs and may begin selling these SPF mother brood to other hatcheries interested in selling only SPF PLs. This should rapidly increase the numbers of SPF PLs. In the fall of 2018, another private lab was able to bring brood shrimp from Thailand that are SPF and is marketing those PLs in Khulna. Other hatcheries are expected to follow.

Besides SPF PLs, DOF law states that only PCR-tested (negative) PLs can be sold, though many of the hatcheries have not responded despite DOF serving notice of even closure of those hatcheries (DoF 2010). The number of PLs caught in Cox's Bazaar or the Sundarbans is declining. Previous to obtaining SPF PLs, those PLs caught in the wild were considered healthier, exhibiting less mortality. Price is an indication: wild PLs do not receive the premium a price as before while SPF PLs are much in demand because of their ability to resist diseases and higher grow out rate. The brood shrimp imported from Hawaii had its genetic background from the Bay of Bengal, which was also an important factor.

Tiger Prawn is a national heritage and can be branded as such by astute processors/exporters. Attempts by GIZ (German Society for International Cooperation) and a local NGO to grow and market organic shrimp failed in its early years. Certification is very difficult to do in Bangladesh. However, "lessons learned" have been published so that many can learn from their efforts (Paul and Vogl 2012).

[11]https://www.worldfishcenter.org - an international, nonprofit research organization that harnesses the potential of fisheries and aquaculture to strengthen livelihoods and improve food and nutrition security.

6.4 Socio-economic and Environmental Factors in Bangladesh Shrimp Production

Despite the importance and positive influence on local and national economy, the shrimp sector has restructured the climate-society interactions in southwest Bangladesh. It creates opportunities, but also at the same time introduces major social, ecological, and economic challenges. Some of these impacts are discussed below.

(a) *Mangroves at Risk.*
Worldwide transformations of forestlands into shrimp farms are considered as one of the major reasons of declining mangrove forests. Local perception about the declining mangrove forest also supports this argument (Shameem et al. 2014). However, due to the shrinking mangrove forest in southwest coastal Bangladesh, further salinity intrusion reaches the inland areas and makes more lands unproductive for crops. This again forces more landless people and makes them more vulnerable to various other climate risks. The massive adoption of shrimp farming in the region produces volumes of waste that harm ecologically sensitive zones, contributing to the destruction of coastal mangrove forests, loss of local biodiversity, and rampant pollution through the discharge of shrimp pond effluents into already fragile ecosystems (Cons and Paprocki 2014; Hossain and Hasan 2017).

In addition, the Sundarbans also act as the natural defense of the region. Almost every year, tropical cyclones are formed in the Bay of Bengal, and most of them reach the mainland with immense destructive force. Mangroves act as natural barriers and slow down the intensity of the cyclones, which saves human lives and properties. Decrease of mangrove forests is also a major challenge for millions of people, whose livelihoods are concentrated on various forest-dependent activities and resources, such as honey collection, crab collection, firewood collection, and fruit collection. Mangroves are also a haven for biodiversity. At this moment, nearly half of the amphibians, reptiles, mammals, and birds living in Sundarbans mangrove forests are threatened with extinction. These animals, in most part, are not found anywhere else. They are only available in the Asian and Australian mangrove forests (Bales 2016). Removal or decrease of mangroves from coastal Bangladesh will force millions of people further into disaster vulnerability and will take a large toll on local biodiversity and ecosystem services.

(b) *Integration into Global Markets.*
With the growing reliance on shrimp farming as a livelihood option, local farmers and gher households experience a new form of dependency, which in turn translates

to a new type of livelihood risk. The shrimp export market is organized by large-scale export companies with capital-intensive cold storage facilities that provide access to local industry among North American and European buyers. Engagement with these international companies involves a complex supply chain of prawn/shrimp buyers at different scales of operation. However, local small-scale shrimp farmers tend to sell to local middleman in transactions of small volumes. Unfortunately, local shrimp farmers have little access to effective price information, which puts them in a disadvantaged position to bargain with the middleman. Besides, lack of information and capacity to interact with international buyers including middle man force the small shrimp farmers from the area to remain at the margin of any perceived benefit.

Furthermore, since the income of local shrimp farmers is dependent on the value of the shrimp they can sell, the well-being of their households and communities are not anymore dependent within their spatial boundary of their communities. Local communities have a growing reliance on the outside market to meet all their basic/daily needs. Their income and well-being are now merged with the global shrimp industry. Previously they used to feed and maintain their families not just by the rice they grew, but also, they could exchange foods with other products. Now they must buy everything from their nearby towns or markets in exchange of cash (Cons and Paprocki 2014). If for some reasons, they become out of cash, they will remain in hunger. Therefore, traditional shrimp farming societies are at higher risk of hunger, poverty, and food insecurity, which of course contradict with the traditional perception of shrimp farming communities.

Conversely, it is also true that integration into global food chain through commercial shrimp farming brought some benefits to the local communities. In some local families, it has contributed to increased household consumption, contributed to increase the level of education and access to health, as well as helped the local communities to connect with the outside world in some ways. Due to their increased access to information and communication technology by cell phones (157 million SIM cards),[12] many local farmers now get early warning before any cyclone with sufficient lead time. Increased income contributed to female education, gender equality, as well as maternal and children's health. Some of the gher areas are remote and any government support barely reached those communities. Due to the shrimp farming revolution, people are exposed to various exogenous influences that create both opportunities and new costs. This has been seen in other local, mostly rural systems, which participate in these global markets for food supply, interlocking in new ways the food-water-climate dynamics. A recent study measured the well-being of shrimp vs. prawn producers and found divergence with prawn producers more positive and shrimp producers more negative (Belton 2016). The paper examined the resulting changes in food security status, access to land, gendered labor relations, and their implications for the social well-being of those involved rather than just economic gain.

(c) *Social Impacts.*

Shrimp farming requires a lot less human labor than rice farming. In a study in southwest Bangladesh, Gurung et al. (2016) reported that even though aquaculture is profitable for local people in some cases, it engages almost 20% less labor than rice farming operations, which means that people who do not own their own lands are more likely to be left without any work and income opportunity and so they will face food insecurity and deepening poverty unless they find supplemental income sources. Shiva (1995) also found similar results in India. He reported that rice farming on 40 ha of land engages approximately 50 laborers, however, shrimp farming in the same amount of land engages only five workers. In addition to labor engagement, conversion from rice farming to shrimp farming has reduced the opportunities for other traditional dry season activities, such as homestead gardening (Alam et al. 2005).

Because of the rapid and uncontrolled expansion of shrimp farming, local people are also losing their access to common property resources such as mangroves, marshes, etc. Many of their livelihoods are dependent on their access to these resources. This is quite common that smallholder farmers often face increasing challenges to pursue their conventional farming in the area. This deprivation of resources and opportunities forces many of the local smallholder farmers to consider migration as one of their options to cope with the changing pattern of resource access and livelihood opportunities. Cons and Paprocki 2014; Hossain et al. 2013). Some of these migrations are seasonal or temporary in nature and some are permanent. Since most of these migrated people do not have any major skill beyond farming, many will end up working as daily-wage laborers with lower wage and will be living in slums/informal unhygienic settlements. With increased uncertainty with income and opportunities, many of them will be in the loop of the poverty cycle and face even further impoverishment and marginalization (Ahmed 2018).

[12] http://www.btrc.gov.bd/content/mobile-phone-subscribers-bangladesh-december-2018

Continued marginalization and persistent poverty in some of the areas are a regular phenomenon. These forms of impoverishments could cause social tensions and even violence (Stonich and Vandergeest 2001). In addition, in the salinity-prone southwest Bangladesh, where certain levels of salinity are favorable for shrimp farming, salinity intrusion is not always solely a hydrological phenomenon. Sometimes the encroachment of salt water and pressure from large and powerful landowners are supported by both hired goons and local politicians (Ahmed 2018). In some instances in the past, they had forced local small-scale farmers along with many other landless farmers to sell or lease their lands to large landowners for local shrimp production. This transformation made a huge structural shift in social relations and produced a range of confrontations and conflicts at various scales during the 2000s, but after a rampant disease hit shrimp farms, lands reverted to small landholders (Guhathakurta 2008).

Since shrimp aquaculture involves saline water, it creates more salinization in local aquifers (Deb 1998; Paul and Vogl 2011; Primavera 1997). In the southwest coastal Bangladesh, shrimp farming has further contributed to the processes of de-peasantization and out-migration by making small-scale farming more difficult in most active shrimp cultivation zones (Guimaraes 1989). In contrast, Sen (2010) indicated because of the increased profitability of shrimp cultivation over rice, the Sundarbans is less pilfered by the poor for its resources than previous decades.

(d) *Changing Patterns of Social-ecological Systems.*
Current patterns of livelihood systems in southwest coastal Bangladesh are creating long-term social-ecological traps. Studies on resilience evolved from an original focus on multi-stable states in ecological systems (Holling 1973), to the study of nested cycles of adaptive changes in social-ecological systems in which persistence and novelty are intertwined, and finally to transformations that can cascade up scales when small, fast events trigger big, slow ones (Holling et al. 2002). Human interventions can substantially alter feedbacks and drivers of social-ecological systems, which can lead to regime shifts. Such changes can alter ecosystems capacity to provide services for human well-being, which can also trigger various social responses. Social phenomena such as de-peasantization can interact with ecosystem dynamics in mutually reinforcing ways and can generate vulnerable pathways of development and follow toward undesirable states that are not resilient. Any perceived social-ecological trap in southwest coastal Bangladesh indicates a situation where social and ecological feedbacks mutually reinforce each other and maintain or push a social-ecological system towards an undesirable state (Ahmed 2018). For the majority of people, particularly those who are dependent on farming or fishing and have limited resources to cope with any structural shifts, they are likely to face social-ecological traps and deepening poverty with increased food insecurity. Without large-scale structural change, social innovations, policy interventions along with opportunities for co-production of knowledge that link to livelihoods, sustainability, and resilience, for small-scale farmers in southwest coastal Bangladesh, it is almost impossible to get away from these unsustainable trajectories (Ahmed 2018). Over the years, shrimp farming forces marginalized farmers to migrate, and the process is very gradual. It is difficult to estimate that how many people have migrated from the region after losing their lands and livelihoods. However, Baird and Quarto (1994) estimated that the expansion of shrimp farming in Satkhira region in Bangladesh has displaced approximately 120,000 smallholder farmers from their farmlands. Many of these displaced farmers opted for migration. Two major factors shape these migration outcomes: available resources and social network that facilitate migration, and environmental stressors, such as loss of arable lands (Bernzen et al. 2019). In situations of extreme poverty, a subset of those displaced farmers was not even able to migrate due to the lack of required resources or social networks that could have contributed to their migration decisions. In literature, these people are now known as "trapped population" (Ayeb-Karlsson et al. 2018). So far there is no estimate of the trapped population in coastal Bangladesh. However, with decades of 6 + % economic growth rate, such structural changes, social innovations, and policy interventions that affect opportunities for improved livelihoods have a greater chance of occurring.

6.5 Conclusion

Shrimp farming in Bangladesh has provided mixed outcomes. It began as a grower-driven adaptation to saline conditions caused by the polderization of coastal plains. It is currently the fourth largest export from Bangladesh and has created a well-developed value chain industry. Because the GOB has chosen not to allow the whiteleg species into the country, Bangladesh is unique among the world's shrimp industry to only allow its native species, the Black Tiger shrimp, to be cultivated and exported, though much lower yielding than the whiteleg shrimp grown mostly throughout the world. The industry remains sustainable as well as profitable because this shrimp has a niche market and higher price in the international markets. Laws and policies have assisted the industry to remain robust. There certainly has been a rocky start where the larger landowners capitalized the market in the beginning of the industry in the 1990s. However, currently both the large and small growers are able to sell to the processors through a system that works for all of them. At the same time, evidence suggests that the benefits of shrimp farming were not equally distributed among all social groups. It presents risks to the mangrove forests, and peoples' dependency on volatile global market has increased. Social outcomes are also mixed. Present evidence on the changing pattern of social-ecological systems indicates an unsustainable long-term trajectory. Since the country is trying to achieve national and international development goals, such as the United Nations Sustainable Development Goals, it is critical to engage people from all social and economic classes in a sustainable manner, so that Bangladesh's resource-base is not compromised both now and for the future. New innovations such as e-traceability, third-party certification of hatcheries/feed/processing factories will also create a better environment for marketing the unique Black Tiger shrimp of Bangladesh.

References

Ahmed S (2018) Shrimp farming at the interface of land use change and marginalization of local farmers: critical insights from southwest coastal Bangladesh. J Land Use Sci 13(3):251–258

Ahmed N, Demaine H, Muir JF (2011) Freshwater prawn farming in Bangladesh: history, present status and future prospects. Aquac Res 39(8):806–819

Alam SMN, Lin CK, Yakupitiyage A, Demaine H, Phillips H (2005) Compliance of Bangladesh shrimp culture with FAO code of conduct for responsible fishes: a development challenge. Ocean Coast Manage 48:177–188

Ayeb-Karlsson S, Smith CD, Kniveton D (2018) A discursive review of the textual use of "trapped" in environmental migration studies: the conceptual birth and troubled teenage years of trapped populations. Ambio 47:557–573

Baird IB, Quarto A (1994) The environmental and social costs of developing coastal shrimp aquaculture in Asia In: Trade and environment: prospects for regional cooperation Nautilus Institute, Berkeley, pp 188–214

Bales K (2016) Blood and Earth: modern slavery, ecocide, and the secret to saving the world. Spiegel & Grau, New York

Belton B (2016) Shrimp, prawn and the political economy of social wellbeing in rural Bangladesh. J Rural Stud 45:230–242

Bernzen A, Jenkins JC, Braun B (2019) Climate change-induced migration in coastal Bangladesh? A critical assessment of migration drivers in rural households under economic and environmental stress. Geosciences 9:51. https://doi.org/10.3390/geosciences9010051

Cons J, Paprocki K (2014) Brackish waters and salted lands: the social cost of shrimp in Bangladesh. Food First Backgrounder, Institute for Food and Development Policy 20(3). https://foodfirst.org/publication/brackish-waters-and-salted-lands/

de Guimaraes JP (1989) Shrimp culture and market incorporation: a study of shrimp culture in paddy fields in Southwest Bangladesh. Dev Chang 20:653–682

Deb AK (1998) Fake blue revolution: environmental and socioeconomic impacts of shrimp culture in the coastal areas of Bangladesh. Ocean Coast Manage 41:63–88

Department of Fisheries, GOB (2015) https://dsiweb.cse.msu.edu/demo/Bangladesh_aquaculture_data_portal_seven_districts/

DoF (2010) Fisheries resource development initiatives. Department of fisheries/Ministry of Fisheries and Livestock, Dhaka, p 120

Guhathakurta M (2008) Globalization, class and gender relations: the shrimp industry in southwestern Bangladesh. Development 51(2):212–219

Gurung K, Bhandari H, Paris T (2016) Transformation from rice farming to commercial aquaculture in Bangladesh: implications for gender, food security, and livelihood. Gend Technol Dev 20(1):49–80

Holling CS (1973) Resilience and stability of ecological systems. Annu Rev Ecol Syst 4:1–24

Holling CS, Gunderson LH, Ludwig D (2002) In quest of a theory of adaptive change. In: Gunderson LH, Holling CS (eds) Panarchy: understanding transformations in human and natural systems. Island Press, Washington, D.C., pp 3–22

Hossain MAR, Hasan MR (2017) An assessment of impacts from shrimp aquaculture in Bangladesh and prospects for improvement. FAO Fisheries and aquaculture technical paper #618, Dhaka

Hossain MS, Uddin MJ, Fakhruddin ANM (2013) Impacts of shrimp farming on the coastal environment of Bangladesh and approach for management. Rev Environ Sci Biotechnol 12:313–332

Ishtiaque A, Sangwan N, Yu D (2017) Robust-yet-fragile nature of partly engineered social-ecological systems: a case study of coastal Bangladesh. Ecol Soc 22(3):5. https://doi.org/10.5751/ES-09186-220305

Islam MR (2004) Coastal embankment improvement project. Draft final report, Sept 2012

Karim M, Meisner CA, Phillips M (2014) Shrimp (Penaeus monodon) farming in the coastal areas of Bangladesh: challenges and prospects towards sustainable development. In: Delany CA (ed) Shrimp evolutionary history, ecological significance and effects on dietary consumption. Novo Publisher, New York, pp 57–88

Muir JF (2003) The future for fisheries: economic performance. fisheries sector review and future development study. Commissioned with the association of the World Bank, DANIDA, USAID, FAO, DFID with the cooperation of the Bangladesh Ministry of Fisheries and Livestock and the Department of Fisheries, Dhaka, p 172

Nishat A (1988) Review of the present activities and state of art of the coastal areas of Bangladesh. In: National workshop on coastal area resource development and management. Part II CARDMA, Dhaka, pp 23–35

Parvin GA, Ali MH, Fujita K, Abedin MA, Habiba U, Shaw R (2017) Land use change in southwestern coastal Bangladesh: consequence to food and water supply. In: Banba M, Shaw R (eds) Land use management in disaster risk reduction. Disaster risk reduction (methods, approaches and practices). Springer, Tokyo

Paul BG, Vogl CR (2011) Impacts of shrimp farming in Bangladesh: challenges and alternatives. Ocean Coast Manage 54:201–211

Paul BG, Vogl CR (2012) Key performance characteristics of organic shrimp aquaculture in Southwest Bangladesh. Sustainability 4:995–1012

PDO-ICZMP (2004) Areas with special status in the coastal zone. PDO-ICZMP project, Dhaka

Primavera JH (1997) Socio-economic impacts of shrimp culture. Aquac Res 28:815–827

Sen SG (2010) Conservation of the Sundarbans in Bangladesh through sustainable shrimp aquaculture. Policy analysis paper by Harvard Kennedy School. https://www.innovations.harvard.edu/conservation-sundarbans-bangladesh-through-sustainable-shrimp-aquaculture

Sereno G (1981) A reconnaissance study of changes in settlement and related non-agricultural land use in Bangladesh. FAO/UNDP land use policy project. Ministry of Agriculture, Dhaka

Shameem MIM, Momtaz S, Rauscher R (2014) Vulnerability of rural livelihoods to multiple stressors: a case study from the southwest coastal region of Bangladesh. Ocean Coast Manage 102:79–89

Shiva V (1995) The damaging social and environment effects of aquaculture. Third World Resourg 59:22–24

Stonich S, Vandergeest P (2001) Violence, environment, and industrial shrimp farming. In: Peluso NL, Watts M (eds) Violent environments. Cornell University Press, Ithaca, NY

Swapan MSH, Gavin M (2011) A desert in the delta: participatory assessment of changing livelihoods induced by commercial shrimp farming in Southwest Bangladesh. Ocean Coast Manage 54(1):45–54

Talukder MA (1991) Current status of land reclamation and polder development in coastal lowlands of Bangladesh, polders in Asia, Economic and Social Commission for Asia and the Pacific (ESCAP), Atlas Urban Geology, vol. 6, UN

Winrock (2017) SAFETI project funded by USAID. https://www.winrock.org/project/safeti/

World Bank (1999) World Bank review on shrimp farming and the environment 2, and an April 1999 meeting on shrimp management practices hosted by NACA and WWF in Bangkok. 1 FAO. 1998. http://library.enaca.org/Shrimp/Publications/WBfinal.pdf

Part II

Economic Transformations

Inland, Coastal, and Offshore Fishery Resources: Solution for Mass Protein in Bangladesh

7

M. Gulam Hussain

Abstract

Bangladesh's extensive interior, coastal, and marine water resources make it one of the top aquaculture nations in the Asia-Pacific area. In terms of inland aquaculture production, Bangladesh now ranks in the top 10 nations in the world (5th in the ranking). The average growth rate for fisheries over the last 15 years has been 5.27%, whereas the average growth rate for aquaculture has been above 8.2%. The potential for the growth of mariculture has been significantly increased since the resolution of marine border conflicts with neighboring nations, such as Myanmar and India. In order to start and implement aquaculture of a few chosen marine finfish, shellfish, and non-traditional marine species, efforts must be done. The nation would be able to accomplish the projected bulk protein solutions and realize substantial economic benefits if these opportunities could be successfully implemented in the coastal, coastal, and offshore sectors.

Keywords

Inland · Coastal · Marine · Fishery resources · Mass protein · Bangladesh

7.1 Introduction

Bangladesh is blessed with exceptional inland and marine water resources. With around 820 species of finfish, shellfish, and non-traditional species in fresh, brackish, and marine waterways, the nation has Asia's third-largest aquatic biodiversity, behind China and India (Mazid 2002). The biggest flooded wetland in the world, the Bengal Delta, and three major river systems that flow from the Himalayan Mountains into the Bay of Bengal—the Ganges, Brahmaputra, and Meghna—have been credited with the species variety. Additionally, the nation possesses a great amount of coastal and marine water resources, including millions of tiny ponds, seasonal ditches, borrow pits, flooded paddy fields, extensive floodplains, and vast coastal and marine water resources, all of which have a good chance of increasing fish yields through improved freshwater and marine aquaculture operations.

Seven of the top 10 producers of aquaculture worldwide now, including Bangladesh, are from the Asia-Pacific area (FAO 2016a). Bangladesh, which produces 2.65% of the world's aquaculture, was rated fifth in 2014. Vietnam came in fourth, followed by India, Indonesia, and 6.61%, 5.82%, and 14.6%, respectively. Bangladesh's GDP is boosted by aquaculture alone by more than 2.81%, more than any other nation in Asia-Pacific outside Vietnam (FAO 2017). Inland aquaculture has been cited as one of the leading sub-sectors over time, accounting for 56.4% of Bangladesh's overall fish production in 2016–17 (DoF 2017). Aquaculture has risen by more than 8.2% over the past 15 years compared to fisheries'average growth rate of 5.27% (DoF 2017). Therefore, inland aquaculture has a huge impact on the nation's productivity, protein supply, economic growth, and way of life for millions of people. The potential for developing marine aquaculture opportunities in coastal, coastal, and offshore areas has recently increased thanks to the country's vast coastal and seawater resources, which have recently been enriched following the end of the final settlement of maritime border disputes with neighboring countries, namely Myanmar and India.

This chapter reviews the country's overall aquaculture trends as well as the state and trends of aquaculture in the Asia-Pacific area. Due to Bangladesh's need for mass protein, it specifically addresses the potentials, prospects, and limitations of aquaculture in the available inland, coastal, and marine water areas.

M. G. Hussain (✉)
Feed the Future Fish Innovation Lab, Mississippi State University, Mississippi State, MS, USA
e-mail: hussain.mgulam@gmail.com

R. Ahmed et al. (eds.), *Transforming Bangladesh*, World Regional Geography Book Series,
https://doi.org/10.1007/978-3-031-45093-8_7

7.2 Review on Aquaculture Status and Trends in Asia-Pacific Region

Aquaculture output has historically been dominated by the Asia-Pacific region. In 2014, the region produced 65.2 million tons, or 88% of the world's aquaculture production (FAO 2016b). The value of regional aquaculture production in the Asia-Pacific in that year was US$127 billion, or 79% of the global total. The overall aquaculture production in the globe in 2014 was 73.8 million tons, valued at an estimated US$ 160.2 billion, according to FAO (2016a). China produced 45.5 million tons of fish from aquaculture in 2014, or more than 60% of the world's total production, and 69.7% of the Asia-Pacific region's production, with the remaining four countries—India, Indonesia, Viet Nam, and Bangladesh—producing the remaining 22.3% (FAO 2017). In 2014, a list of the top 10 worldwide aquaculture producers was shown (Table 7.1).

Aquaculture of marine fish is based on the harvest of fish seeds and young fish from wild fisheries in tropical and subtropical Southeast Asia. As wild fish stocks fall, production rises, and environmental limits are placed on gathering wild species for stocking cages, there is a progressive shift away from marine fish farming to hatchery aquaculture throughout the region (Marine Aquarium Council/The Nature Conservancy 2004). The marine aquaculture industry has not grown as rapidly as freshwater fish and crustaceans. The primary causes are (a) insufficient hatchery-produced seed, (b) a lack of readily available, cost-effective formulated feeds, (c) insufficient investments, and (d) a lack of high-level knowledge transfer, government priority, and patronage (FAO 2017). China, Indonesia, Malaysia, Taiwan Province of China, Thailand, Singapore, and Vietnam are the nations advancing fish aquaculture in the Asia-Pacific region. As one of the key obstacles to extending offshore marine aquaculture, increasing the usage of hatchery-grown seed is a priority (FAO/NACA 2011).

Table 7.1 Top 10 aquaculture producers of the world in 2014 by volume and value

By volume		By value	
Country	Tons (thousand)	Country	US$ (Million)
China	45,471	China	73,296
India	4881	India	10,768
Indonesia	4298	Chile	10,276
Viet Nam	3397	Indonesia	8915
Bangladesh	1957	Viet Nam	7173
Norway	1332	Norway	7068
Chile	1215	Bangladesh	4853
Egypt	1137	Japan	3728
Myanmar	962	Thailand	2636
Thailand	935	Egypt	2025
Other	8247	Other	29,645
Total	**73,832**	**Total**	**160,383**

Source: FAO (2016b)

7.3 Inland Aquaculture Status and Trends in Bangladesh

7.3.1 Water Resources Potentials and Ecosystem Diversity

Bangladesh has abundant water resources that are both inland and coastal in nature (Table 7.2). There are 4,700,795 hectares of inland water resources in the nation, of which 83% are open waters (3,906,434 ha), and 17% are closed water bodies (794,361). Floodplains (2,691,910 ha); rivers and estuaries (853,863 ha); beels and haors (114,161 ha), etc., are the main inland open water capture fisheries. The Kaptai Lake, the sole reservoir, has an area of 68,800 acres. Ponds (372,397 ha), ox-bow lakes (5488 ha), and coastal shrimp farms (275,583 ha) make up the inland closed-water fishery zones. Bangladesh has established sovereign rights to 30,772,448 hectares (i.e., 118,813 km^2) of marine water resources, extending up to 12 nautical miles of territorial sea, a further Exclusive Economic Zone (EEZ) of 200 nautical miles into the sea, and a Continental Shelf extending up to 354 nm (MoFA 2014). These rights follow the recent demarcation of maritime boundary issues with Myanmar and India. According to Chowdhury (2014), Fig. 7.1 depicts Bangladesh's current marine border/area.

Table 7.2 Statement of water resources in Bangladesh (2015–2016)

Water resources	Area (hectares)
A. Inland open water bodies	
Rivers and estuaries	853,863
Sundarbans	177,700
Beels and haors (natural depressions)	114,161
Kaptai lake	68,800
Floodplains	2,691,910
Total inland open water bodies	**3,906,434**
B. Inland closed water bodies	
Ponds	372,397
Seasonal water bodies	133,330
Baors (Oxbow lakes)	5488
Shrimp/prawn farms	275,583
Pen culture water bodies	7553
Cage culture water bodies	10
Total inland closed water bodies	**794,361**
Total inland water bodies (A + B)	**4,700,795**
C. Marine water bodies	**30,772,567**[a]

Source: DoF (2017)
[a]MoFA (2014)

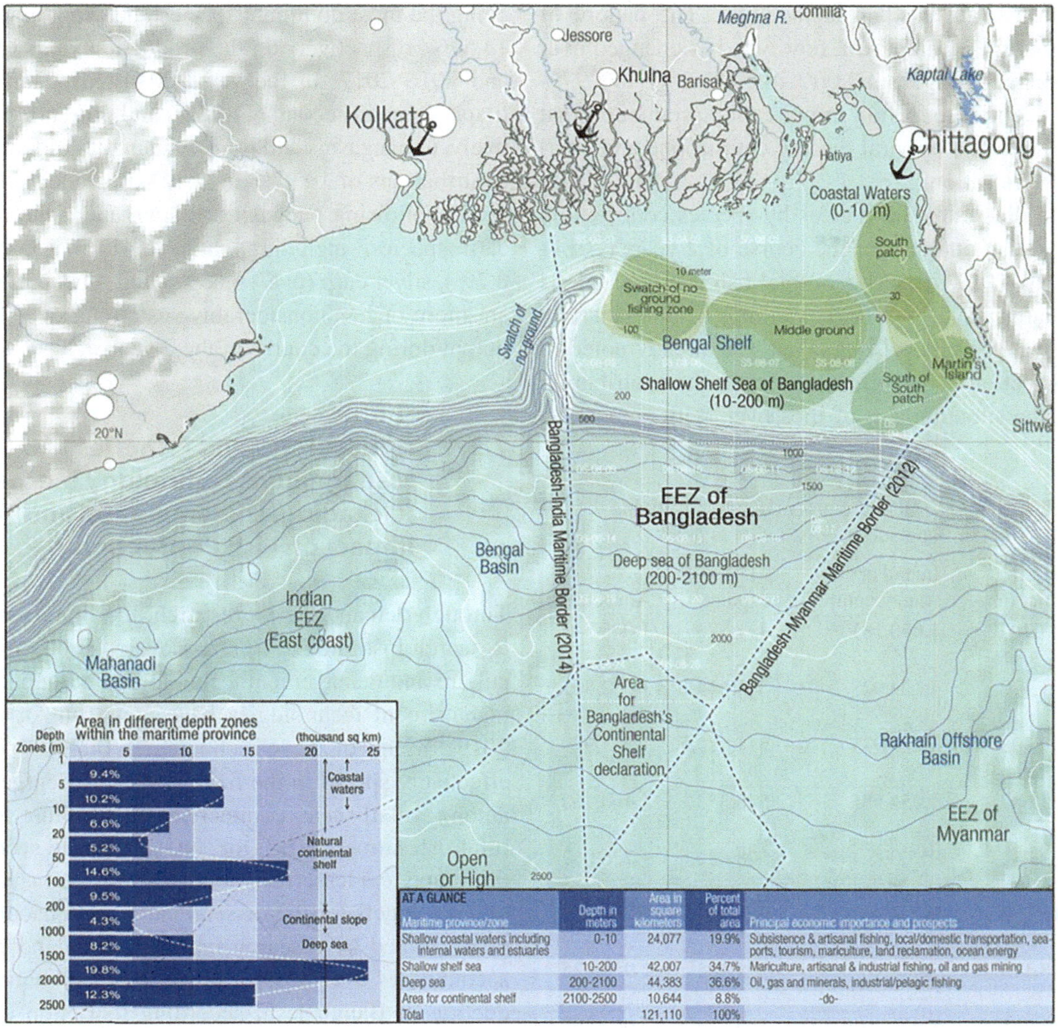

Fig. 7.1 Maritime area of Bangladesh. (Chowdhury 2014)

7.3.2 Trends in Fish and Aquaculture Production by Volume

The nation's overall fish production grew during the past 10 years, rising from 2,563,296 tons in 2008 to 4,134,434 tons in 2017 (Table 7.3). Inland capture fisheries account for 28.14% of total fish production between 2016 and 2017, followed by aquaculture at 56.44% and marine fisheries at 15.42%. The fishing sector currently contributes 3.65% of GDP (DoF 2017). Bangladesh's GDP is boosted by aquaculture alone by more than 2.81%, more than any other country in the Asia-Pacific area outside Vietnam (FAO 2017). Aquaculture output in Bangladesh has expanded significantly in recent years, much like it has in many other nations in the Asia-Pacific area. Among the top 10 nations in the world for inland aquaculture production, Bangladesh is one of the leaders (fifth in ranking) (FAO 2016a).

7.3.3 Group and Species-Wise Contribution to Total Fish Production

Table 7.4 provides a summary of the 10-year group-wise trends in total fish output (inland aquaculture + inland capture + sea fish production). Carp species—both native and alien carps, as well as other carp species—dominated the overall fish production in every year. This trend revealed that major indigenous carps, major exotic carps, and other carps (i.e., minor carps) contributed 9.91%, 2.44%, 31.98%, and 19.63%, respectively, to the overall carp population in 2016–17. In terms of volume and percentage of overall fish production, pangasius (12.34%), tilapia (8.95%), and shrimp/prawns (5.97%) are in second and third place, respectively, behind carps. Due to a sustained increase in pangasius and tilapia production in recent years, the contribution of all carps decreased from 33.37% in 2005–06 to 31.98% in 2016–17 (DoF 2017). Only shrimp and prawns from marine fish and shellfish are used in aquaculture. The pattern of support from shrimp and prawn, however, has changed over the past 10 years, falling from 9.06% in 2005–2006 to 5.97% in 2016–17 due to a reduction in per capita production that was greatly impacted by viral disease catastrophe and environmental deterioration.

Figure 7.2 illustrates the species-by-species contribution to total inland fish output over the course of 2 years (2014–2015 and 2016–2017). In terms of total fish and total inland fish output, carp species—all of them—play a dominant role, followed by pangasisus and tilapia, as was already noted in the previous section. However, if we look at the contribution per species, the situation sort of flips over, with pangasisus being the most dominant species and providing 13.19% and 14.59%, respectively, in the production years of 2014–2015 and 2016–2017. In the years 2014–2015 and 2016–2017, tilapia came in second (11.28%) and third (10.58%), respectively, to pangasius and rohu. As a matter of fact, Rohu made contributions of 10.97% in 2014–2015 and 10.6% in 2016–2017. Following pangasius, tilapia, and rohu, other notable carp species, including catla (6.41–6.7%), mrigal (6.2–6.7%), silver carp (6.47–6.63%), and other carps (4.76–5.0), are more or less considerably sustaining overall inland fish output during the course of the 2 years.

Table 7.3 Total fish and aquaculture production trends during the last 10 years (2008–2017) in Bangladesh (*Production in tons*)

Year	Total fish production	Inland open-water capture	Marine capture	Aquaculture
2007–08	2,563,296	1,060,181	497,573	1,005,542
2008–09	2,701,370	1,123,928	514,644	1,062,801
2009–10	2,899,198	1,029,937	517,282	1,351,979
2010–11	3,061,687	1,054,585	546,333	1,460,769
2011–12	3,261,782	957,095	578,620	1,726,067
2012–13	3,410,254	961,458	588,988	1,859,808
2013–14	3,548,115	995,805	595,385	1,956,925
2014–15	3,684,245	1,023,991	599,846	2,060,408
2015–16	3,878,324	1,048,242	626,528	2,203,554
2016–17	4,134,434	1,163,606	637,476	2,333,352

Source: DoF (2017)

7.3.4 Aquaculture and Fisheries Management Technology Innovation, Packaging, Dissemination and Their Impact on Fish Production

Through demand-driven research, the Bangladesh Fisheries Research Institute (BFRI) has developed 60 improved aquaculture and management technologies. 53 aquaculture and 7 management technologies have come out of those (BFRI 2019). Many of these technologies (Table 7.5) have been effectively spread in the field to varying degrees, especially among small-scale commercial aquaculture farmers and rural fish farmers. The DoF, NGOs, private sector organizations, and to a lesser extent NARS institutions have been the main channels for the dissemination of aquaculture innovations created by research institutions. Inland pond culture accounts for over 86% of all documented aquaculture production in Bangladesh, according to Belton et al. (2011), making it the industry's mainstay. Carp production dominates pond culture, while the next two most significant species raised in ponds are tilapia and pangasius catfish. In Bangladesh, tilapia farming has advanced significantly between 2005 and 2016 (Fig. 7.3). According to FAO

Table 7.4 Group-wise fish production (tons) and contribution (percent) to total fish production from 2005–06 to 2016–17 (DoF 2017)

	2005–06		2009–10		2014–15		2016–2017	
Group	Tons	Percent	Tons	Percent	Tons	Percent	Tons	Percent
Indigenous carps	509,995	21.90	692,597	23.89	755,074	20.50	811,588	19.63
Exotic carps	257,739	11.07	3,776,006	12.97	363,737	9.87	409,801	9.91
Other carps	9303	0.40	64,359	2.22	80,997	2.20	100,730	2.44
Pangas	–		–		406,818	11.04	510,097	12.34
Other catfish	34,104	1.46	208,972	7.20	64,537	1.75	66,646	1.61
Tilapia	–		–		347,801	9.44	370,017	8.95
Live fish	60,292	2.60	101,368	3.50	133,512	3.62	127,120	3.07
Snake head	101,309	4.35	113,989	3.93	69,305	1.88	72,991	1.77
Other inland fish	634,829	27.26	575,620	19.85	542,711	14.73	598,923	14.49
Hilsa	277,123	11.90	313,753	10.82	387,211	10.51	496,417	12.01
Shrimp/prawn	211,010	9.06	186,418	6.43	230,244	6.25	246,774	5.97
All other marine fish	232,841	10.00	266,116	9.19	302,296	8.21	323,332	7.28
Total production	**2,328,545**		**2,899,198**		**3,684,245**		**4,134,434**	

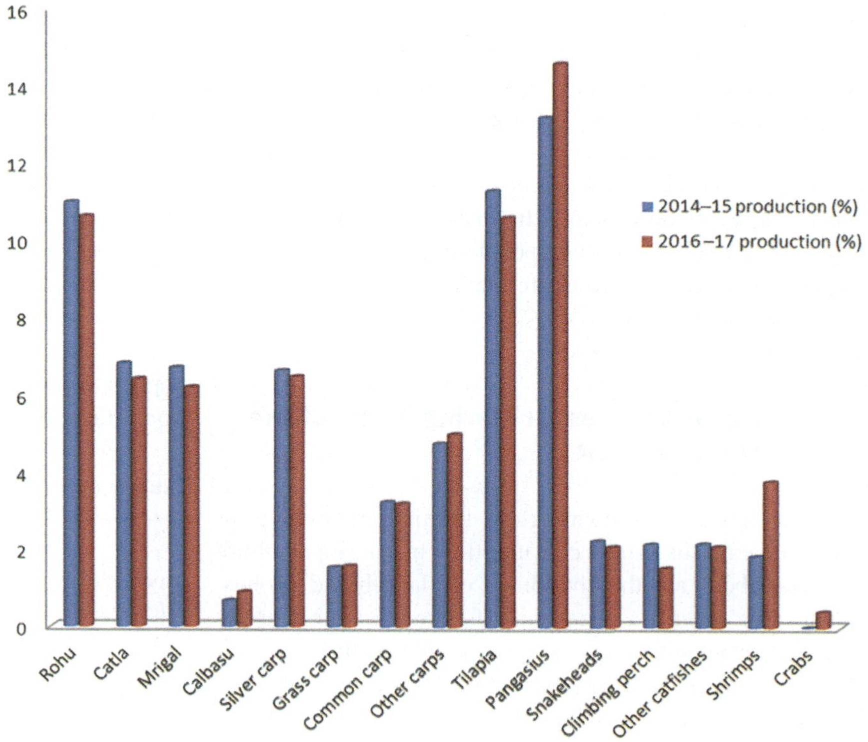

Fig. 7.2 Species contribution to annual fish production of inland water 2014–2015 and 2016–2017 in Bangladesh. (FRSS 2017; DoF 2017)

Table 7.5 List of major aquaculture technologies disseminated to the rural fish farmers and commercial aquaculture farming entrepreneurs in Bangladesh

Sl. #	Name of the aquaculture technologies
1.	Carp breeding and hatchery management
2.	Improved nursery management of carps
3.	Monosex seed production of BFRI super GIFT strain of tilapia
4.	Mass seed production technique of climbing perch (*Anabas testudinius*)
5.	Mass seed production of pangasius catfish
6.	Improved nursery management of carp, catfish, tilapia, and prawn
7.	Culture of silver barb in seasonal ponds
8.	Mixed and mono sex culture of GIFT strain of tilapia in seasonal ponds
9.	Polyculture of endemic and Chinese major carp
10.	Pond culture of Pangasius catfish
11.	Breeding and culture of seed production and culture of indigenous catfish species
12.	Integrated rice-fish and floodplain fish farming
13.	Fish culture in pens
14.	Improved extensive culture of shrimp in enclosures
15.	Crab fattening
16.	Fish health management

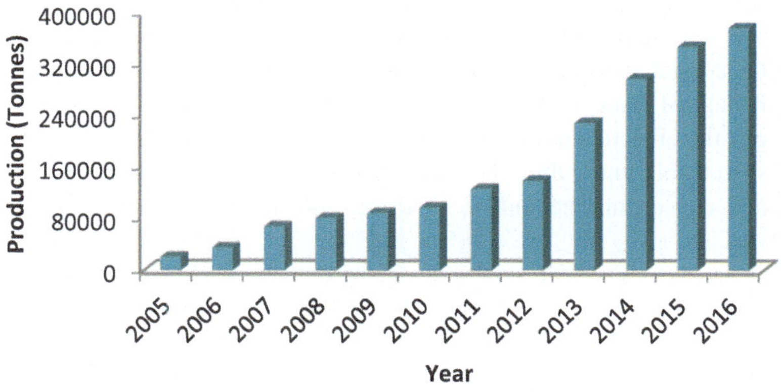

Fig. 7.3 Tilapia production over the last 12 years (2005–2016) in Bangladesh. (Hussain et al. 2017c)

(2016a), Bangladesh was one of the top eight tilapia-producing nations in the Asia-Pacific in 2014, trailing only China and Indonesia. According to Hussain et al. 2017a, production climbed from 19,268 tons in 2005 to 377,346 in 2016.

Many stakeholders rely on this sub-sector because it has grown to be a significant industry in the nation. According to Hossain (2014), 14.7 million people in Bangladesh are currently employed in aquaculture, including fish farmers and prawn/shrimp farmers.

7.4 Opportunities for Marine Aquaculture Development

For the purpose of commercial farming, including for financial gain for human consumption, marine aquaculture is described as the building of land-based ponds, enclosures/cages in coastal and offshore areas for the rearing of marine aquatic species/groups such as finfish, shellfish, and algae. With the exception of the tiger shrimp, *Penaeus monodon*, which is raised in the coastal regions using both conventional and more advanced extended methods, there have been little initiatives over the past 40 years to increase marine aquaculture in Bangladesh. The nations are located in the Asia-Pacific area. Using hatchery-produced seeds and formulated feeds, China, Indonesia, Malaysia, China's Taiwan Province, Thailand, Singapore, and Vietnam are growing the finfish mariculture industry. The government's plan for developing the blue economy can be applied to these nations. Bangladesh might quickly start up marine aquaculture employing the identified and readily available fish, shellfish, and non-traditional species. According to Hossain et al. 2017b, the shallow sea shelf and the shallow coastal waters (which include inland waters and estuaries) make up around 20% and 30% of the total marine waters, respectively. The potential and opportunity for the development of marine aquaculture are considerably increased by these facilities. As a result, brackish and marine species as well as non-traditional marine species have a lot of potential for mariculture. The nation may attain the anticipated bulk protein solutions and long-lasting advantages in the economic sectors if these opportunities could be successfully realized in the coastal, coastal, and offshore sectors. For the long-term benefit of Bangladesh's economy, environment, and society, the objective in this respect is to build a profitable and sustainable mariculture business that produces acceptable and chosen marine finfish, shellfish, and non-traditional species.

7.4.1 Potential Species for Marine Aquaculture

A recent joint regional workshop on the Blue Economy in South East of Bangladesh in Chittagong, Bangladesh, organized by Chittagong University, the European Union Delegation (EUD), Dhaka, and the Ministry of Foreign Affairs highlighted the potentials and opportunities of marine aquaculture: Major Opportunities and Constraints discussed in detail matters, government. from Bangladesh (Hossain et al. 2017a). In Bangladesh, the blue economy idea has been applied to a number of prospective marine aquaculture locations (Table 7.6).

Table 7.6 Marine aquaculture species/group-wise opportunities in coastal and marine waters of Bangladesh

Species/group	Main locations/areas for mariculture
Marine fin fish breeding and farming	
Sea bass (*Lates calcarifer*), Grey mullet (*Mugil cephalus*), Green back mullet (*Chelon subviridis*), Pomfret (*Pampus argenteus*), Hilsa (*Tenualosa ilisha*), saline tolerant tilapia, and other species	Land-based brackishwater pond culture, cage culture, and pen culture inshore and offshore areas, i.e., Moheshkhali-kutubdia channel, Sonadia island, Dubla island
Marine shrimp broodstock domestication and SPF seed production and farming	
Broodstock domestication, breeding, and farming of *Penaeus monodon, P. indicus,* and *P. merguiensis*	Cox's Bazar, Satkhira, Khulna
Crab breeding and farming	
Mud crab, *Scylla serrata*	Shamnagar, Shatkhira; Moheshkahi; Cox's Bazar
Mussel breeding and culture	
Clams (Anadara)	Chittagong, Moheshkhali, Kutubdia coast
Mussel (green mussel, *Perna viridis*)	St. Martin's Island, Moheshkhali, Kutubdia
Clam (*Meretrix meretrix*) and oyster (*Crassostrea madrasensis*)	Cox's Bazar, Sundarban mangrove
Pearly oyster (blacklip, nei)	Cox's Bazar, Patuakhali
Culture of marine aquatic Plants	
(a) Seaweeds	
Sargassum sp., *Hypnea* sp., *Caulerpa* sp. *Ulva* sp., and others	St. Martin's, Cox's Bazar and Khulna
(b) Marine microalgae	
Marine micro alagae viz. *Skeletonema costatum, Thalassiosira* sp., *Chaetoceros gracilis, Tetraselmis* sp., *Nanochloropsis oculata, Chlorella* sp., etc.	Chittagong, Cox's Bazar, Dhaka, Khulna (can be used as live feed and biodiesel production by marine biotechnology)
Culture of marine coral and other non-traditional species	
Marine corals, sea cucumber, sea urchin, etc.	St. Martin's Island

7.4.2 Essential Requirements and Challenges of Marine Aquaculture

In Bangladesh, there is no practice of marine aquaculture and no inshore or marine fin fish farming (Hossain 2014). Breeding specific marine species is a brand-new field called mariculture. In this approach, the mariculture industry can be developed to its maximum potential by taking lessons from both recent changes in other nations and our own inland aquaculture. The following sections provide a succinct description of the necessary conditions for a successful implementation of mariculture.

7.4.3 Ensure Suitable Site Selection for Mariculture

A good location in the coastal, coastal, and offshore zones is crucial for the construction of the necessary marine aquaculture facilities. Bangladesh's coastal regions and surrounding areas are appropriate for a variety of marine aquaculture types and forms due to the country's geographic location and climatic circumstances. Priority should be given to thorough feasibility studies that take into account the necessary environmental and ecosystem criteria before choosing locations for a certain mariculture crop. All ideal locations should have easy road access, necessary value chain infrastructure, ice factories, and market access, whether they are seed-producing hatcheries, nurseries, or commercial farms.

7.4.4 Secure Marine Fish and Shellfish Seed Production and Supply by Establishing Hatcheries

Hatcheries are a crucial component of the infrastructure since they produce large quantities of seeds that are necessary for the development of mariculture. Without sufficient quality, quantity, and consistency of marine species seeds or hatchlings, the mariculture sector would not prosper (Anon 2018). Sadly, there are no sea fin fish, crab, clam, or seed-producing seaweed breeding areas in the nation. To support current shrimp farming, tiger shrimp seeds are flown to the Khulna and Shatkhira regions from a number of shrimp hatcheries in the Coxs Bazar region. However, aside from a relatively small number of these outdated infrastructures, the majority of these shrimp hatcheries have outdated interiors. It is critically necessary to create a number of fish and crab hatcheries in the proper places in Khulna, Shamnagar, and Shatkhira. Regions of Cox's Bazar, Kutubdia, and Moheshkahi. These hatcheries should be equipped with all contemporary indoor and outdoor seed-rearing apparatus, including sufficient saltwater supply systems, egg incubators and jars, larval troughs and tanks, juvenile tanks and ponds, etc. On the coast of Saint Martins Island and other similar areas, specialized hatcheries from South Korea, Japan, and Australia might be taken over in tandem to establish hatcheries for the production of mussel and seaweed seeds.

7.4.5 Design and Build Mariculture Farming Infrastructure and Facilities

Establishing commercial mariculture agricultural infrastructures for raising marine fish, shellfish, and other relevant species is the next job after ensuring adequate seed and juvenile production and availability. Given that the following systems must be built in order of priority:

(a) *Coastal marine fish farms having earthen ponds*

Large expanses of low-lying coastal lands are typically used for traditional shrimp farming in Bangladesh, mostly in the South West region, Khulna and Shatkhira districts. To start semi-intensive shrimp and crab farming as well as finfish farming, such as sea bass, grey mullet, green back mullet, pomfret, etc., better farming infrastructure can be built in those areas, especially in the greater Khulna region as well as in Chakuria, Kutubdia, and Moheshkhali of Cox's Bazar region. Better yield and output in this system will require artificial feeding using premium floating or sinking pelleted feeds. A well-designed farm should have a size of between 5 and 25 hectares, a sufficient supply of salt water, and suitable water depths.

(b) *Inshore raceways using net enclosures*

There are numerous excellent coastal locations to build inshore raceways, where net enclosures can be erected at the coastal lagoons and canals at various lengths and required depths. With bamboo, PVC, or steel poles, polyethylene/nylon net materials with mesh sizes >2–5 cm can be fixed. With supplemental feeding and water quality control, marine fin fish juveniles, such as hilsa, sea bass, grey mullet, etc., can be actively stocked and nurtured to grow to marketable size.

(c) *Offshore floating net cages*

Bangladesh will launch a new program to cultivate marine finfish species in offshore net cages. It is a typical practice in various countries of the Indo-Pacific region, including Indonesia, China, Norway, Japan, Brazil, and Mexico. At the artisanal level, net cage farming can be implemented in the nation with a straightforward design, small size, and depths of 4–5 m (Hossain et al. 2017b). They suggested that using locally accessible materials like bamboo, wooden boards, steel or PVC pipes, and nylon nets, local farmers and fishers might construct their own cages. In the future, Sonadia Island, the Andermanik rivers, and other suitable locations could be used to begin the improved construction of net cages. Marine finfish juveniles like hilsa, sea bass, grey mullet, seabream, and other species can be stocked densely and fostered with consistent, high-quality artificial feeding to grow in net cages.

(d) *Hanging racks and floating rafts for mussels culture*

Mussels are a common non-traditional marine species found in our marine waters' coastal zone. These creatures are grouped together on different substrates and feed on the phytoplankton and other debris that is present in the ocean. The most effective source of tasty and nutrient-rich animal protein is found in mussels, according to Aypa (1980). In order to cultivate gathered mussel spats into sizes that may be sold, both floating rafts and hanging racks in the shallow area are typically used. The Moheshkhali, Kutubdia, and Sonadia islands of the Cox's Bazar and Teknaf regions offer particularly good potential for the production of green mussel, clam, and oysters (Shahabuddin et al. 2010).

(e) *Setting reef flats, mangrove stakes/nets, and long line culture methods of seaweed culture*

This nation might start a widespread seaweed culture to produce food products and earn foreign exchange. Seaweeds that are naturally growing can be found in Bangladesh from October to April in the littoral and sub-littoral zones from St. Martin's Island to the Sundarbans Mangrove forest area (Islam and Aziz 1987; Islam 1998). The St. Martin's Island is a superb location for naturally occurring seaweeds due to its water quality parameters and rocky substratum (Khan 1990; Tomascik 1997; Zafar 2005). According to Sarkar et al. (2016), Bangladesh is home to 19 commercially significant seaweed species out of a total of 193 species. The two methods of long line and reef flat cultivation of seaweeds are used most frequently in this country for commercial seaweed cultivation. The mangrove stakes are pushed into the substrate, and the twine is connected to seedlings and strung between them. At a depth of 1 m, this off-ground technique is still among the most significant ones now in use. Deeper water, at a depth of around 7 m, can be employed for new longline farming techniques. Algae farming is reportedly capable of contributing to biological carbon sequestration.

(f) *Adoption of integrated multi-trophic aquaculture*

One adaptation strategy being used in the present to address the problem of climate change is integrated multi-trophic aquaculture (IMTA). An innovative method of producing different finfish and shellfish with algae in an integrated farm would be open-water IMTA on the Bangladeshi coast (Ahmed and Glaser 2016). According to Hussain et al. (2017b), IMTA is an ecosystem-based adaptation technique to climate change that may have positive effects on both the environment and the economy. Over 40 nations, including Canada, Chile, China, Japan, the USA, and several European nations, participate in IMTA on an experimental and commercial basis (Chopin 2011). Recent study and development on variety production in Bangladesh was taken into consideration by IMTA (Kibria 2016). Hossain et al. (2017b) observed that although regions like the Cox's Bazar—Teknaf coast and the islands of St. Martin's, Moheshkhali, and Sonadia can be thought of as potential locations for the deployment of IMTA, no attempts have yet been made to build an MTA system in Bangladesh.

7.4.6 Ensure Domestication of Mariculture Species

In inland aquaculture systems around the world, including Bangladesh, domesticating wild fish species for use in the production of artificial seed is a prevalent practice. To start off accurately for a few marine finfish and shellfish species in Bangladesh, similar techniques won't differ. Therefore, for a few months in breeding tanks, raceways, or ponds under ideal salinity, wild-collected juveniles or brood stocks must be accustomed to and domesticated in order to improve their gonadal and sexual maturity before seed production can begin in marine hatcheries. The next phase is to choose the best-sized, leanest, and mature finfish and shellfish for hatchery breeding. For growing mollusks and other non-traditional species, a similar scenario might need to be followed. For some inbred and chosen breeding stocks, selective breeding technique(s) could be used to prevent genetic degeneration.

7.4.7 Ensure Production and Availability of High-quality Feeds

For some species, such as tiger shrimp and predatory finfish like sea bass, intensive feeding with high-quality feeds is necessary, and it may be preferable in the case of other marine species to enhance production, shorten grow-out times, and accelerate the return on investment in ponds or cages. Dry pelleted formulated feeds have become more popular in Bangladesh, where there are several well-known feed mills, such Saudi Bangla Fish Feed, Nusrish, Quality Feed, and Paragon Feed Mills. Science and economics have shown that high-quality meals produce high and reliable performance. They greatly lower labor expenses and are simple to handle and store. To ensure an optimal standard of such feeds, it is necessary to take into account the simultaneous import of high-quality fish meal or other necessary materials from overseas in order to maximize feed costs vs. quality.

7.4.8 Secure and Promote Investment in Mariculture

Gaining sufficient assets to support mariculture operations in Bangladesh is still a struggle for many interested parties and businesspeople because it is a new endeavor and a risky task. The construction of new infrastructure is necessary for marine hatcheries, commercial land-based farms, inshore raceways, offshore cages, etc. No one will be interested unless and until a foreign donor or a state-owned bank from within the country steps forward to assist initiatives like the costs of infrastructure construction and the operation of mariculture systems. It will be wiser to create public-private partnerships to manage the mariculture sector for the long term. For the advantage of the industry, the relevant government departments and ministries should take appropriate action.

7.4.9 Build Public Understanding and Support for Mariculture

Building public awareness and support for one of the recently emerging sectors of the blue economy, namely mariculture, is essential to the development of mariculture in Bangladesh, unlike other fisheries and agricultural sectors. No amount of public or private investment can result in execution and success without the enthusiastic support of the affected population and the ensuing political approbation (Anon 2018). It is clear that mariculture's sustainable function will assist coastal towns with limited resources and, eventually, the nation's environment and economy.

7.4.10 Develop Mariculture Markets and Products

There may not be adequate demand for some processed products and products from the marine sector, such as carbohydrates, mussels, and seaweeds. The producers and government departments/ministries will therefore be concerned about how to export those mariculture products to domestic and foreign markets. To develop a workable solution in this situation, domestic and international marketing organizations will need to work together. Research and development in marketing and product development will aid in boosting production as a nation's mariculture of shellfish and aquatic plants expands, creating more opportunities and dependable revenue streams for the sector and the nation.

7.4.11 Build Public Institution(s) for Mariculture Research and Technical Education

There is just one technical institution in Bangladesh that does apply marine aquaculture research and training, and that is the Marine and Technology Station of the Bangladesh Fisheries Research Institute (BFRI). For conducting need-based marine fisheries/aquaculture and oceanographic research, the BFRI's marine research station in Coxs's research facilities is insufficient (Hussain et al. 2017c). There is potential for an equitable and mutually beneficial partnership between European and other international universities and Bangladeshi research institutions/universities in this area to create high tech. education programs pertaining to mariculture. On the other hand, a new initiative can be launched soon to establish a fully flagged new mariculture research institution in the Cox's Bazar coastal belt with all labs and other facilities for marine fish/shellfish and non-traditional species breeding/propagation, biotechnology, culture, health management, and product development.

7.5 The Way Forward

The following is the top advice from this thorough assessment of prospective interior, coastal, and offshore fisheries resources to guarantee solutions for mass protein in Bangladesh:

- Ongoing inland aquaculture effort supported by technology to increase fish output in current freshwater bodies
- Establish contemporary hatcheries and seed-rearing facilities to ensure enough seed production and supply of mariculture species
- Develop new aquaculture strain(s) and new technologies through demand-driven and biotechnological research. Ensure domestication of potential mariculture of finfish and shellfish species
- Create mariculture farming infrastructure and facilities
- Ensure production and availability of high-quality feeds
- Build public understanding and support for mariculture
- Build public institution(s) for mariculture research and technical education
- Ensure equitable dissemination of newly innovated adequate inland freshwater and mariculture technologies to boost national fish production and economy.

References

Ahmed N, Glaser M (2016) Can "integrated multi-trophic aquaculture" adapt to climate change in Coastal Bangladesh. Ocean Coast Manag 132:120–131

Anon (2018) Alaska mariculture development plan. State of Alaska, 81p. http://www.adfg.alaska.gov/Static/fishing/pdfs/mariculture/ak_mariculture_devplan_06-29-18.pdf

Aypa SM (1980) Factors affecting recovery and growth rate of transplanted mussels, Perna viridis (Linneus). Master thesis submitted to U.P

Belton B et al (2011). Review of aquaculture and fish consumption in Bangladesh. Studies and reviews 2011–53. The World Fish Center, November 2011

BFRI (2019) BFRI at a glance. Bangladesh Fisheries Research Institute, Mymensingh, 40p

Chopin T (2011) Progress of the integrated multi-trophic aquaculture (IMTA) concept and up scaling of IMTA system towards commercialization. Aquacult Eur 36:5–12

Chowdhury SR (2014) Maritime province of Bangladesh (map). University of Chittagong, Chittagong

DoF (2017) Yearbook of Fisheries Statistics of Bangladesh 2016–17. Fisheries Resources Survey System (FRSS), Department of Fisheries, vol 34. Director General, DoF, Pilkhana, 129p

FAO (2016a) 2016 global aquaculture production dataset 1950–2014 (Fishstat). Available at: www.fao.org/fishery/statistics/software/fishstatj/en

FAO (2016b) The state of world fisheries and aquaculture 2016. Contributing to food security and nutrition for all. FAO, Rome, 200pp. FAO, Fish Stat J Database

FAO (2017) Regional review on status and trends in aquaculture development in Asia-Pacific – 2015, FAO fisheries and aquaculture circular no. 1135/5. Rome, Italy, 45pp

FAO/Network of Aquaculture Centres in Asia-Pacific (NACA) (2011) Regional review on status and trends in Aquaculture Development in Asia-Pacific – 2010, FAO fisheries and aquaculture circular. No. 1061/5. Rome, FAO, 89 pp

FRSS (2017) Yearbook of fisheries statistics of Bangladesh, vol 33. Fisheries Resources Survey System (FRSS), Department of Fisheries, Dhaka, 124 p

Hossain MAR (2014) An overview of fisheries sector of Bangladesh. Res Agric Livest Fish 1(1):109–126. https://doi.org/10.3329/ralf.v1i1.22375

Hossain MS, Failler P, Hussain MG (2017a) Regional workshop on blue economy in south east of Bangladesh: major opportunities and constraints, 8 May 2017. Chittagong, Bangladesh, 20p

Hossain MS, Chowdhury SR, Sharifuzzaman SM (2017b) Blue economic development in Bangladesh: a policy guide to marine fisheries and aquaculture. Institute of Marine Sciences and Fisheries, University of Chittagong, Chittagong, 32p

Hussain MG, Kohinoor AHM, Moshiur Rahman MM, Zillur Rahman M, Abdullah Al Masum AA, Nguyen NH (2017a) Genetic improvement and effective dissemination improved Nile tilapia strain in Bangladesh. Aquacult Asia Pac Mag 13(6):49–50. November/December issue

Hussain MG, Failler P, Karim AA, Alam MK (2017b) Review on opportunities, constraints and challenges of blue economy development in Bangladesh. J Fish Life Sci 2(1):45–57

Hussain MG, Failler P, Karim AA, Alam MK (2017c) Major opportunities of blue economy development in Bangladesh. J Indian Ocean Reg. https://doi.org/10.1080/19480881.2017.1368250

Islam AKMN (1998) The seaweed resources of Bangladesh. In: Critchley A, Ohno M (eds) Seaweed resources of the world. Japan International Cooperation Agency, Nagai, pp 106–109

Islam AKMN, Aziz A (1987) Addition to the list of marine algae of St. Martin's Island, Bangladesh. III. Redalage (rhodophyceae). Nova Hedwigia 45:211–222. 16

Khan G (1990) Status of production and utilization of seaweeds in Bangladesh. In: Report of the regional workshop on the culture & utilization of seaweeds. FAO, Rome

Kibria ASM (2016) Studies on Integrated multi-trophic aquaculture (IMTA) systems in freshwater ponds, PhD thesis. Department of Aquaculture, Bangladesh Agricultural University, Mymensingh

Marine Aquarium Council/The Nature Conservancy (2004) International standards for the trade in live reef food fish. Hawaii. July 2004, 18p

Mazid MA (2002) Development of fisheries in Bangladesh: plans and strategies for income generation and poverty alleviation. Dhaka, Bangladesh, 176p

MoFA (2014) Press release: press statement of the Hon'ble Foreign Minister on the verdict of the arbitral tribunal/PCA. Ministry of Foreign Affairs, Dhaka. Retrieved from http://www.mofa.gov.bd/PressRelease/PRDetails.php?txtUserId=&PRid=854

Sarkar MSI, Kamal M, Hasan MM, Hossain MI (2016) Present status of naturally occurring seaweed flora and their utilization in Bangladesh. Res Agric Livest Fish 3(1):203–216

Shahabuddin AM, Wahab MA, Miah MI, Salam MA (2010) Abundance, distribution and culture potentials of three commercially important mollusks species along the Coast of Bay of Bengal. Res J Agric Biol Sci 6(6):754–762

Tomascik T (1997) Management plan for coral resources of Narikel Jinjira (St. Martin's Island). In: Final report, national conservation strategy implementation project-1. Ministry of Environment and Forest, Government of Bangladesh, Dhaka, p 125

Zafar M (2005) Seaweed culture in Bangladesh holds promise. INFOFISH Int 1:18–10

Commercial Fisheries and Aquaculture Shrimp in Bangladesh: Production, Export, and Local Supply Chain

8

Md. Saidul Islam

Abstract

The chapter provides an overview of Bangladesh's landscapes, emphasizing its topography, climate, population, and sources of revenue, as well as its levels of poverty and standard of living. The chapter begins with a brief description of Bangladesh as a geographical location before examining Bangladesh's fisheries sector, emphasizing its water area, types of fisheries, production, and yields, income and employment contribution to the economy, institutional arrangements, and policies. Shrimp is primarily regarded as an indispensable component of Bangladesh's fisheries sector because of this. The research gaps are then identified after a thorough examination of the prior literature on shrimp aquaculture in Bangladesh. Finally, the chapter gives a broad overview of Bangladesh's shrimp industry, concentrating especially on its production, export, and local supply chain.

Keywords

Fisheries · Aquaculture · Shrimp · Commodity chain · Department of fisheries · Bangladesh

8.1 Introduction

In some areas, traditional export commodities like coffee, tea, sugar, tobacco, cocoa, and others have fallen over the past few decades, and "high-value foods" (HVFs), as Goodman and Watts (1997: 10) refer to them, have taken their place. Shrimp, which is largely cultivated in tropical nations in the Global South, is one of these "high-value foods". Bangladesh's coastal regions have evolved into an essential geographic site for shrimp production since grown shrimp require a special type of tropical environment. By producing this high-value food item, Bangladesh entered the global supply chain for agro-food commodities (Islam 2014).

The process of local or regional integration into the global markets was furthered by the "first green revolution" (McMichael 2004). In the 1960s, the "second green revolution" was described as the application of green revolution technologies to luxury foods, sometimes known as "high-value foods" (Sanderson 1986; Raynolds et al. 1993; Raynolds 1994). Ten years hence, the global agro-food system could see the "blue revolution," or an analogous change in aquaculture. Cultured shrimp emerged as the "pinnacle of the blue revolution's achievement" (Public Citizen 2005: 9), and Bangladesh eventually entered the group of countries known as the "new agricultural countries" (NACs). Despite the fact that Bangladesh's subsistence farmers have been harvesting shrimp for centuries without substantially impacting society or the environment, the country's integration into the global food system has altered its agrarian and environmental landscapes, creating complex intersections and occasionally tense situations (Islam 2009, 2014).

A World Bank structural adjustment program (SAP) loan to Bangladesh totaling $1.76 billion from 1979 to 1996 was one of the main factors influencing this transformation (Battacharya et al. 1999). As this program was being put into action, "policies that had limited trade were replaced with new policies that encouraged exports," and "the changes created an environment in which private investments in shrimp culture, shrimp processing, and shrimp exports flourished" (Roheim 2004: 277). The crucial World Bank financing and subsequent political developments led to a major expansion in commercial farming in Bangladesh's coastal regions. With time, Bangladesh started to become a major role in the world agro-food system. As foreign investment increasingly strengthens the local economy, many rice farmers in Bangladesh were pushed to use low-salinity culture systems that rely on seawater or salt farm effluent trucked inland and

Md. S. Islam (✉)
Division of Sociology, Nanyang Technological University Singapore (NTU), Singapore, Singapore
e-mail: msaidul@ntu.edu.sg

R. Ahmed et al. (eds.), *Transforming Bangladesh*, World Regional Geography Book Series,
https://doi.org/10.1007/978-3-031-45093-8_8

blended with freshwater. When they started converting paddy fields to shrimp ponds, the amount of land used for shrimp farming increased, leading to major changes in Bangladesh's agricultural and natural landscapes (Islam 2009).

The chapter begins with a brief description of Bangladesh's geography before moving quickly through an analysis of the sector's water area, types of fisheries, production and yields, economic contribution in terms of income and employment, and institutional structures and policies. This is because shrimp is widely regarded as an essential part of Bangladesh's fisheries sector. The research on shrimp aquaculture in Bangladesh is thoroughly analyzed in the section of the chapter that follows, and common trends and trajectories are discovered. The chapter concludes with a general overview of the shrimp business in Bangladesh, focusing particularly on its privatization policies and regional supply chain.

8.2 Overview of the Fisheries Sector in Bangladesh

8.2.1 Water Area and Types of Fisheries

Bangladesh's water bodies are the most valuable resource for the freshwater ecology and are perfect for fish farming. After China and India, Bangladesh is the third-largest producer of inland fish worldwide. More over 60% of the population's per capita animal intake comes from fish in Bangladesh, making it the country's main source of animal protein. Approximately 5.3 million hectares of water are used for inland fishing, of which 92.27% is inland open water and 7.73% is inland closed water (DOF 2002). In the past, fish were fished out of rivers, canals, and other natural sites. Aquaculture became the main fish source in the 2000s, accounting for more than 40% of the nation's total value fish output (Karim et al. 2006: 2). The Ministry of Fisheries claims that this development was brought about by three main factors. First, as more extensive use of the common property increased pressure on the property's natural and environmental boundaries, the likelihood of further boosting fish productivity by exploiting open water bodies drastically decreased. Second, the demand for fish was substantially increased by the growing dietary requirements of a rapidly expanding population with greater purchasing power. This led to a significant increase in the real price of fish and improved the profitability of fish farming compared to other food crops. With regard to shrimp in particular, the opening of a worldwide market considerably increased demand and revenue. In order for farmers to analyze the costs and benefits of various commodities and technologies and decide whether aquaculture was more profitable than other rice-producing alternatives, less complex and less expensive technologies were made available to them (Karim et al. 2006; Khatun 2004). The following table lists the types of waters and areas that are appropriate for fisheries and aquaculture (Table 8.1).

According to the Ministry of Fisheries, Bangladesh has 36 shrimp species, 25 prawn species, 475 marine fish species, and about 260 freshwater fish species in the early 2000s. Hilsha, carp, catfish, prawn, and other fish species are among Bangladesh's top five fish species (DOF 2004). During monsoon floods, all fish and shrimp species enjoy population increase, filling every opening in the expanded aquatic environment that the floods generate. 4.92 million ha are used for inland capture (open water), including polder; 0.44 million ha are used for inland culture (closed water, including brackish water); 0.14 million ha are used for brackish water; and 16.61 million ha are used for marine water. The Ministry of Fisheries (DOF 2004) determined these four groups. Over the previous 10 years (2004–2005 to 2013–2014 F.Y.), the growth of the fishing industry has been more or less stable, averaging 5.38% per year and varying from 7.32% growth in 2009–2010 to 4.04% growth in 2013–2014 (Bangladesh Economic Review 2014; Shamsuzzaman et al. 2017).

8.2.2 Production and Yield

According to Khatun (2004), inland fisheries account for over 78% of total fish production in Bangladesh, making them the country's main source of fish. The entire area of inland fisheries was projected to be 5.3 million ha in 2002, with a total catch of 1475 thousand metric tons (mt), or 278 kg per hectare. Compared to open water capture, closed culture water contributes more to inland fisheries. Inland waterways' open-water fisheries are natural systems. Pond

Table 8.1 Water types and areas in Bangladesh

	Water type	Area
(a)	Closed water body (culture-based)	7,97,851 ha
	(i) Pond & ditches	3,91,753 ha
	(ii) Oxbow lake	5488 ha
	(iii) Shrimp farm	2,58,681 ha
(b)	Open water body (capture-based)	39,27,142 ha
	(i) River & estuaries (without Sundarban)	8,53,863 ha
	(ii) Beel	1,14,161 ha
	(iii) Kaptai lake	68,800 ha
	(iv) Flood plain:	27,12,618 ha
(c)	Marine fisheries	
	(i) Territorial water	2680 sq. n. miles
	(ii) Exclusive economic zone	41,040 sq. n. miles
	(iii) Continental shelf	24,800 sq. n. miles
	(iv) Coastline	710 km.

Source: DOF (2022)

fishing and marine aquaculture are the main sources of the inland confined water fisheries. The production of brackish water shrimp is the primary element of coastal aquaculture. Due to its production for sale to the European Union, the United States, and Japan, farmed shrimp is only occasionally consumed locally. Locals rely on local freshwater fish instead. About 98% of all fish exports go to the aforementioned countries, with the remaining 2% going to Southeast Asia and the Middle East (DOF 2022). In comparison to the production from wild capture (1.023 million metric tons) in 2016, aquaculture production increased from 712,640 to 2,060,408 metric tons between 2000 and 2016 (Shamsuzzaman et al. 2017: 145). Table 8.2 provides a concise summary of the value and productivity of fisheries.

8.2.3 Income and Employment

Many people rely on the fishing industry as a source of income and employment, particularly in Bangladesh's rural areas. The Ministry of Fisheries of Bangladesh has divided fishing fishermen into the following four categories: There are four different categories of fishermen: (i) professionals who rely solely on fishing for a living; (ii) part-timers who occasionally fish for subsistence rather than income; (iii) subsistence fishermen who occasionally fish for subsistence rather than income; and (iv) fish farmers who own ponds and manage fishing operations. The sector employs more than two million people full-time, making up 7% of all occupations in Bangladesh (DOF 2002).

The sector's overall income increased from 65.82 billion taka in 1994 to 137.49 billion taka in 2002. Value added by the fishing industry as a percentage of GDP climbed from 5.04% in 1994 to 5.24% in 2002. Fish exports have grown to be one of Bangladesh's most significant sources of foreign exchange gains. In terms of export earnings, fish and fish products rank first among non-traditional goods. Despite the sector's share of export earnings declining from 6.91% in 1991 to 4.76% in 2002, the amount of fish sent has almost doubled (41,482 mt) from 22,080 in 1992. The principal fish exports are frozen fish, frozen frog legs, and frozen shrimp (Khatun 2004). Shamsuzzaman et al. (2017: 149) estimate that the export value was around $443.1063, 598.4771, 611.5135, 560.6393, and 636.7686 million USD in 2010, 2011, 2012, 2013, and 2014.

Table 8.2 Export of fish and fish products

	Source-wise production		Other fish products		Total
Year	Quantity (M.T.)	Value (crore taka)[a]	Quantity (M.T.)	Value (crore taka)	Value (crore taka)
2017–2018	36,167.77	3527.07	32,767.93	782.87	4309.94
2016–2017	39,705.85	3682.26	28,599.83	605.38	4287.64
2015–2016	40,726	3598.67	34,612	684.15	4282.82
2014–2015	44,278	3937.60	39,246	723	4660.60
2013–2014	47,635	4118.80	29,693	658.12	4776.92

Source: DOF (2022)
[a](1 core = 10 millions)

8.2.4 Institutional Arrangements and Policies

A number of factors, according to Bangladesh, make effective fishery management essential, including (i) the availability of cheap protein, (ii) the reduction of poverty through the creation of jobs and income, (iii) the maximization of economic yield, (iv) the maintenance of the quality of fish exports, and (v) the efficient and equitable distribution of benefits among the fishing community. Due to the interconnected nature of the challenges pertaining to the sector, Bangladesh's institutional structure for fisheries is extremely large, involving a number of ministries, directorates, and institutes. Even though the Ministry of Fisheries and Livestock is in charge of overall administration, management, extension, production, research, and training, other ministries also have some responsibilities. An overview of Bangladesh's institutional structure for fisheries (which includes shrimp farming) may be seen in Table 8.3, which was modified from Khatun (2004), Gammage et al. (2006), and Islam (2014).

8.3 Critical Research on Aquaculture in Bangladesh

The initial goal of commercial aquaculture's introduction in Bangladesh was to increase foreign investment and local employment opportunities, but it also has a number of social and environmental consequences. Since the early 1990s, numerous national and international NGOs, researchers, and others have voiced serious concerns regarding the negative externalities and effects of commercial shrimp farming in countries like Bangladesh that produce shrimp. A wealth of literature on the subject, spanning from production to consumption in various parts of the world, has linked the emergence and growth of shrimp culture to a number of social and environmental issues (Shiva 1995; Philips et al. 1993; CAP 1995; Clay 1996; Sernbo and Kloth 1996; Boyd and Clay 1998; Vandergeest et al. 1999; Flaherty et al. 1999; Scott 2000; Clay 2004). Some of these problems include the removal of natives frequently as a result of violence and con-

Table 8.3 Institutional setup of shrimp culture in Bangladesh

	Ministry/institutions	Activities
1.	*Ministry of Fisheries and Livestock*	
	Directorate of Fisheries	Administrative, management, development, extension and training
	Bangladesh Fisheries Development Corporation	Training, production and development
	Fisheries Research Institute	Research and training
2.	*Ministry of Local Government, Rural Development and Cooperative*	
	Rural Development Board	Fisheries component of integrated rural development
	Directorate of Cooperatives	Registration and supervision of fisheries cooperatives
	Bangladesh Jatioy Matshyajibi Samabaya Samity Ltd.	Development of fisheries cooperative, operation of ice plant, and import of gear
	Bangladesh Samabaya Bank Ltd.	Financing fisherman cooperatives
	Upazilla Administration	Management of water bodies less than 20 ha
3.	*Ministry of Land*	
	Land Administration and Land Reform Division	Leasing of public water bodies
4.	*Ministry of Irrigation, Water Management and Flood Control*	
	Bangladesh Water Development Board	Leasing of reservoir and irrigation canals
5.	*Ministry of Commerce*	
	Department of Commerce	Leasing of fish processing plant
	Export Promotion Bureau	Export promotion of shrimp, fish, and fish products
6.	*Ministry of Industry*	
	Bangladesh sugar and food industries corporation	Processing of shrimp and fish
7.	*Ministry of Shipping*	
	Mercantile Marine Department	Registration of fishing boat
8.	*Ministry of Youth*	
	Youth Development Training Centers	Extension and the training of unemployed young people and fish farmers.
9.	*Ministry of Planning*	
	Fisheries Section	Planning and overall coordination of all development activities on fisheries
10.	*Ministry of Foreign Affairs*	
		Negotiation of exclusive economic zone

flict, the destruction of mangrove trees and other species in coastal belts, and tensions in the countryside. Others include the invasion of salt into paddy crops and its attack on the surroundings. There is still a lot of debate over the sustainability of shrimp culture, despite the fact that several initiatives have been taken to solve these problems (Bene 2005; Vandergeest et al. 1999; Gammage et al. 2006; USAID 2006; Roheim 2004; Bush et al. 2013). As a result of this situation, numerous environmental and social justice NGOs are still actively engaged in this area. According to reports, more than 60% of Asia's mangroves, including the Sundarban, the world's biggest mangrove forest, have already been converted into aquaculture farms, primarily for the production of shrimp. Although mangrove forests may not always be the ideal setting for shrimp farming, environmental and social justice NGOs have made a strong case that shrimp cultivation in tropical regions is negatively correlated with mangrove degradation. Since shrimp is such a big export for its farmers and has such a clear negative impact on their environment, numerous environmental, social justice, and human rights NGOs are actively involved in this industry, either fighting against or promoting it (Islam 2014). Bangladesh is undoubtedly a top option for inquiry as one of the world's leading producers of shrimp. Numerous studies on shrimp aquaculture in Bangladesh conducted since the early 1990s have documented detrimental effects brought on by the industry, particularly on the nation's social and environmental landscapes (Nijera Kori 1996; Rahman 1995; Manju 1996; Adnan 1991; Thamina 1995; Ahmed 1996; Deb 1998; Battacharya et al. 1999). The identified impacts, according to Battacharya et al. (1999), are "well-articulated if not always well evidenced" (p. 19). The majority of studies concluded that Bangladesh's growing shrimp culture was primarily unplanned, uncoordinated, unmanaged, and unregulated. As a result, Deb (1998) referred to it as a "fake blue revolution" (p. 63). The general impact of this unregulated expansion, as documented by the researchers, were, among many other things: (a) overfishing in the rivers and estuaries (as the industry is mainly dependent on wild sources of fry); (b) the loss of shellfish and finfish (due to the over catching of wild tiger shrimp fry); (c) the destruction of fragile mangrove ecosystems in the coastal regions (which provide many other benefits to society, ecology and other species); (d) soil acidity and salt incursion with flow-on effects on other coastal agricultural activities, including paddy farming, (e) contribution to both inorganic and organic coastal pollution (as intensive shrimp farming also requires large amounts of freshwater, approximately 25–30 million liters to raise one ton of shrimp, and shrimp pond effluent such as ammonia, nitrate, phosphate, bacteria); and (f) adverse socio-economic effects as shrimp cultivation included the erosion of the traditional resource-use rights of coastal people, the displacement of the local population, and rural conflict (Islam 2014).

The majority of these studies, which were conducted in the middle of the 1990s, also revealed that Bangladesh's shrimp aquaculture industry altered land-use patterns, which had a considerable impact on traditional agricultural practices and endeavors (Nijera Kori 1996). Large shrimp growers were out-of-town company owners who had no long-term commitment to or interest in the development of Bangladesh's coastal community, according to Manju (1996) and Nijera Kori (1996), who also stated that the issue grew worse as a result. The Bangladeshi government issued permits to a range of business owners so they could cultivate shrimp in Bangladesh's coastal districts when shrimp aquaculture initially got started. Because most of the businesspeople were immigrants to those regions, the locals labeled them as "outsiders." Sharecroppers were subsequently denied access to the primary productive resource and opportunity, which resulted in a significant rise in unemployment. Sharecroppers had previously depended on renting and leasing cultivable land for their living. Aside from that, traditional economic activities like social forestry, domestic gardening, small-scale fishing, and poultry farming began to decline. Shrimp gradually became unattainable in many areas where they had been produced for a long time.

The negative outcomes weren't ignored or uncontested. The shrimp growers, who were mostly owners of outside businesses, and the locals who were affected by the new industry regularly got into violent fights. The conflict frequently led to poor law and order situations, violence, serious human rights violations, and occasionally even fatalities, according to Manju (1996), Nijera Kori (1996), and Tutu (2004). According to the report, the local authorities, police, and other law enforcement agencies frequently sided with the shrimp growers. In light of this, NGOs such as Nijera Kori (1996) called for a total ban on shrimp farming in Bangladesh or, at the very least, for limitations to be placed on shrimp farming in areas where there is clear evidence of extremely negative externalities and impacts (Islam 2014).

In order to counteract the negative externalities brought on by commercial aquaculture, a separate discourse was also present that emphasized the positive consequences of the industry. Shrimp business owners, shrimp farm owners, and the government (especially the Department of Fisheries) prepared documents focusing on the industry's potential role in the economy, indicating its growing significance as the country's second-largest non-traditional export-earning activity, after ready-made clothing (see, Aftabuzzaman 1996; Battacharya et al. 1999; Karim 2000). These publications claimed that any short-, medium-, or long-term negative externalities, such as the growth of the local economy, more jobs, and the generation of income for the local communities, are much offset by the positive externalities. The materials also asserted that the underlying advantages will soon have a major multiplier impact, benefiting not just entrepreneurs but also the entire local population (Islam 2014).

In order to promote entrepreneurship in the shrimp industry, the government of Bangladesh's support for shrimp farming and institutional push in that direction were crucial. The government, through its various ministries, provided critical assistance to the industry's entrepreneurs during the early stages of this commercial aquaculture through land acquisition, the leasing of khas (state-owned) land to shrimp farmers, and the provision of fiscal and financial incentives in a variety of forms for the production and processing of shrimp. The government, however, launched a number of initiatives and measures to contain and mitigate these negative effects as they gradually started to manifest in the shrimp farming regions and ecological-environmental concerns sparked discussions and disagreements. One of them was the establishment of Shrimp Culture Steering Bodies at the local (thana), regional, and national levels (Aftabuzzaman 1996). These regulations included those that dealt with land leasing, provided rules for the construction of shrimp farms, demanded the approval of adjacent farmers before the construction of shrimp farms, and demanded the approval of regional farmers. Although government policies and initiatives were put into place on schedule, they have come under fire for being insufficient, not being strictly enforced, and being indifferent to environmental concerns (Nijera Kori 1996; Manju 1996; Habib 1998).

We can distinguish at least three important perspectives about Bangladesh's export-focused shrimp farming after carefully reading the aforementioned literature: First, the negative outlook: It highlights how shrimp aquaculture is an environmentally unsustainable practice in Bangladesh's coastal ecosystem. This viewpoint contends that adverse externalities are not artificial but endemic, systemic, and long-lasting, calling for a complete ban on shrimp farming in Bangladesh's coastal regions. Second, a positive outlook: It highlights the potential economic benefits of the sector, such as significantly better earnings, employment opportunities, and foreign exchange earning potential. This viewpoint contends that the benefits much outweigh the costs associated with any potential negative social and environmental repercussions of shrimp cultivation. The Reformist View emphasizes that negative environmental and social externalities of shrimp aquaculture can be successfully addressed by a successful combination of policies, technologies, and tools (Islam 2014), without disputing or questioning the externalities themselves. After 2000, literature largely absorbed the reformist perspectives, and commercial aquaculture is now widely accepted without much controversy.

8.4 Private Regulations Governing Aquaculture

As the debate about potential gains and losses in the sector's production site continued, the shrimp business faced yet another challenge from the buyers to ensure the "quality" of the shrimp. Customers in the European Union forbade the export of shrimp from Bangladesh in 1997 because they thought the product was "sub-standard." Other large importers, such as the United States and Japan, put emphasis on the implementation of Hazard Analysis Critical Control Point (HACCP) to maintain shrimp freshness and quality (Pokrant and Reeves 2003). Seven guiding principles for freshness and safety comprise the HACCP system: The procedures are as follows: (a) conduct a hazard analysis; (b) identify the critical control points (CCPs); (c) establish critical limits, target levels, and tolerances; (d) establish a monitoring system through scheduled testing or observations; (e) establish corrective actions to be taken when a CCP is not within tolerance; and (f) establish verification procedures, which include appropriate supplementary testing (Islam 2010).

Due to pressure for "quality shrimp" emanating from the shrimp commodity chain, which includes demands from buyers, consumers, and environmental groups, the top shrimp processing facilities in Bangladesh have consistently had to adjust toward new regulations and reorganize local institutions. Significant local repercussions resulted from this, such as increased expenses and ongoing vulnerability. The EU embargo was lifted once Bangladesh began implementing HACCP in a number of institutions. However, due to poor product quality, Bangladesh used to receive at least 10% less in price for its shrimp products on the international markets. According to US and EU food safety requirements, the Shrimp Seal of Quality (SSOQ), a third-party certifier, was founded in Bangladesh in 2002 (SSQO 2002; Gammage et al. 2006). In order to achieve higher pricing in the international frozen food markets, the SSOQ began certifying shrimp processors and hatcheries in February 2005 (The Independent, February 18, 2005).

As part of a USAID program, the SSOQ has been in use in the country since early 2002, and it took nearly 3 years to begin the certification process. To make sure that shrimp production adheres to global norms of conduct, the SSOQ, a private, autonomous, non-profit organization, is in charge. It involves respect for human rights, moral labor standards, hygienic food handling, and effective environmental management. The SSOQ also monitors production standards and markets finished items on international markets. The business also provides services for technical support, lab analysis, market research, and product development. The SSOQ was mostly ineffective because of the Department of Fisheries' opposition, and Bangladeshi shrimp that isn't certified fetches substantially lower prices on the world market (Islam 2008).

Shrimp exported from Bangladesh is also checked in Singapore to make sure it complies with significant importer standards or codes of conduct before being supplied to the buyers. The pre-shipment inspection is carried out by international organizations SGS (Société Générale de Surveillance), Switzerland, and Lloyds, United Kingdom, prior to shipment to the buyers. The EU delegated Government bodies, buyers' chosen quality assurers, and appointed quality inspectors for processors and exporters also conduct some in-country inspections in Bangladesh (Gammage et al. 2006; Islam 2014).

8.5 Aquaculture Commodity Chain

The restaurants, grocery stores, seafood businesses, and purchasing organizations in Europe, North America, and Japan control the global shrimp supply chain. Various environmental organizations with operations ranging from the local to the global have an impact. Bangladesh's shrimp farming is at the extraction, production, and processing end of this lengthy commodities chain. Although vertical integration is still relatively limited in this business, it has grown since the mid-1990s. Incorporating shrimp farming into their operations has proven difficult for shrimp factories or processors, who operate as the primary conduit between domestic shrimp growers and foreign markets. As a result, they would rather leave the risky business of cultivating shrimp to the thousands of shrimp farmers, which they consider to be risky from an economic, political, ecological, and environmental perspective. The processors, working through company agents, independent suppliers, shrimp depot owners or parties, and commission agents (arotdars), effectively purchase shrimp from independent shrimp farmers directly and through advanced payment arrangements (dadon). In turn, feed mills, hatcheries, and feed supply brokers sell shrimp farmers shrimp fry and other inputs. The supply chain for the Bangladeshi commercial shrimp business is briefly depicted in the following graphic, which was modified from Khatun (2004) and the author's ethnography (Fig. 8.1).

Less emphasis has been paid to freshwater prawn farming (golda) than brackish water shrimp farming (bagda). This is due to the fact that the former typically involves large sums of money and causes social and environmental concern. The majority of small-scale farmers that raise freshwater prawns, on the other hand, have recently started growing rice and shrimp after converting en masse from rice farming to prawn farming in the 1990s. The local economy of Bangladesh has grown significantly over the last 10 years, but more recently, management and environmental concerns have begun to threaten the profitability of agricultural enterprises. The structure of the regional supply chain is also changing as a result of the adoption of global norms for hygienic practices and food safety (Islam 2014). Climate change and disease outbreaks both have a big impact on this industry. When these tendencies are combined, small-scale shrimp growers are disadvantaged more and more. Despite challenges, if the right technology and sustainable policies are implemented, Bangladesh's fisheries industry has the potential to earn over US$ 2 billion.

Fig. 8.1 Supply chain of cultured shrimp

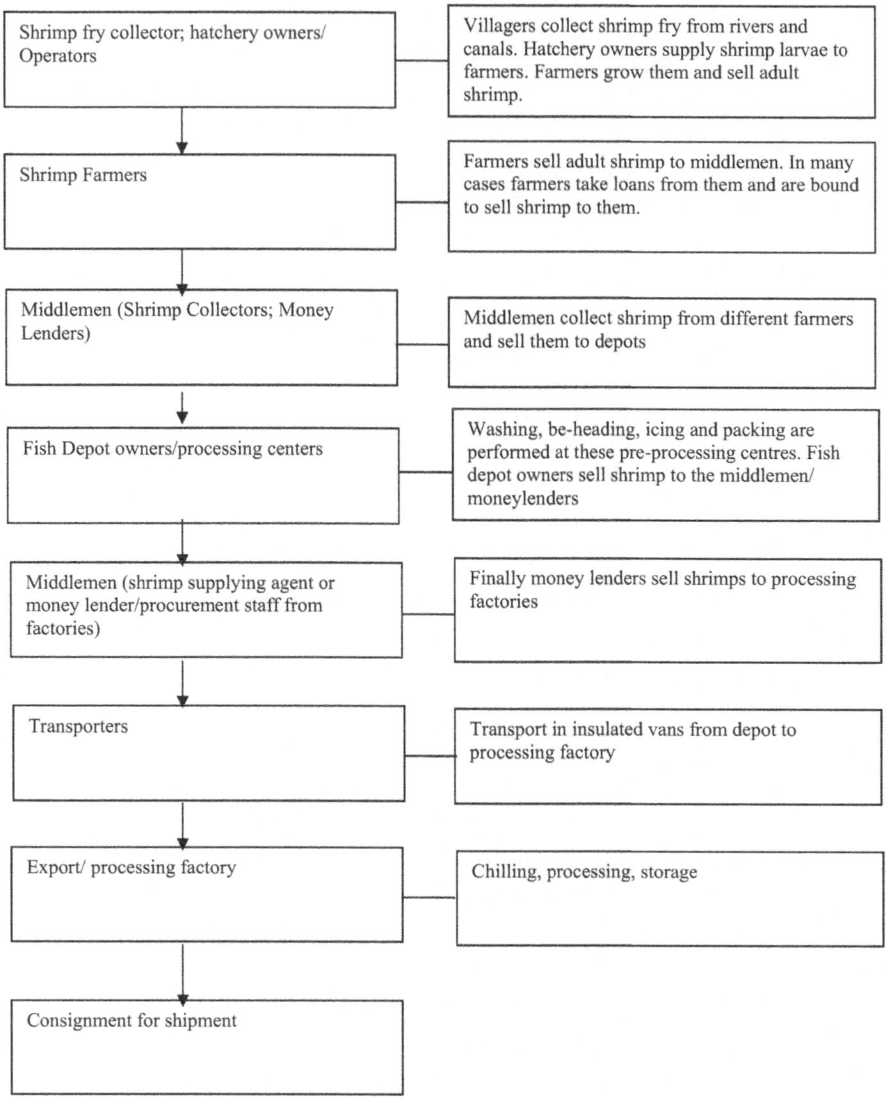

8.6 Conclusion

The chapter provides a brief overview of Bangladesh's fisheries and aquaculture sector and illustrates how, while being a country that struggles with poverty, Bangladesh is ideally situated to produce lucrative fisheries and shrimp for rich consumers in the North. The chapter demonstrates how fragile and unstable Bangladesh's shrimp sector is right now, despite having a large economic impact on the nation. The supply chains for shrimp and other fisheries are being forced to restructure their agricultural institutions as a result of the constraints posed by the current global market, as was previously mentioned. Small farmers consequently have limited access to capital, which ultimately has an impact on the ownership models of shrimp farms. The development of purchasing regulations, the consequences of climate change, and the spread of different diseases are some additional key challenges the industry is currently experiencing.

As inland fisheries and aquaculture are seen as alternatives to protein loss brought on by the global depletion of marine fisheries, the market for these products will grow. This industry will either flourish or fail depending on how Bangladesh responds to the problems and seizes the opportunities. Strong political will, technological know-how, solid buyer networks, successful sustainable policies, and pro-poor initiatives are the key success factors.

References

Adnan S (1991) Minority view of appraisal Mission for phase-III of DDP regarding project policy towards shrimp culture: an alternative approach. Report prepared for the DDP phase III appraisal mission, October, 1991, Dhaka

Aftabuzzaman (1996) Swimming through troubled waters: Bangladesh perspective. Paper prepared for Bangladesh Frozen Foods Exporters Association (BFFEA), Dhaka

Ahmed F (1996) In defense of land and livelihood. Sierra Club, Ottawa
Bangladesh Economic Review (2014) Economic Advisers Wing, Finance Division, Ministry of Finance, June
Battacharya D, Rahman M, Khatun FA (1999) Environmental impacts of trade liberalization and policies for the sustainable management of natural resources a case study on Bangladesh's shrimp farming industry. Center for Policy Dialogue, Dhaka
Bene C (2005) The good, the bad and the ugly, discourse, policy controversies and the role of science in the politics of shrimp farming development. Dev Policy Rev 23(5):585–614
Boyd CE, Clay J (1998) Shrimp aquaculture and the environment. Sci Am 278(6):59–65
Bush S, Belton B, Hall D, Vandergeest P, Murray F, Ponte S, Oosterveer P, Islam MS, Mol APJ, Hatanaka M, Kruissen F, ThuHa TT, Little D, Kusumawati R (2013) Global food supply: certify sustainable aquaculture? Science 341(9 September):1067–1068
CAP (Consumers Association of Penang) (1995) Aquaculture: problems and issues. Dossier Compiled by Consumers Association of Penang, Penang
Clay JW (1996) Market potentials for redressing the environmental impact of wild cultured and pond produced shrimp. Report Prepared for World Wildlife Fund, Geneva
Clay JW (2004) World agriculture and the environment. Island Press, Washington, DC
Deb AK (1998) Fake blue revolution: environmental and socio-economic impacts of shrimp culture in the coastal areas of Bangladesh. Ocean Coast Manag 4(1998):63–88
DOF (The Department of Fisheries) (2002) Annual report of Department of Fisheries. Government of the People's Republic of Bangladesh, Dhaka
DOF (The Department of Fisheries) (2004) DOF annual report 2004. Department of Fisheries, Ministry of Fisheries and Livestock, People's Republic of Bangladesh, Dhaka
DOF (The Department of Fisheries) (2022) Fisheries resources and information on Bangladesh. Government of the People's Republic of Bangladesh, Dhaka
Flaherty M, Vandergeest P, Miller P (1999) Rice paddy or shrimp pond: tough decision in rural Thailand. World Dev 27(12):2045–2060
Gammage S, Swanburg K, Khandkar M, Islam MZ, Zobair M, Muzareba AM (2006) A gendered analysis of the shrimp sector in Bangladesh. Greater Access to Trade and Expansion, USAID, Dhaka
Goodman D, Watts MJ (1997) Agrarian questions, global appetite, local metabolism, nature, culture and industry in fin-de-siecle agro-food system. In: Goodman D, Watts M (eds) Globalizing food, agrarian questions and global restructuring. Routledge, London, pp 1–32
Habib E (1998) Legal aspects of shrimp cultivation. Paper presented at the workshop on environmental impact of structural adjustment policies in Bangladesh. The Centre for Policy Dialogue, Dhaka, 17 May 1998
Islam MS (2008) From sea to shrimp processing factories in Bangladesh: gender and employment at the bottom of a global commodity chain. J South Asian Dev 3(2):211–236
Islam MS (2009) In search of 'white gold': environmental and agrarian changes in rural Bangladesh. Soc Nat Resour 22(1):66–78
Islam MS (2010) Regimes of environmental regulations and governance: opportunities and challenges for shrimp aquaculture in Bangladesh. J Bangladesh Stud [Penn State Univ] 12(1):44–62
Islam MS (2014) Confronting the blue revolution: industrial aquaculture and sustainability in the global south. University of Toronto Press, Toronto/Buffalo/London
Karim M (2000) Preserving shrimp quality. Agribus Bull 61, 25 February
Karim M, Ahmed M, Talukder RK, Taslim MA, Rahman HZ (2006) Policy working paper: dynamic agribusiness-focused aquaculture for poverty reduction and economic growth in Bangladesh. World Fish Center, Ministry of Fisheries and Livestock, and Bangladesh shrimp and Fish Foundation, Dhaka
Khatun F (2004) Fish trade liberalization in Bangladesh: implications of SPS measures and co-labeling for the export-oriented shrimp sector, FAO project PR26109. United Nations, Rome
Manju TH (1996) Political economy of shrimp culture in Bangladesh, Report prepared for Grameen Trust. Grameen Bank, Dhaka
McMichael P (2004) Development and social change: a global perspective, 4th edn. Pine Forge Press, Thousand Oaks
Nijera Kori (1996) The impact of shrimp cultivation on soils and environment in Paikgacha region, Khulna (limited to polders 20, 21, 22, 23 and 29). Report prepared for Nijera Kori by Environment Study Centre, Dhaka
Philips MJ, Lin CK, Beveridge MCM (1993) Shrimp culture and the environment: lessons from the world's most rapidly expanding warmwater aquaculture sector. In: Pullin RSV, Rosenthal H, Maclean JL (eds) Environment and aquaculture in developing countries. ICLARM, Bangkok
Pokrant B, Reeves P (2003) Work and labour in Bangladesh brackish-water shrimp export sector. South Asia J South Asian Stud XXVI(3):359–389
Public Citizen (2005) Fishy currency: how international finance institutions fund shrimp farms. Public Citizen, Pennsylvania
Rahman A (1995) Shrimp culture and environment in the coastal region, Working paper new series, no. 8, Dhaka, BIDS
Raynolds L (1994) Institutionalizing flexibility: a comparative analysis of Fordist and post-Fordist models of third world agro-export production. In: Gereffi G, Korzeniewicz M (eds) Commodity chains and global capitalism. Praeger, Westport, pp 143–161
Raynolds L, Myhre D, McMichael P, Carro-Figueroa V, Buttel FH (1993) The "new" internationalization of agriculture: a reformation. World Dev 21(7):1101–1121
Roheim CA (2004) Seafood, trade liberalization and impacts on sustainability. In: Aksoy MA, Beghin JC (eds) Global Aquacultural trade and developing countries. World Bank Publications, Washington, DC, pp 275–295
Sanderson S (1986) The emergence of "world steer": internationalization and foreign domination in Latin American cattle production. In: Tullis FL, Hollist WL (eds) Food, the state and international political economy. University of Nebraska Press, Lincoln
Scott D (2000) The environmental and social impacts of commercial shrimp farming. Reports, November 8
Sernbo K, Kloth J (1996) Prawn farming: a study of intensive aquaculture. Report prepared for Swedish Society for Nature Conservation, Stockholm
Shamsuzzaman MM, Islam MM, Taniaa NJ, Al-Mamun MA, Xua PPBX (2017) Fisheries resources of Bangladesh: present status and future direction. Agric Fish 2(4):145–156
Shiva V (1995) The damaging social and environmental effects of aquaculture. Third World Resurgence 59:22–24
SSOQ (Shrimp Seal of Quality Organization) (2002) The seal of quality program: questions and answers on the seal of quality program. SSOQ, Dhaka
Thamina Q (1995) Profit by destruction. International workshop on ecology, politics and violence of shrimp cultivation for export. Dhaka, Bangladesh
The Independent (2005) U.S. firm starts shrimp certification. February 18. http://independent-bangladesh.com/. Retrieved 25 Feb 2005
Tutu A-U-A (2004) Commercial shrimp farming in Bangladesh: in respective of environment, land rights and labour rights. Coastal Development Partnership, Khulna
USAID Bangladesh (2006) A pro-poor analysis of the shrimp sector in Bangladesh. Greater Access to Trade Expansion, Development & Training Services, U.S. Agency for International Development, Arlington
Vandergeest P, Flaherty M, Miller P (1999) A political ecology of shrimp aquaculture in Thailand. Rural Sociol 64:573–596

Recent Trend of Foreign Trade of Bangladesh: Opportunities, Challenges, and Issues Forward

9

Rejaul Karim Bakshi

Abstract

International trade is considered as the engine of economic growth for developing countries like Bangladesh. Bangladesh has produced sustained high economic growth in the past two decades thanks to export and remittance growth. However, there are particular challenges ahead. The import growth rate of Bangladesh remained equally higher, resulting in the continuous trade deficit. The country's growing trade deficit with India and China poses to be alarming. Geopolitical issues like Brexit and the US-China trade war are also likely to change the status quo in trade relationships and trade patterns. The country, therefore, needs to redesign its trade and macroeconomic policies carefully.

Keywords

Foreign trade · Economic growth · Readymade garments · Remittance · Bangladesh

9.1 Introduction

Trade is considered as the engine of economic growth, especially for developing countries like Bangladesh. For developing countries, succession from lower income to middle and higher income country depends mainly on gains from trade in the global market (Balassa 1978; Dodaro 1993; Ramos 2001; Howard 2002). Developing countries like Bangladesh can benefit from international trade in various ways (Love 1994; Michaely 1977; Tyler 1981). Trade is the source of foreign currency to keep country's balance of payment; helps financing import bill of essential food and raw materials; expands of export industries to create employment; contains inflation through import; and accelerates development through innovation and technological transformation (Hossain and Salim 2009; Kavoussi 1984). Export earnings are essential to meet import expenditure of necessary items, including food, medicine, and raw materials (Williamson 1978; Feder 1983). Foreign exchange earnings earned from exports also help make up for import bill for types of machinery and equipment necessary for industrial transformation of the country (Ram 1985; Jung and Marshall 1985; Chow 1987). However, Bangladesh export is facing stiff competition at the world market due to various issues. This paper, in this context, explores the performance, trends, composition, challenges, and policy options for foreign trade of Bangladesh. The paper postulates Bangladesh's export and import scenario relative to other nations over the past years. We also discuss the sources and directions of Bangladesh's imports and exports and the dynamic changes over the years. The paper also highlights significant issues and challenges that Bangladeshi products face globally and suggests some policy reforms to address those challenges.

9.2 Composition and Performance of Exports of Bangladesh

9.2.1 Export and Export Growth

The export sector of Bangladesh witnessed tremendous growth in the post-independence years, especially after 1990. The export volume heats the USD 5 billion mark in 1997–1998, propelled to USD 15 billion in 2007–08 and to USD 35 billion in 2015–16 (Fig. 9.1).

9.2.2 Export Basket Diversification

The country's rapid expansion of export earnings comes with strong performance of readymade garments industries

R. K. Bakshi (✉)
Department of Economics, University of Rajshahi,
Rajshahi, Bangladesh
e-mail: rkbakshi@ru.ac.bd

R. Ahmed et al. (eds.), *Transforming Bangladesh*, World Regional Geography Book Series,
https://doi.org/10.1007/978-3-031-45093-8_9

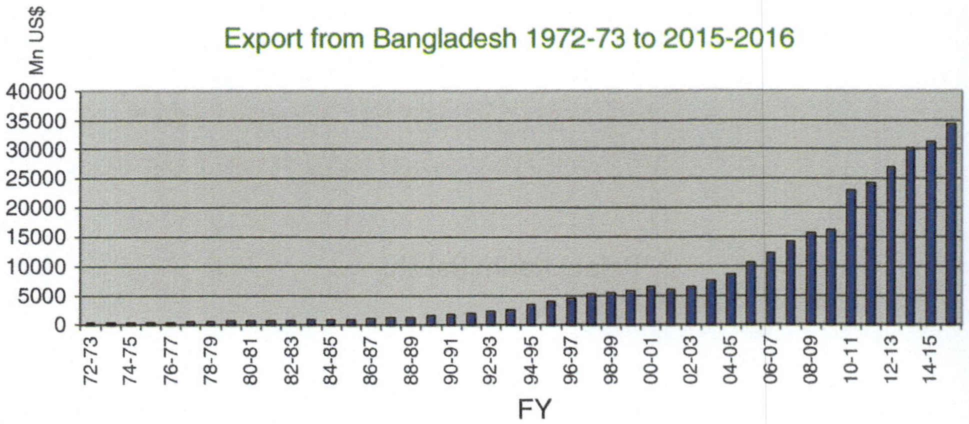

Fig. 9.1 Export earnings and export growth
Data Source: BBS (2018)

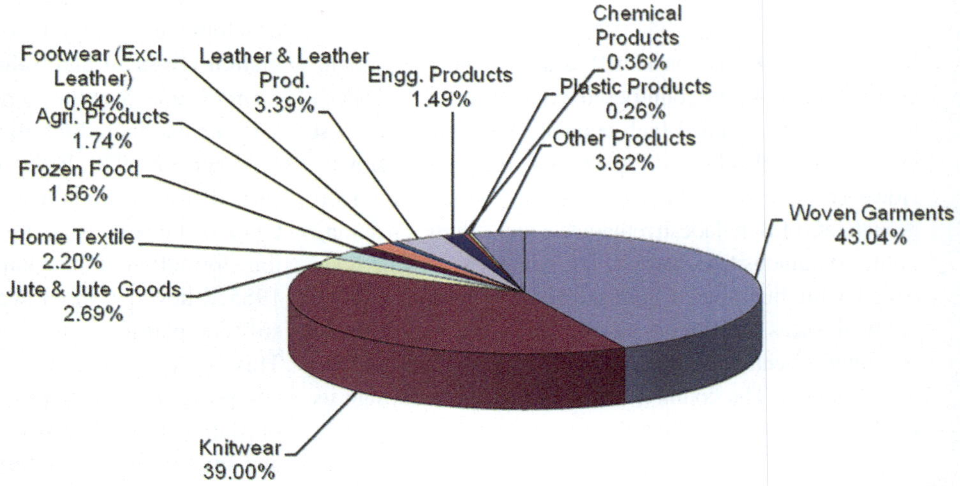

Fig. 9.2 Export earnings by commodities in 2015–16
Data Source: BBS (2018)

(RMG). As seen from Fig. 9.2, woven garments contribute 43.04% of the total export earnings, followed by Knitwear with another 39% export share in 2015–16. Among the non-RMG export baskets, the highest contributor is Leather & Leather products with a mere 3.39% export share, and Jute & Jute products with 2.69%.

9.2.3 Contribution of Readymade Garments in Export

Another essential feature of Bangladesh export sector is the rapid expansion of RMG industries' contribution compared to other industries. The share of non-RMG sectors in the export volume is decreasing over time. This can be seen in Table 9.1. In the 1970s and 1980s, export sector was dominated by non-RMG primary sectors including Jute, Leather, and Tea. But the situation reversed after 1990. This is encouraging as conventional wisdom suggests that manufacturing sector can grow faster and also earn more value from the world market.

9.2.4 Export Market Diversification

Bangladesh's export destination is dominated by industrial countries of the European Union and the USA. As Fig. 9.3 shows, EU countries provided 54.58% of total export in 2015–16, followed by the American region with 22.71%. The major destinations are the USA, the UK, Germany, France, Spain, the Netherlands, Italy, Denmark, Belgium, Japan, Turkey, China, and Australia. In comparison, it can be seen from Fig. 9.4 that Asia (28.4%), Africa (17.9%), Middle East (10.5%), and East Europe (10.1%) were dominating in export volumes in 1981–82. Thus there is a categorical shift not only in export commodities but also to export destinations.

9.2.5 Export, Import, and the Trade Balance

The country's export earnings have increased consistently in the past three decades, similar to the growing trend of import.

Table 9.1 Export growth of RMG and non-RMG sectors (in million USD)

FY	RMG	Non-RMG	Total	RMG share (% in total)
1992–93	1445.03	937.86	2382.89	60.65
1995–96	2547.13	1335.29	3882.42	65.61
1998–99	4020.23	1292.63	5312.86	75.67
2000–01	4860.56	1606.74	6467.3	75.16
2003–04	5686.09	1916.9	7602.99	74.79
2006–07	9211.23	2966.63	12177.86	75.64
2009–10	12496.72	3707.93	16204.65	77.12
2010–11	17914.46	5013.76	22928.22	78.14
2011–12	19089.73	5212.17	24301.9	78.56
2012–13	21515.73	5511.63	27027.36	79.61
2013–14	24491.88	5694.74	30186.62	81.14
2014–15	25491.4	5717.54	31208.94	81.68
2015–16	28094.16	6163.02	34257.18	82.01

Data Source: BBS (2018)

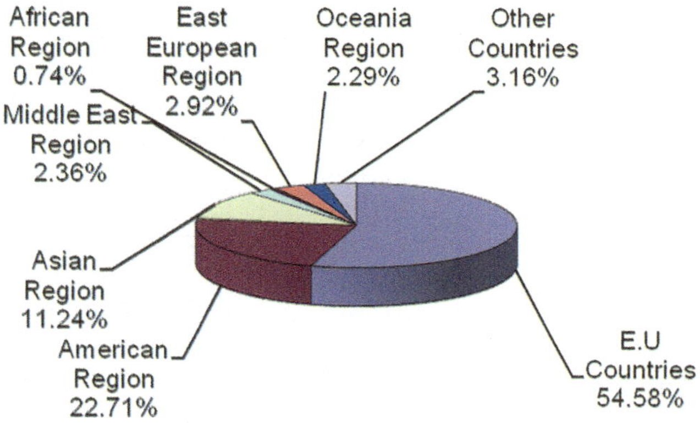

Fig. 9.3 Export by countries/regions in 2015–16
Data Source: BBS (2018)

As shown in Table 9.2, the trade balance is negative and the deficit has been widening in total terms as import for industrial machineries, raw materials, essential medicines, and food staples are increasing to meet. However, the last column of Table 9.2 shows that the trade deficit as % of total export earnings has declined from 166.84% in 1975–76 to just 16.87% in 2015–16.

9.3 Major Issues, Opportunities, and Policy Options

Though Bangladesh's export has witnessed tremendous growth in recent time, the export market is heavily concentrated and readymade garments predominantly dominate the export basket. The reasons for such market concentration of Bangladesh export are manifold. Export basket of the country is too narrow. It is essential to encourage export diversification through appropriate fiscal and monetary policies. Furthermore, the price of readymade garment products remains low while safety issues also remain a critical hindrance for this sector's long-term prospect. Some recent phenomenon also deserves due attention for the interest of the country's trade.

9.3.1 BREXIT and Bangladesh Trade

Britain's decision to leave the European Union (EU), popularly known as Brexit, is set to pose significant economic implications at its domestic level and for the rest of the world, including Bangladesh. EU is the major trading partner for Bangladesh and so is Britain.

Bangladesh is a current beneficiary of the preferential treatment of Britain under EU's Everything But Arms (EBA) provision as a least developed country (LDC). Bangladesh has gained eligibility for LDC graduation in March 2018, and is expected to finally graduate in 2024. According to EU policy, Bangladeshi goods will enjoy duty-free market for another 3 years after graduation (till 2027). So far, the UK has committed to offering a duty-free market to Bangladesh as a LDC country (till 2024) but did not commit to its preferential treatment after graduation.

UK is a major trading partner for Bangladesh. The bilateral relationship encompasses a wide range of areas, including trade, remittances, FDI, aid, and educational development. Bangladesh's export volume to the UK stands more than $4 billion and is about 11% of its total exports—only third to the USA and Germany. Apart from export earnings, Bangladesh's remittance inflows from expatriates living in the UK are over $1.2 billion and are about 7.5% of the total remittances for Bangladesh, while FDI inflow stands at $1.60 billion (roughly 11.25% of the total FDI inflow). Of the total exports of Bangladeshi goods to the UK, over 92% comes from readymade garments. Leather goods, frozen foods, and light engineering goods constitute the rest of the export

Fig. 9.4 Export by countries/regions in 1981–82Data Source: BBS (2018)

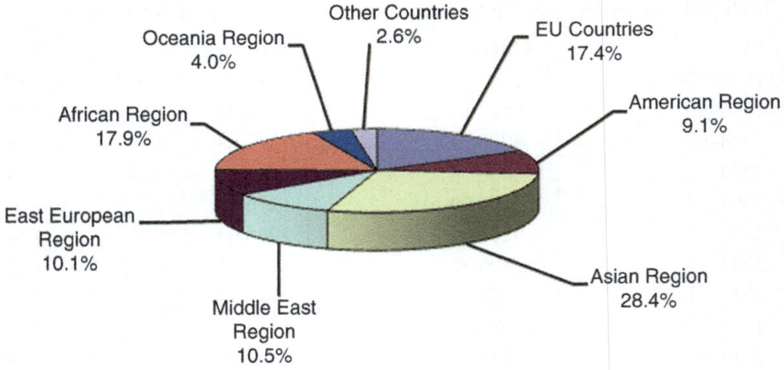

Table 9.2 Export, import, and the trade balance (in million US dollars)

FY	Export	Import	Trade balance	Trade balance as % of export
75–76	380	1014	−634	−166.84
81–82	626	1931	−1305	−208.47
84–85	934	2641	−1707	−182.76
90–91	1718	3511	−1793	−104.37
93–94	2534	4191	−1657	−65.39
96–97	4418	7150	−2732	−61.84
99–00	5752	8403	−2651	−46.09
02–03	6548	8661	−2113	−32.27
05–06	10,526	14,746	−4220	−40.09
08–09	15,565	22,507	−6942	−44.60
10–11	22,928	33,657	−10,729	−46.79
11–12	24,302	35,516	−11,214	−46.14
13–14	30,187	40,693	−10,506	−34.80
14–15	31,209	40,639	−9430	−30.22
15–16	34,257	40,036	−5779	−16.87

Data Source: BBS (2018).

basket. If Bangladesh is losing preferential treatment due to Brexit, Bangladesh goods are to face duties as under MFN tariffs of EU, the exporters may lose as high as 11.7% of the export earnings (Stevens and Kennan 2016). This will result in serious loss in export competitiveness for Bangladesh. The loss can be triggered further as some other EU countries often import Bangladeshi goods via the UK with EU common market facility. After Brexit, such import via the UK may drop as the phenomenon of common market may no longer exist in the first place.

Secondly, due to Brexit, the value of Bangladeshi currency, Taka, may show some appreciation as Pound is likely to lose value as a worldwide payment standard. Such currency appreciation means Bangladeshi goods to appear costlier in Britain—reducing import from Bangladesh. This effect, however, is not likely to be great as Bangladeshi goods are not subject to domestically produced goods in Britain and currency appreciation is likely to occur in almost all trading rivals.

The third effect of Brexit is likely to be felt in migration and remittance flow. Since a key reason for Brexit is securing local employment controlling migration, this is going to hit Bangladesh migration to Britain. But to remain globally competitive, Britain will still require low-wage labor and manpower exports from densely populous countries like Bangladesh may remain high as a result. Flow of remittance is critical for Bangladesh as a significant portion of its GDP is coming from remittance, hovering about 8–10% of GDP in the past few years (BBS 2018).

It is thus important for Bangladesh to remain engaged with Britain in order to retain its business interest in the post-Brexit era.

9.3.2 Free Trade Agreement: Opportunities and Challenges

Trade openness of Bangladesh has risen significantly in recent years. The share of import and export to GDP has crossed more than 50% in the recent time. Another important feature of Bangladesh trade is that from 50 of its top trading partners, the country is experiencing trade surplus with 25 countries in 2018 and deficit with others (BBS 2018). Proponents of the FTAs advocate that in the absence of a generalized free trade arrangement, bilateral and multilateral arrangement of free trade among countries with mutual trading interest can benefit substantially at least in the short and medium term. As argued by the conventional Heckscher-Ohlin theory, successful FTAs require that the partner countries are enjoying comparative advantage across different sets of commodities. The prospect of successful FTAs also depends on the level of tariffs and the preferential market access currently enjoyed by the member countries.

At present, Bangladesh is enjoying Preferential Market Status fully in 21 of its major 50 trading countries namely Germany, the UK, France, Italy, Spain, Belgium, the Netherlands, Sweden, Norway, Finland, Denmark, Switzerland, Slovakia, Greece, Poland, Austria, Turkey, Portugal, Czech, Canada, and India (UN 2018). The country also privileged with preferential market access partly to five major world economies including the USA, Ireland, Japan, Hong Kong, and Taiwan. Thus, over half of its top 50 country trading partners, Bangladesh has already preferential trading agreement fully or partly. It is however, important to note that Bangladesh is enjoying preferential trading arrangement with these countries as a LDC country. The scenario will change once the country graduates from LDC scheduled in 2024 and the preferential trading arrangement ends with a 3-year grace period in 2027.

Bangladesh's exports are dominated by textiles and apparels. About 70% of its total exports are coming from apparels. These major export products are overwhelmingly exported to major Western countries. USA has suspended its GSP facilities and hence apparel exports from Bangladesh are subject to tariff. High tariffs are also seen in other major markets including Russia, China, Turkey, Ukraine, Croatia, Serbia, Argentina, Taiwan, Korea, Malaysia, and Singapore. Such tariff rates vary from 5% to 45% of export value. Even the EU countries are currently imposing 15% tariffs on Bangladeshi exports with preferential market access. Thus, there are opportunities to trading agreements reducing tariffs on Bangladeshi exports specially textile and apparels.

At present, Bangladesh is enjoying duty-free access or preferential tariff advantage with 23 EU countries. The natural choice for potential free trade deals from trade perspective is countries with significant trading volume as well as with higher level of trade restrictions including tariffs. The import-export basket of Bangladesh combined with prevailing tariff rates show that among the top trading partners of the country, Bangladesh faces substantial tariff in the market of the USA (0–55%), Argentina (14–54%), Brazil (8–35%), Russia (0–33%), Mexico (0–30%), Malaysia (0–20%), and Korea (0–13%) (UN 2018). Thus, Bangladesh should emphasize on these countries while pursuing for FTAs.

9.3.3 Revision of Trade Policies of Bangladesh

9.3.3.1 Revision of Tariff

The country's tariff structure should be revised addressing sectoral needs imperative to enhance competitiveness of strategic sectors within a specific time period. Tariffs should not be seen as a means to garner hefty revenue for the government rather as a tool to protect different sectors on a targeted basis for a stipulated time limit. Furthermore, it is often common practice in Bangladesh to impose different forms of supplementary duties on "ad hoc basis" which does not help establishing confidence among the investors. There should be coordination between tariff, tax, VAT, and supplementary duties applied to various products.

9.3.3.2 Improving Governance in Trade-Related Issues

Import of the country often faces unfair competition due to unscrupulous practices including under-invoicing, custom valuation, false declaration about imported items, and dumping. On the other hand, export products face non-tariff barriers because of anti-dumping duties, bureaucratic complexities, unnecessary security, and safety concerns means to curb trade volumes. Such practices should be correctly specified and need to address these issues to boost trade volume. The country should enhance technical capacity and economic diplomacy to successfully tackle these issues, particularly in tariff classification, customs standardization, product labeling, and identifying dumping and anti-dumping cases.

9.3.3.3 Coordination of Macroeconomic Policies

The country's trade policy needs to be coordinated with appropriate fiscal, monetary, and exchange rate policies. Better coordination of different macroeconomic policies is necessary to foster targeted outcome from a strategic trade policy. Policy instruments designed by central bank and relevant authorities dealing with fiscal, monetary, and exchange rate policies need to be inter-linked suitably to achieve the stipulated goal of the trade policy. Monetary and fiscal policies should facilitate and direct market forces in a direction that are helpful in trade balance. Macroeconomic policies should identify and target specific sectors with potential for world market space. In this perspective, strong coordination is imperative between different agencies of the government and policy-making bodies.

9.3.3.4 Diversification of Export Basket and Export Market

Readymade garments and textile products heavily dominate Bangladesh's export. In addition, country's export market is also concentrated with a limited number of countries including the USA and EU. In this context, it is important to devise strategies to diversify export baskets of the country as well as export destinations. Market and product diversification are critical in order to reduce the risk of market volatility such global financial crisis (GFC). Policies should be designed enabling diversification of export products and export markets.

Product designing, finishing, labeling, marketing, and show-casing should be synchronized with diverse demand perspectives from various corners of the world. Institutional arrangement should be done in this aspect.

9.3.3.5 Improving Port and Transport Facilities

Transport and transaction costs for commodities from and to Bangladesh borders are very high, affecting trade potential negatively. Chattogram Port, the key gateway of the country, is one of the most inefficient in handling containers. Ships cannot get birth in Bangladeshi ports without waiting 3–6 days. Due to huge congestion and other difficulties, Chattagram port is labeled among the five worst ports of the world by the body of the Word Merchant Marine (Bright Hub 2002). Trade policy should address efficiency of ports of the country, container handling, hassles, bribe, and strike-related issues at port and border customs stations. Automation system in cargo handling can be of great help in this regard. Furthermore, road and rail transport between the capital city Dhaka and the port city Chattogram should be given priority in order to facilitate trade. High-speed electric train and expressway for uninterrupted road communication can significantly reduce transport cost as well as lead time of export products that are essential to improve market competitiveness at the global level.

9.3.3.6 Focusing on Trade Diplomacy

Trade in this competitive world has also become a matter of bilateral and multilateral diplomacy. Country successful in trade diplomacy and negotiation can harness significant access to market, tariff and non-tariff protection. As a developing country, Bangladesh is in a position to negotiate favorably in these issues in line with WTO agreement. Trade diplomacy of Bangladesh should also prioritize searching for and extending new destinations for its apparel products in addition to its established market in the EU and North America. Countries with huge potentials of apparel export include China, India, Russia, Turkey, Brazil, and Japan. The country can frame beneficial trade agreement along with free trade agreement within SAARC, ASEAN, BIMSTEC, BRICS, and Commonwealth forum.

9.4 Summary and Conclusion

Bangladesh's export sector has been performing well and has been working as the engine of sustained economic growth in the past three decades despite certain structural and infrastructural limitations. The country experienced higher export growth than most of the developing countries of the world. But the import growth rate of Bangladesh remained equally higher resulting in continuous trade deficit. The imports essentially constitute industrial machineries, raw materials, food grain essentials, petroleum, and other intermediate inputs. Another positive sign is that the trade deficit as a percentage of export earnings and trade deficit ratio to GDP has declined in the recent time. But the country's growing trade deficit with India and China poses to be alarming and needs careful thoughts.

It is seen that that the export composition of Bangladesh has changed in recent years. Bangladesh used to export primary commodities in the 1970s and 1980s. But today, the export basket is dominated mainly by manufactured commodities. North America now dominates the export market compared to traditional European destination. The recent trend also indicates huge export potentials in Asian markets including Korea, Japan, China, Indonesia, Malaysia, India, and Australia.

It is important that country must emphasize on export diversification, market expansion, and quality improvement. Product diversification and new markets may be catalyst for the stability of the export sector. The forward and backward linkage industries of readymade garments such as yarn and fabric manufacturing should be given priority. Government should also work closely with its trading partners to formulate preferential and free trade arrangements. Country should also negotiate with India and China to remove tariff and non-tariff barriers in order to surge the alarming trade deficit with these countries. Finally, there should be coordination in formulating monetary policy, fiscal policy, and foreign exchange policy to boost investment, employment, output, and export.

References

Balassa B (1978) Exports and economic growth: further evidence. J Dev Econ 5(1):181–189

BBS (2018) Bangladesh economic review. Bangladesh Bureau of Statistics, Dhaka

Bright Hub (2002) https://www.brighthubengineering.com/seafaring/122948-best-and-worst-ports-in-the-world/

Chow PCY (1987) Causality between export growth and industrial performance: evidence from the NICs. J Dev Econ 18(1):55–63

Dodaro S (1993) Exports and growth: a reconsideration of causality. J Dev Areas 27:535–542

Feder G (1983) On export and economic growth. J Dev Econ 12(1):59–73

Hossain MA, Salim RA (2009) The linkages between export and income: further evidence from Bangladesh. Singap Econ Rev 56(1):79–95

Howard M (2002) Causality between exports, imports and income in Trinidad and Tobago. Int Econ J 16(4):97–106

Jung WS, Marshall PJ (1985) Exports, growth and causality in developing countries. J Dev Econ 18(1):1–12

Kavoussi RM (1984) Export expansion and economic growth: further empirical evidence. J Dev Econ 14(2):241–250

Love J (1994) Engines of Growth: the exports and government sectors. World Econ 17:203–218

Michaely M (1977) Export and growth: an empirical investigation. J Dev Econ 4:49–53

Ram R (1985) Export and economic growth: some additional evidence. Econ Dev Cult Chang 33(2):415–425

Ramos FF (2001) Exports, imports and economic growth in Portugal: evidence from causality and cointegration analysis. Econ Model 18:613–623

Stevens C, Kennan J (2016) Trade implications of Brexit for commonwealth developing countries, Commonwealth trade hot topics 133. The Commonwealth, London

Tyler WB (1981) Growth and export expansion in developing countries. J Dev Econ 9(1):121–130

UN COMTRADE (2018) www.comtrade.un.org

Williamson RB (1978) The role of exports and foreign capital in Latin American economic growth. South Econ J 45:410–420

Foreign Remittance and Its Effect on the Economy

10

Nazmus Sadat Khan

Abstract

Remittances are defined as foreign currency transferred by people working in foreign countries to people or groups of people in his/her home country. After the liberation war in 1971, the country was left with very few resources. The primary resource Bangladesh had at that time was "human resources". As a result, discussion on making the best use of this resource became common in policymaking circles. The possibility of exporting workforce to other countries gained momentum soon and new opportunities opened up during the oil boom of 1970s. As the transaction of money became more accessible and more convenient, this transfer of cash started to play an ever-increasing role in the economy of many developing countries, including Bangladesh. Since then, Bangladesh has successfully exported human resources to several countries, and the remittances sent by these hard-working people have become an essential part of the economy. In the latest fiscal year, remittance earning was USD 14.98 billion, much larger than the foreign aid and foreign direct investment. Therefore, this chapter explores how this sector works and where its future lies.

Keywords

Remittance · Liberation · Human resources · Investment

10.1 Introduction

For Bangladesh, remittances have emerged as an essential source for foreign currency and economic growth. Remittances are defined as foreign currency transferred by people working in foreign countries to people or group of people in his/her home country. After the liberation war in 1971, the country was left with very few resources. The primary resource Bangladesh had at that time was "human resource". As a result, discussion on how to make the best use of this resource became common in policymaking circles. The possibility of exporting manpower to other countries gained momentum soon and new opportunities opened up during the oil boom of 1970s. As the transaction of money became easier and convenient, this transfer of money started to play an ever-increasing role in the economy of many developing countries, including Bangladesh. Since then Bangladesh has successfully exported manpower to several countries and the remittances sent by these hard-working people became an important part of the economy. In the latest fiscal year, remittance earning was USD 14.98 billion, much larger than the foreign aid and foreign direct investment. It is therefore very important to know how this sector works and where its future lies.

There are two types of channels through which remittances can be discussed: channels through which remittances get affected and the channels through which remittances affect the overall economy. As the importance of remittance in Bangladesh's economy grew over time, these channels became more important. Remittance is directly related to migration of people from their home country to another. Therefore, any factor that affects migration can influence remittance inflow as well. Bangladesh has always been a large country in terms of population. It has an area roughly equivalent to Greece but a population larger than Russia. Therefore, in spite of being the 43rd largest economy in the world in terms of nominal GDP, it ranks 146th in terms of per capita GDP based on the 2018 data. The country always struggled to provide its citizen with basic food, housing, medical care, and employment. For low-skilled workers with very little or no education, it was even more difficult. As a result, the penchant to migrate was always there. However, a large number of unemployed population with very few formal employment opportunities also meant that the labor

N. S. Khan (✉)
Economist, The World Bank, Dhaka, Bangladesh

R. Ahmed et al. (eds.), *Transforming Bangladesh*, World Regional Geography Book Series,
https://doi.org/10.1007/978-3-031-45093-8_10

was cheap. In a world that was getting more globalized every year, this meant it was a matter of time when other countries with low population growth and booming economies take advantage of this cheap labor. When the domestic pressure to migrate converged with the international demand for cheap labor, migration started to happen. Migration has a long history in the region. Some earlier migration from Bangladesh in the middle of the twentieth century was concentrated to the UK, when the UK was in labor shortage after the Second World War. However, the numbers were relatively low. Oil boom in the 1970s opened up real opportunities for low-skilled workers from Bangladesh. The increasing need for cheap labor in the Middle East then led to a rapid growth of international labor migration from Bangladesh. This created a win-win situation for both the labor-importing and exporting countries. Bangladesh was quick to realize the potential for future labor export. It included labor migration as part of the overall development plan in 1976 and started to promote international labor migration. About fourteen thousand people migrated to Middle East in 1976. Soon the country became a major source country of migrant workers. Labor migration was later expanded to the newly industrialized countries of South East Asia, particularly Malaysia.

Over the years, several international factors affected migration from Bangladesh and hence the remittances. The need for labor is directly related to the growth prospect of the countries. Since most of the migration from Bangladesh occurred in The Gulf Cooperation Council (GCC) countries and the economy of these countries is highly dependent on oil, any major change in the oil price is likely to affect remittance through migration. Different geopolitical developments and wars also affected migration to these GCC countries. For example, migration to GCC countries faced a tough time during the Gulf War of 1990–1991. Migration to Libya stopped and most of the emigrants there returned home during the Arab Spring. Economic downturns also affect migration greatly. For example, the Asian financial crisis of 1997–1999 slowed down migration to Malaysia significantly. Bi-lateral relation between Bangladesh and some labor-importing countries also affected migration to those countries. In spite of these occasional setbacks, demand for cheap labor from Bangladesh persisted in the last 40 years and remittance earning continued to grow.

These remittances helped Bangladesh's economy tremendously. Remittance can affect a country's economy through several channels. First of all, these are money that are directly transferred to the households living in villages who are usually poor. This direct cash transfer to the family members of the migrant workers empowers them in the society. It helps them to consume and save more. Higher demand of goods helps the rural economy to gain momentum. This often leads to higher employment, higher wages, and have a multiplier effect. It also works through the investment channel. Since the households can save more now, they often use their savings to invest in businesses, which in turn helps the rural economy thrive. This helps to reduce income inequality between rural and urban residents and indirectly scale down migration from villages to cities.

Secondly, remittances are foreign money. For a developing country like Bangladesh, this is crucial. Foreign currency is required to make import payments and other necessary payments to the foreigners. This means it increases supply of foreign currency in the economic system and helps boost foreign currency reserves. It also helps the central bank to invest in gold and other relatively safe interest-bearing financial instruments.

Most importantly, adequate foreign currency reserve has helped the central bank to maintain a stable exchange rate. Since, Bangladesh follows a "managed float" exchange rate regime, with sufficient foreign exchange reserves, central bank gains the ability to keep the exchange rate at a desired level. Sufficient foreign currency reserves also provide confidence to the economic system. It assures the stakeholder that the central bank can take action in times of emergency to protect their investment and make sure the country will meet its external obligations. All these played an essential part in maintaining a healthy GDP growth rate for Bangladesh in the last 40 years.

10.2 Literature Review

The impact of remittances on Bangladesh's economy has attracted the interest of economists both locally and globally. An early quantitative study by Qubria (1997) on the impact of remittances on macroeconomic variables in Bangladesh compared to other Asian countries showed that the steady flow of remittances eased foreign exchange restrictions, improved the balance of payments, and increased national savings. Murshid et al. (2002) examined the prospects and challenges of official routing of remittances by national and multinational banks. They note that an increase in remittances of one million taka (local currency of Bangladesh) would increase national income by 3.33 million taka. A study by Khan (2008) used data from the 2005 Household Income and Expenditure Survey to draw conclusions about the status of poverty in Bangladesh, and estimated that income from remittances resulted in an approximately eighteen percent reduction in poverty. Positive effects of remittances on GDP and the balance of payments were also observed by Faruque and Sarkar (2016).

There is some evidence of a U-shaped relationship between remittances and growth. That is, the growth effect of remittances is initially negative, but later becomes positive. Hassan and Shakur (2017) found evidence of such a nonlinear relationship. They showed that the unproductive use of remittances was initially widespread when received by migrant families, but better social and economic investments led to more productive uses of remittance earnings in later periods. Similar U-shaped relationship was also observed by Hassan et al. (2016). Using single-equation cointegration methods, they found that remittances' effect on long-run growth in Bangladesh is negative and falling until the remittances-to-GDP ratio is roughly eight percent. The benefits of remittance receipts outweigh their costs and their net effects start to become positive when the ratio exceeds 14 percent. Similar finding was also observed in Siddqui (2004). He showed that in the 1990s the migrant families effectively utilized the remittances they received by investing in nutritious food for the family members, health, education, land purchase, and financing migration of other family members.

Causal relationship between remittances and other variables is investigated in some studies. Kumar et al. (2017) find bidirectional causality between remittances and total factor productivity growth, which implies TFP growth and remittances are mutually reinforcing. Hatemi-J and Uddin (2013) applied a newly developed method based on simulations to show that causality nexus of poverty and remittances is bi-directional. They also found that the causal impact of poverty reduction on remittance is stronger than the reverse impact. Using the augmented Solow framework with the autoregressive distributed lag bounds procedure, Kumar et al. (2017) examine the cointegration relationship, the short-run and long-run effects, and the causality nexus between remittances per worker, capital per worker, and the output per worker. The results showed that remittances have a mixed effect in the short-run, however, a momentous positive effect in the long run (0.11%), on the output per worker. From the Granger causality assessment, they found a bidirectional causality between remittances and output (in per-worker terms) and a unidirectional causation from capital to remittances (in per-worker terms). Mamun et al. (2016) investigated the possibility of Dutch disease through remittances. Using annual data for 1982–2013 and standard testing methodologies, they showed that Dutch disease via remittances does not exist and remittance flow significantly improves domestic labor productivity in Bangladesh in the long run.

10.3 Recent Trends

Bangladesh is one of the top ten remittance recipient countries in the world. Figure 10.1 (a) shows the growth rate of remittance inflow since 1992.[1] Throughout this 27-year period, remittance inflow experienced an average growth rate of 12.1% per year. This impressive double-digit growth rate reflects how remittances became an important part of Bangladesh's economy. Growth rate was particularly higher in early 2000. However, since the financial crisis, there has been a decreasing trend in this rate. It even went into a negative territory in 2014, 2016, and 2017. Particularly, a steep 14.5 percent decline in 2017 raised many concerns among policymakers and economists. However, remittance earning picked up since then and saw an increase of 17 percent in 2018. Figure 10.1 (b) shows the share of remittance in total GDP. Remittance was as high as 9.6 percent of GDP in 2012 and 2013. However, the share of remittance decreased since then and currently it is below 6 percent.

To understand the evolution of remittance from different countries, Table 10.1 shows the growth rate of remittance inflow from some of the major countries. 2017 has been a disappointing year in terms of remittance earning from most countries except Qatar growth which was 33.4 percent. However, the numbers picked up again in 2018.

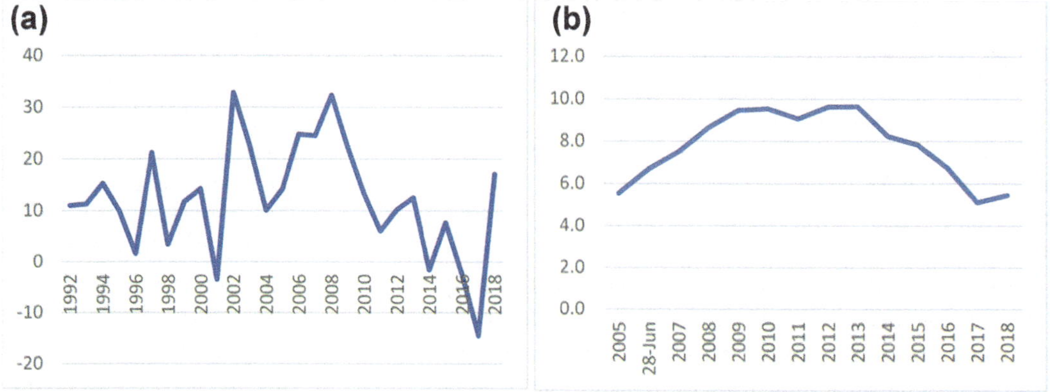

Fig. 10.1 (**a**) Remittance growth rate. (**b**) Remittance as a % of GDP

[1]Years represents fiscal years and not calendar years.

Table 10.1 Growth rate of remittances by country

Years	KSA	U.A.E.	UK	Kuwait	USA	Libya	Malaysia	Oman	Qatar
2006	12.35	26.95	47.89	21.53	36.49	−33.33	−18.38	25.59	28.76
2007	2.23	43.35	59.60	37.68	22.30	1350.00	−43.13	19.12	32.75
2008	33.98	41.04	1.04	26.89	48.34	−86.21	680.74	12.30	24.28
2009	23.01	54.78	−11.88	12.39	14.14	247.22	205.30	31.46	18.49
2010	19.87	7.59	4.79	4.99	−7.83	16.80	108.03	20.35	5.11
2011	−4.00	5.94	7.50	5.55	27.32	256.16	19.87	−4.23	−11.52
2012	11.99	20.08	11.00	10.63	−17.96	148.08	20.43	19.92	4.98
2013	3.94	17.60	0.44	−0.27	22.61	346.67	17.62	52.10	−14.42
2014	−18.56	−5.06	−9.13	−6.74	24.95	24.90	6.81	14.96	−10.24
2015	7.26	5.19	−10.67	−2.61	2.80	−36.47	29.77	30.55	20.43
2016	−11.51	−3.87	7.08	−3.73	1.07	−73.14	−4.15	−0.41	39.16
2017	−23.41	−22.89	−6.26	−0.43	−30.03	−81.84	−16.66	−1.51	33.47
2018	14.30	15.36	36.74	15.66	18.21	5.83	0.32	6.63	46.28

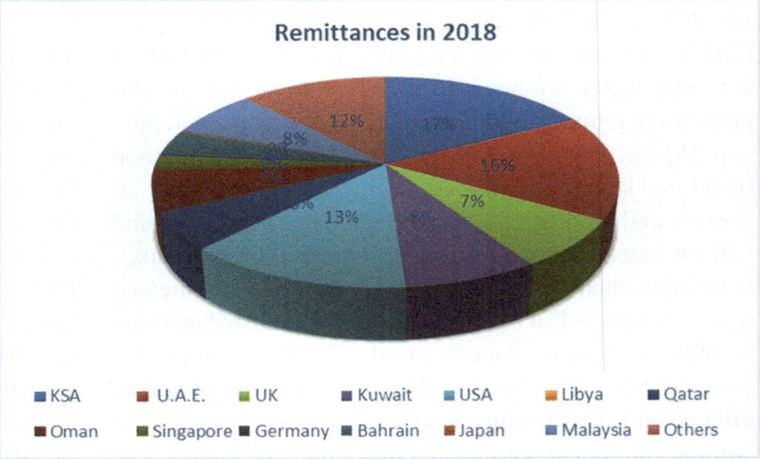

Fig. 10.2 Remitances in 2018

Remittances from Libya experienced very high growth rate until 2013 but then it fell steeply due to Arab Spring. Negative growth rate in 2014, 2016, and 2017 from Saudi Arabia, the largest remittance sender for Bangladesh, also raised concerns. Remittance from Malaysia increased significantly during 2008–2010, but growth has been negative or very low for the last three years. In fact, in 2018, when most of the countries experienced strong growth, remittance from Malaysia increased only by 0.32 percent. Figure 10.2 shows the remittance earning by government in the latest.

Fiscal year. According to the figure, Saudi Arabia remains the leading country for remittance earning for Bangladesh. 17 percent of total remittance came from this country. However, UAE became close second with 16 percent. The UK, Kuwait, and the USA were the three other major countries.

10.4 Conclusion and Policy Implications

There is no doubt that remittances have played a crucial role in Bangladesh's economy since its independence. If these large number of migrant people stayed at home, put extra pressure on the economy, and the government missed out on the foreign currency they received over the years, Bangladesh would have been economically weaker economically. However, as Bangladesh climbs up the income ladder and aspire to become a middle-income country, it is time to rethink seriously on the future of this growth model based on cheap labor export. In future, exporting labor can become more difficult due to various factors. As income rises, labor will not always be cheap and the traditional labor-importing countries may not have the need for them either. In January 2018, to provide Saudi citizens more job opportunities in the private sector, the Saudi government issued a decree to limit work in 12 sectors for foreign workers. There are recent reports of Bangladeshi workers getting deported from Saudi Arabia in spite of having valid work permit. Malaysia also stopped hiring Bangladeshi workers from September 2018.

Bangladesh government should take policies based on future needs. Firstly, new destinations for labor exports should be sought. Currently, many developed countries are facing low or declining population growth and the need for labor in these countries may rise in the future. Steps should be taken now to negotiate with those governments to understand their need. Secondly, local population aspiring to go abroad should be properly trained to acquire the skills needed for those particular countries. The technical training centers currently run by the Bureau of Manpower, Employment and Training (BMET) do a very poor job in terms of building the necessary skills. The certificates they provide are also often not recognized by foreign employers. Bangladeshi workers also earn less than workers from other countries due to lack of skills. Thirdly, migration from Bangladesh is concentrated on certain parts of the country. Steps should be taken so that residents of other districts with higher poverty rate also become aware of the opportunities. Fourthly, large amount of remittance still arrives in this country through informal channels. Steps should be taken to make the transfer of money through formal channels more easy and attractive. Some steps have already been taken in this regard, but more needs to be done.

Finally, Bangladesh should take long-term plans based on the possibility that remittances based on cheap unskilled labor are likely to face a downward trend in the future. It is very much expected that a lower number of low-skilled workers will immigrate (as they will have better opportunities at home). The remittances coming from these workers will also gradually slow down. Bangladesh's economy cannot thrive on remittances forever. However, it will take time. In the short to medium term, the country is still likely to remain one of the most densely populated countries globally, and it is unlikely that the country will be able to provide good quality jobs for all of its citizens. Therefore, the drive to immigrate will be there among the people. The government now needs to concentrate on the qualitative nature of this immigration as well. More concentration should be given on how the citizens of this country are treated by employers when they go abroad. Workplace safety and security should be ensured. Proper negotiation should be conducted with relevant country officials to stop physical torture, sexual abuse, and wage-related irregularities – even if it comes at a cost of lesser migrants to those countries. Also, the migration should be made smoother so that the recruiting agencies do not exploit the migrants. Dependency on unskilled labor should go down gradually and even if remittance plays an important role in the country's economy in the long run, it should come from skilled labor. The country should take policies accordingly.

References

Faruque O, Sarkar U (2016) Impact of remittances to the economic development of Bangladesh. J Business Stud 9:213–230

Hassan GM, Chowdhury M, Bhuyan M (2016) Growth effects of remittances in Bangladesh: is there a U-shaped relationship? Int Migr 54(5):105–121

Hassan GM, Shakur S (2017) Nonlinear effects of remittances on per capita GDP growth in Bangladesh. Economies 5(3)

Hatemi-J A, Uddin GS (2013) On the causal nexus of remittances and poverty reduction in Bangladesh. Appl Econ 46(4):374–382

Khan MWR (2008) The micro level impact of foreign remittances on incomes in Bangladesh: a measurement approach using the propensity score. Centre for Policy Dialogue, Dhaka

Kumar RR, Stauvermann PJ, Kumar NN, Shahzad SJH (2017) Revisiting the threshold effect of remittances on total factor productivity growth in South Asia: a study of Bangladesh and India. Appl Econ 50(26):2860–2877

Mamun A, Shohag K, Samargandi N, Yasmeen F (2016) Does remittance fuel labor productivity in Bangladesh? The application of an asymmetric non-linear ARDL approach. Appl Econ 48(50):4861–4877

Murshid K, Iqbal K, Ahmed M (2002) A study on remittance inflows and utilization. International Organization for Migration (IOM), Geneva

Qubria MG (1997) International migration, remittances and income distribution in the source country: a synthesis. Bull Econ Res 49:29–46

Siddqui T (2004) Efficiency of migrant workers' remittances: the Bangladesh case. Asian Development Bank, Metro Manila

11 Garment Industries: Its Impact on Economic Growth and Prevailing Challenges Towards Sustainability

María Gabriela Costa and Gregor C. Falk

Abstract

The Bangladeshi garment industry is a result of a trade quota system that enabled a degree of economic growth, and also implies questionable social and environmental costs. To further explain the phenomenon of the Bangladeshi garment industry arousal as one of the leading ready garment manufacturers in the world, we will describe what conditions helped them to obtain the role they perform, and associated profitability in the global production network. Further, we will portray the evolution of the exports and its significance for economic growth. We will unveil some global criticisms to the fast fashion production under the lenses of expected sustainability, to finalize describing the general supply management challenges that commonly organizations face to implement successful changes.

Keywords

Bangladesh · social and environmental · sustainability · exports

M. G. Costa
Faculty of Environmental and Natural Resources, Albert Ludwig Freiburg University, Freiburg im Breisgau, Germany

Facultad Jurídica, Social y Administrativa, Universidad Nacional de Loja, Loja, Ecuador
e-mail: maria.g.costa@unl.edu.ec

G. C. Falk (✉)
Institute for Geography and Geography Education, University of Education Freiburg, Kunzenweg, Freiburg, Germany
e-mail: gregor.falk@ph-freiburg.de

11.1 The Entrance of Bangladesh to the Global Production Network of Garments

Trade is one of the oldest economic activities, within it textiles and garments have one of the most globalized paths. In 1945 an earth system governance aroused through the establishment of the United Nations (UN), related international organizations, and agreements (Biermann and Boas 2010). Under this, the degree of participation in international trade is decided through agreements, trade quotas, tariffs, currency revaluation, preferential access to overseas providers and technology availability (Gereffi 1999).

The World Trade Organization Agreement on Textiles and Clothing (WTO-ATC) in 1995 used to be the main policy framework for garments, until the approval of the Multifiber Agreement (MFA). The MFA prepared a phased out of quota restrictions in 2004. This was seen as a potential threat to Bangladesh, Pakistan and Nepal garment and apparel trade due to the preferred quota they had compared to China and India (Lips et al. 2003).

Besides having advantage in the international quota system to enter overseas countries' market, the reduction of labour costs became an imperative to disperse production *(or triangle manufacture)* to quota-free nations, that usually are low-wage nations (Christerson and Appelbaum 1996). Indeed, policies that emerged to protect labour force in developed countries triggered the first shift of garment production from North America and Western Europe to Japan in 1950. In the 1970s a second shift occurred from Japan to the Newly Industrializing Economies (NIE), namely South Korea, Taiwan, Hong Kong and Singapore (Gereffi 1999). In the 1980s Bangladesh entered into the global supply chain in an assembly role of imported inputs, which further developed into a full-package supply, also known as Original Equipment Manufacturing (OEM) production (Gereffi 1999).

Open international trade enabled a global integrated production system, where firms master specific functions and a

R. Ahmed et al. (eds.), *Transforming Bangladesh*, World Regional Geography Book Series,
https://doi.org/10.1007/978-3-031-45093-8_11

role in the supply chain, which results in certain profitability (Gereffi 1997). Trade liberation improved cooperation and performance based on horizontal and vertical linkages (Bair and Gereffi 2001). The global supply chain of garments is a typical buyer-driven chain, where retailers, brand-named merchandizers and trading companies exercise a decentralized production network on which do not own any production facility but merchandize their designs (Barrientos et al. 2011).

The turn of Bangladeshi garment manufacturers to exporters was based on the world preference for the most inexpensive commodities, turning them in a low-cost supplier of standardized goods in the global sourcing to large volume discount stores (Gereffi 1994). The most renown clients include mass merchandizers as Sears Roebuck, Montgomery Ward, J.C Penney, Woolworth and among discount chains are Wal-Mart, Kmart and Target (Gereffi 1994). United States (US) apparel sourcing imports rose from \$17.3 to \$41.7 billion between 1986 and 1996, on which Bangladesh share rose from 1.0–1.9% to 2.0–3.9% of total imports (Gereffi 1999).

Firms are encouraged to upgrade their process along the supply chain or within their own role. Upgrading differs depending if the impact is social, economical, product based, or an industrial improvement (Gereffi 1999; Bair and Gereffi 2001, 2004; Barrientos et al. 2011; Rossi 2013; Selwyn 2013; Prybyla 2014). An example of industrial upgrading is through mastering higher value activities, for example, from export-oriented assembly to a more integrated manufacture with marketing, and a full package production, known as *Original Equipment Manufacturing* (OEM), or to an *Own Brand name Manufacturing* (OBM) (Bair and Gereffi 2004). The upgrading depends on entrepreneurs capacity to master the dynamics within the commodity chain such as establishing linkages with leading firms or brands to reach a coordinating role and to move upwards in their specialization role (Gereffi 1997, 1999; Gibbon 2001; Rossi 2013).

The profitability depends on the level of specialization of a company, the closeness to consumer, and its role (coordinator or subcontractor) along the supply chain (Gibbon 2001). The profit from a buyer-driven commodity such as garments is derived from design, marketing, sales, and financial services, whereas in a producer-driven supply chain is based on scale economies where transnational corporations or industrial enterprises control the production system of a capital and technology-intensive industry like automobiles, machinery, computers, aircraft, etc. (Gereffi 1994).

The highest profitable activities in the garment supply chain are design and marketing, normally a role from buyers such as fashion bureaus and OBM which are closer to the final consumer, and mainly domain a process of triangle manufacture, instead of owning factories or hiring workers (Gibbon 2001), see Fig. 11.1 representing different profitability depending on the functions developed in the garments supply chain.

Measured consequences of the Bangladeshi role in the garment supply chain due to the MFA imply a structural change in labour force (*mainly benefiting female workers*), on which 6% of labour force shifted from agriculture to export-oriented services such as clothing in 1991–1992. The share of business and commerce in national hold income increased from 14.8% in 1991 to 25.9% in 2000, and experienced a decrease of income from agriculture from 33.4% to 18.9% in the same period (Ahamed and Ehsan 2005).

Since 1995, Bangladesh became a member of the WTO, and in 2001 signed the non-legally binding EU-Bangladesh Co-operation Agreement, which comprises human rights, good governance to trade, environment, and economic devel-

Position:	Design	Raw material	Spinning waving	Manufacture	Transport	Retailer	Marketing	Profitability from role in supply chain
- Assembly				$				>
- Supplier/ Trader		$$	$$		$$			>
- Fashion Bureau	$$$$$					$$$$$	$$$$$	<
- OBM	$$$$$	$$	$$$	$$$		$$$$$	$$$$$	<
- OEM		$$	$$$	$$$	$$			>

Fig. 11.1 Representation of profit margin depending on the role and typical activities performed in the garment supply chain based on: (Gereffi 1994, 1997, 1999; Gibbon 2001; Bair and Gereffi 2004)

opment. By 2018, Bangladesh still receives special grants as duty-free and quota-free access for all exports, except arms and ammunition as part of the EU's arrangement "Everything but Arms" (European Commission 2018). According to the European Commission portal on trade, it is stated that in 2015, over 90% of EU's total imports from Bangladesh were clothing (European Commission 2018).

11.2 The Initiation of the Bangladeshi Garment Industry and Its Impact on Economic Growth

The development of the Bangladeshi garment industry resulted from a spillover effect of technology transfer on 130 workers of Desh Ltd., who were intensively trained by Daewoo Corporation in South Korea in 1979. By 1990, the trained workers became garment manufacturers and garments traders causing a $2 billion garment increase sale, that rose up to $10 billion by 2008 (Mottaleb and Sonobe 2011). Nowadays, there are around 4000 manufacturers accounting 75% of exports earnings (Mottaleb and Sonobe 2011), and which are important job providers for unskilled and semi-skilled female workers who migrated from rural areas (Bakht et al. 2009).

The benefit on economic growth from the MFA was evident as well as an increase in income inequality. Data shows that Gross Domestic Product (GDP) increased from 3.5% (1980s) to more than 5% (1990s), trade-GDP ratio increased 6% from 1980 to 2000 and 6% more to 2004 resulting in 32%, and a Gini coefficient increased from 0.348 to 0.417 (Ahamed and Ehsan 2005). Data from Bangladeshi Garment Manufacturers Exporters Association (BGMEA) shows the growth of Garment Factories, the contribution to exports from ready-made garments, the employment growth and GDP resulting from the garment industry (Fig. 11.2).

The economic performance of companies in the global economy does not necessarily reflect compliance of workplace human rights standards, such as good working conditions, assurance of labour rights, absence of force and child labour (Islam and McPhail 2011), or means improved gender equality. The underlying assumption that there is causality between companies' profitability and social upgrading is derived from neoclassical economics but not from empirical observation (Selwyn 2013).

It is evident to some degree that the shifts of garment production to developing countries brought benefits to a large female force and rural migrants who could access to job opportunities (Barrientos et al. 2011). The Bangladeshi garment industry is the main provider of work for females (Mottaleb and Sonobe 2011). The difference between time wage rate and piece wage rate can be alarming (Chand 2012). The registered minimum wage in Bangladesh for a manufacturer worker is 30 € to 100 € per month and it has been reported by human rights organizations that it can include 10–16 hours of work per day (AWAJ Foundation 2013).

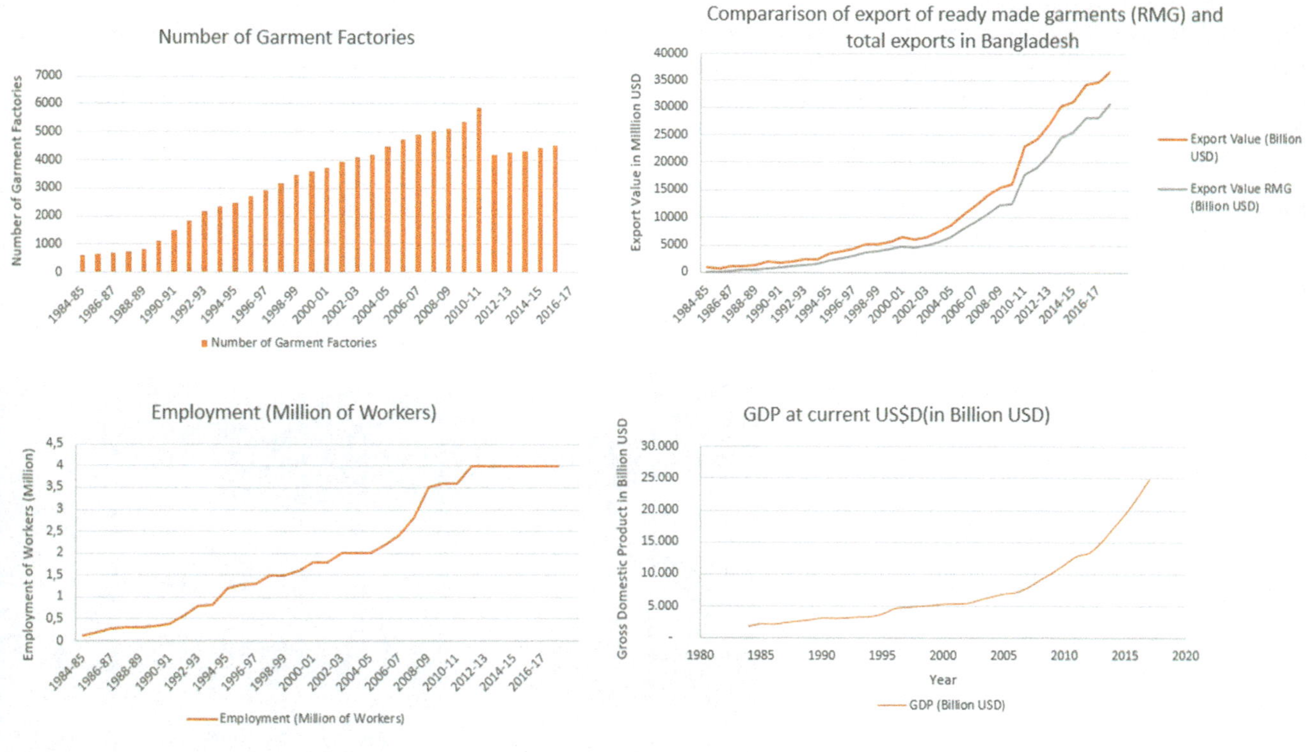

Fig. 11.2 Growth of garments, exports, GDP, and employment

Major problems are the *lead time,* which is a fast-paced production mode, and the profitable competitiveness competition which jeopardizes a real sustainable development, becoming a limiting factor. In the Bangladeshi garment industry offer there are at least 99 different types of products. The highest contributor to the exports earnings were the cotton non-knit shirts, man and boy, accounting for $332 million in 2005 (Haider 2007).

The requirements in trade policy that the Bangladeshi exporters face are barriers such as quotas, 12–15% Most Favored Nation tariffs, Rules Of Origins (ROO), local assembly and yarn forward rule (need local fabrics) (Demidova et al. 2012). ROO constraints depend on the trade quotas or preferences, and if are binding or not. The process is based on documentation (Demidova et al. 2012). Bangladeshi as second most important garment traders (Haider 2007).

The readymade garment industry accounts for almost 80 per cent of the country's export earnings and contributes to more than 10 per cent of the gross domestic product (GDP), employing more than four million workers, the majority women (Ahmed et al. 2014).

11.3 The Quest to a Sustainable Supply Chain

As known, *sustainable development* is a long-term voluntary commitment (1987) adopted in the global governance network of the United Nations' World Commission on Environment and Development (WCED), as a result of the Bruntland Report (Caniato et al. 2012). The Brundtland Report is known as the starting point of a world commitment towards development as Our Common Future (Linton et al. 2007).

The Bruntland Report defined sustainability reflecting upon interlinked future consequences resulted from an extremely polluting global system. Production and consumption modes should consider an optimal management of available resources without compromising the availability of necessary resources for future generations (Linton et al. 2007; Seuring and Müller 2008). A sustainable development comprises an overlapped impact of our decisions in an economic, social and environmental dimension at once (de Brito et al. 2008).

A tipping point between international trade and environmental governance is the analysis of how an earth system governance can be assured an earth system transformation to secure a safe co-evolution considering environmental problems (Biermann 2007). The international sustainability governance has as a guiding principle, the common but differentiated responsibilities, that seeks for responsibility, awareness and a precautionary approach (article 3 of United Nations Framework Convention on Climate Change) (Biermann et al. 2009). Indeed, long-term Agreements on the new anchored Governance Architecture, through its tariffs and barriers cluster of norms, might have resulted in a synergetic fragmentation (Biermann et al. 2009).

Dirty industries are infamous for their environmental impact, and is highly argued a relationship between open trade and a polluting economic development, particularly in developing countries where there are weaker legislations (Mani and Wheeler 1998). Environmental and Social concerns are witnessed at every node of the supply chains, from the excessive use of pesticides, chemicals, CO_2 emissions (not only from unsustainable energy supply but from international transport), as well as lack of management and treatment of water, pre and post-consumption waste (see Fig. 11.3) and human rights violations such as in the Bangladeshi manufacture node.

Fig. 11.3 Textile and garment production waste in Dhaka. (By: MG Costa Garcia)

The garment industry that leads Bangladeshi trade in a globalized production system is fast fashion, characterized by planned obsolescence, which is a trend contrary to the sustainable consumption. Sustainable consumption is embedded in the action plan for sustainable development Agenda 21, adopted in Rio Earth Summit, 1992, and further supported by UN Global Compacts initiatives (Bly et al. 2015), and Global Reporting Initiative (GRI) for accountability and transparency, following corporate social responsibility and environmental management guidelines (Sobhani et al. 2009).

Fast fashion means mass production, that prioritizes cheapness, mass availability and volume purchasing using scales of economics that generate commons costs like increased pollution, resource depletion, climate change costs (Fletcher 2010), and environmental concerns as environmentally friendly practices, environmental compliance, measurement and control of pollution, waste recycling and tree plantation (Sobhani et al. 2009). Indeed, during the 1980s, it was evident the increase and acceleration of pollution and waste in landfills, increasing the cost of landfill disposal from $10 to $45 per ton (Domina and Koch 1997). In Bangladesh pollution of rivers and environment is a concern seen as prone to face an ecological catastrophe (Sobhani et al. 2009).

Alongside, due to the development of communications and transportation aid manufacturing companies to negotiate for the most profitable combination of capital and labour, undermining long-term development by shifting production to countries characterized for having low wages, restrictive collective bargaining rights, and generally weak laws (Christerson and Appelbaum 1996).

Global criticisms to garment industries in developing countries are about human rights abuses such as child and forced labour, accidents from fires, verbal and physical abuse or direct exploitation of workers in factories (Islam and McPhail 2011). Indeed, international criticism led in 1994 to Bangladeshi government and garment industry with support of ILO, to establish a monitoring verification system and a Memorandum of Understanding to eliminate child labour and to provide financial assistance to children under 14 to return to schools (Islam and McPhail 2011).

The globally produced garment industry had reported several cases and misfortunate events related to workers human rights violation such as violence at workplace, threatening to trade unions, low living wages, inadequate working and loss of lives due to safety conditions and over-exploitation due to extended working hours, among others (Seuring and Müller 2008; Sikka 2011; Fink 2014; Ganguly 2015). The ability of Ministry of Labour and Employment to prosecute workers human rights is doubted (Campaign 2016).

A lack of proper working conditions has been recalled on incidents from 2006 to 2010 on which almost 500 workers died due to safety incidents in Bangladeshi garment factories such as Spectrum factory (64 workers died). In 2012, the fire at the Tazreen Fashions factory left 112 workers dead and the catastrophic collapse of Rana Plaza 1100 workers, the latest is considered the second worst disaster in global garment industry after China (Ganguly 2015). Particularly the catastrophic events in 2013 unveiled the extreme production conditions that fast fashion production mode demands, neglecting previous voluntary human rights agreements.

As a result of the Rana Plaza catastrophe there were sanctions imposed by the US such as cut off of trade benefits (Palmer 2013), soft law, hard law and organizational changes taking place such as the Bangladesh Labor Act of 2006, amendment in 2006, Adoption of ACCORD on Fire and Building Safety, the Alliance for Bangladesh Worker Safety (Human Rights Watch 2015), Sustainability Compact with the European Union, on which also participates the USA and Canada (European Commission 2018), the Clean Clothes Campaign (Campaign 2016), IndustriALL Global Union and UNI Global Union (Tula Connell 2014), EU Flagship initiative for a responsible management of the garment supply chain among others aiming to improve labour rights, safety in buildings and business conduct.

The amendments in the Labour Act outlaws numerous "unfair labour practices" in Article 195, such as no threatening, violence, or remove from employment of workers that seeks organizing or to become member or of a trade union (Bangladesh Employers' Federation 2009). After the Labour Law amendments reported in 2014, a prevailing challenge is the assurance of workers collective rights in Bangladesh, even after ratifying labour international conventions such as Conventions 87 and 98 that demand freedom of association and collective bargaining (Human Rights Watch 2014). The estimation suggests that an average of 5000 garment factories in Bangladesh accounts for nearly 4 million or 4000000 workers, from which only less than 10% were registered with trade Unions in the Bangladeshi Ministry of Labor (Human Rights Watch 2015). Before the collapse at Rana Plaza, there were only 2 garment unions between 2010 and 2012 (Tula Connell 2014).

11.4 Prevailing Challenges to Change Towards a Sustainable Production in the Bangladeshi Garment

The interactions in an earth system governance demand the following principles towards sustainable development: the precautionary principle, intergenerational equity, common but differentiated responsibilities and equitable utilization (Harrison 2008), alongside with credibility, stability, adaptiveness, inclusiveness, accountability, and allocation (Biermann 2007). Most nations are committed to increased long-term and far-reaching changes to achieve the sustainable development stated from 1992 in the Rio Declaration (Rennings et al. 2006).

Businesses pursuing sustainable economic development are voluntarily committed to reflect upon externalities such as negative social and environmental effect of economic growth (Banerjee 2008). The determinant factors that encourage environmental innovations in products and processes range from environmental policy, technology push (R&D activities), market push (customer satisfaction), firm characteristics (size and age) (Rehfeld et al. 2007), environmental regulation, environmental management tools, general organizational changes and improvements (Horbach 2008), and actions from the Corporate Social Responsibility (CSR) framed by the World Business Council (Banerjee 2008).

The implementation of sustainable innovations requires a multidisciplinary understanding of supply chain management (Croom et al. 2000), to clarify what is required to include green practices or sustainability (Seuring and Müller 2008; Caniato et al. 2012). Innovations and changes to achieve sustainability range from environmental innovations, technical environmental innovations, environmental product innovations, environmental process innovations, innovations in cleaner production technologies, social innovations, standards, environmental management (Rennings et al. 2006; Rehfeld et al. 2007), and more importantly action regarding direct or indirect costs of respecting environment and people, as part of environmental justice (Oyewole 2001).

The organizational changes towards sustainability face processes that prevent change such as structural process, institutional processes, political processes and learning processes (Hannan et al. 2003), which success depend on leadership and a series of long timing phases that if skipped won't produce satisfying results (Kotter 1995). The expected responses from organizational change, either positive or negative include an emotional, cognitive and intentional dimension (Piderit 2000).

The positive responses to change are obtained when *readiness for change* exists in organizational members' beliefs, attitudes, and intentions, otherwise the obstacle will be *resistance to change* (Armenakis et al. 1993) Kotter suggests eight steps to achieve organizational change: *(1) Establishing a sense of urgency, (2) Forming a powerful guiding coalition, (3) Creating a vision, (4) Communicating the vision, (5) Empowering others to act on the vision, (6) Planning for and creating short-term wins, (7) Consolidating improvement and producing still more change, (8) Institutionalizing new approaches* (Kotter 1995, 2007). The cognitive resistance to change can be reduced with education, participation, involvement, negotiation and agreement (Armenakis et al. 1993).

Complementary, organizations need to enhance their capacity to learn and create, possibly by including in their DNA 5 disciplines to become a learning organization: Systems thinking, personal mastery, mental models, shared vision building and team learning (Senge 1990). A learning organization avoids the creation of silos to ensure a learning flow among the subsystems organization, people, knowledge and technology (Serrat 2017). Other important factors to enable change are allowing experimentation, learn from own and other past experiences and transfer knowledge (Garvin 1993). A challenge is that knowledge capital and technological capabilities that are key, also for research and design (R&D) (Horbach 2008), are limited in developing countries or are obstacle due to cultural factors and power relations, such as in garment organizations of developing countries (Islam and McPhail 2011; Rossi 2013; Selwyn 2013).

Further, the changes that sustainability requires are questioned by the business model profitability and related management decisions on raw material, energy, water, biodiversity, emissions, products and services, compliance, transport and business integration (Caniato et al. 2012). The investments required to adopt green technologies are challenged by the efficiency of investment across firms (Amacher et al. 2004). Generally, changes towards sustainability result in a barrier for products from low-cost countries or of deficient quality such as handmade products (Houe and Grabot 2009).

Industrial ecology takes the perspective of sustainability on the production of by-products (wastes) along a supply chain considering product design, manufacturing by-products, by-product life cycle and recovery processes at the end of life (Linton et al. 2007). Finding new means of recycling solid waste considering a lifecycle analysis potentially could improve post-producer waste and pre-consumer waste (Domina and Koch 1997), and even pre-producer waste and post-consumer waste of garments if upcycling methods get included as implemented in BEXIMCO, a Bangladeshi firm (Aus 2011).

In the case of the garment industry it is assumed that there is little room to innovate technologically in the manufacture node of the supply chain because this is an activity performed mainly by hand, however BEXIMCO has experimented trough up-cycling techniques, which not require monotonous movements but a high degree of creativity and therefore an upgrade of tasks and skills required.

It is argued if the consumers' willingness-to-pay to pay higher prices allows a proper commercial exploitation of environmental innovations (Rehfeld et al. 2007) if the characteristics of the product, or, of its production are informed and transparent (Solaiman et al. 2011). Some of the environmental attributes range from sustainable, recycled, non-toxic, biodegradable, green electricity and cruelty-free (Hamilton and Zilberman 2006), ethical and fair trade (Loureiro and Lotade 2005). Consumers face fraud due to marketing, but often this possibility is perceived as manageable through certification or label that states for reliable reputation (Hamilton and Zilberman 2006).

Eco-labelling is used as an information tool that induces the adoption of green technologies because of its ability to reduce the costs of providing high levels of environmental quality required to obtain or hold ecolabel rights. Some examples of standards for sustainable production are the design for environment ISO 14062 (Houe and Grabot 2009), ISO 14001 and SA8000 for social issues (Seuring and Müller 2008).

The nature of the labelling either as voluntary or mandatory will reflect a degree of compliance such as positive, negative or neutral informing different stages from production, to all lifecycle of the product (Islam 2008). Social issues such as minimum wages are considered a useful tool when government values redistribution to low-wage workers are in place alongside optimal nonlinear taxes or transfers (Lee and Saez 2012).

The expected changes in workers' rights comprise employment, social protection, workers' rights and social dialogue comprised in the International Labor Organization (ILO)'s Decent Work Agenda, as well as others such as improvements to workers' pay and conditions, freedom of association, collective bargaining, and reduction in excessive working hours (Selwyn 2013).

In Bangladesh some companies are implementing sustainability by selecting suppliers within the Fair Trade Network while focusing on green design and working with small producers (Caniato et al. 2012), while others have problems implementing social sustainability because of incoherence between codes of conduct and standards related to hygiene, health care, and working hours. In some cases there is complicity between suppliers, buyers and manufacturers to correct information for audits to avoid sanctions (Anisul Huq et al. 2014). Compared to 1998, when the ILOs workplace human rights were approved, the voluntary disclosure in Bangladesh has engaged in corporate accountability and disclosure (Islam and McPhail 2011).

11.5 Conclusion

The intertwined impacts of our choices and decisions as part of a supply chain are evident, either as suppliers or consumers. Further analysis of strategies set in place to increase dialogue and multi-cooperation in international spheres as coordination mechanisms is required to improve accountability. There is no turning point from a globalized path, however from the fast paced is still possible if a higher price and higher willingness to pay increases. Our purchase decisions and willingness-to-pay for sustainable products pose complex questions and commitments in the global production network that will be translated in transparency in supply chains of labour and resource-intense commodities.

Acknowledgements Walking around Dhaka to learn from the Bangladeshi garment industry was possible thanks to the Müller-Fahnenbergplatz Foundation, that allowed data collection in the field, and the help and warming help from all my Bangladeshi friends that guided me and opened the doors for this research: Khalid Khan, Saamera Zaman, Protyasha Prachurja and Mahbub Rubel. Professors' contribution Benno Pokorny from Freiburg University, Raquib Ahmed from Rajshahi University, and Gregor Falk from PH Freiburg was determinant to keep me motivated to continue researching on the branches of sustainable development. Ferdinand Peters is as always my backbone of encouragement and love. Thanks to all of you.

References

Ahamed SM, Ehsan M (2005) Bangladesh's poverty reduction strategy: an analysis from the labour market perspective. Asian Aff 27:51–66

Ahmed FZ, Greenleaf A, Sacks A (2014) The paradox of export growth in areas of weak governance: the case of the ready made garment sector in Bangladesh. World Dev 56:258–271

Amacher GS, Koskela E, Ollikainen M (2004) Environmental quality competition and eco-labeling. J Environ Econ Manag 47:284–306

Anisul Huq F, Stevenson M, Zorzini M (2014) Social sustainability in developing country suppliers: an exploratory study in the ready made garments industry of Bangladesh. Int J Oper Prod Manag 34:610–638

Armenakis AA, Harris SG, Mossholder KW (1993) Creating readiness for organizational change. Hum Relat 46:681–703

Aus R (2011) Trash to trend-using upcycling in fashion design. Dissertation, Estonian Academy of Arts

AWAJ Foundation (2013) Workers' voice report 2013. https://www.wiwiss.fu-berlin.de/forschung/Garments/Medien/2013_AWAJ-Foundation_Workers_-Voice-Report-Bangladesh.pdf. Accessed 26 Sep 2018

Bair J, Gereffi G (2001) Local clusters in global chains: the causes and consequences of export dynamism in Torreon's blue jeans industry. World Dev 29:1885–1903. https://doi.org/10.1016/S0305-750X(01)00075-4

Bair J, Gereffi G (2004) Upgrading, uneven development, and jobs in the North American apparel industry. In: Labor and the globalization of production. Global Networks, London, pp 58–87

Bakht Z, Yamagata T, Yunus M (2009) Profitability and diversity among knitwear-producing firms in Bangladesh: the prospects of a labor-intensive industry in a least developed country. Dev Econ 47:340–366. https://doi.org/10.1111/j.1746-1049.2009.00089.x

Banerjee SB (2008) Corporate social responsibility: the good, the bad and the ugly. Crit Sociol 34:51–79. https://doi.org/10.1177/0896920507084623

Bangladesh Employers' Federation (2009) A handbook on the Bangladesh Labour Act, 2006. https://www.ilo.org/dyn/travail/docs/352/A%20Handbook%20on%20the%20Bangladesh%20Labour%20Act%202006.pdf. Accessed 25 Sep 2018

Barrientos S, Gereffi G, Rossi A (2011) Economic and social upgrading in global production networks: a new paradigm for a changing world. Int Labour Rev 150:319–340. https://doi.org/10.1111/j.1564-913X.2011.00119.x

Biermann F (2007) Earth system governance'as a crosscutting theme of global change research. Glob Environ Change 17:326–337. https://doi.org/10.1016/j.gloenvcha.2006.11.010

Biermann F, Boas I (2010) Preparing for a warmer world: towards a global governance system to protect climate refugees. Glob Environ Polit 10:60–88. https://doi.org/10.1162/glep.2010.10.1.60

Biermann F, Pattberg P, Van Asselt H, Zelli F (2009) The fragmentation of global governance architectures: a framework for analysis. Glob Environ Polit 9:14–40. https://doi.org/10.1162/glep.2009.9.4.14

Bly S, Gwozdz W, Reisch LA (2015) Exit from the high street: an exploratory study of sustainable fashion consumption pioneers: sustainable fashion consumption pioneers study. Int J Consum Stud 39:125–135. https://doi.org/10.1111/ijcs.12159

Campaign CC (2016) Hazardous workplaces: making the Bangladesh Garment industry safe. https://cleanclothes.org/resources/publications/2012-11-hazardousworkplaces.pdf. Accessed 27 Sep 2018

Caniato F, Caridi M, Crippa L, Moretto A (2012) Environmental sustainability in fashion supply chains: an exploratory case based research. Int J Prod Econ 135:659–670. https://doi.org/10.1016/j.ijpe.2011.06.001

Chand A (2012) At the bottom of a global commodity chain—how different really are hourly wage rates and piece rates? Ind Relat J 43:171–191. https://doi.org/10.1111/j.1468-2338.2012.00668.x

Christerson B, Appelbaum R (1996) Imperatives for profitability in labor-intensive export industries: implications for China's economic development. J Chin Polit Sci 2:23–50. https://doi.org/10.1007/BF02915934

Croom S, Romano P, Giannakis M (2000) Supply chain management: an analytical framework for critical literature review. Eur J Purch Supply Manag 6:67–83. https://doi.org/10.1016/S0969-7012(99)00030-1

de Brito MP, Carbone V, Blanquart CM (2008) Towards a sustainable fashion retail supply chain in Europe: organisation and performance. Int J Prod Econ 114:534–553. https://doi.org/10.1016/j.ijpe.2007.06.012

Demidova S, Kee HL, Krishna K (2012) Do trade policy differences induce sorting? Theory and evidence from Bangladeshi apparel exporters. J Int Econ 87:247–261. https://doi.org/10.1016/j.jinteco.2011.12.008

Domina T, Koch K (1997) The textile waste lifecycle. Cloth Text Res J 15:96–102. https://doi.org/10.1177/0887302X9701500204

European Commission (2018) Countries and regions Bangladesh. Policies Inf Serv. http://ec.europa.eu/trade/policy/countries-and-regions/countries/bangladesh/. Accessed 15 Aug 2018

Fink E (2014) Trade unions, NGOs and transnationalisation: experiences from the ready-made garment sector in Bangladesh. ASIEN 130:42–59

Fletcher K (2010) Slow fashion: an invitation for systems change. Fash Pract 2:259–265. https://doi.org/10.2752/175693810X12774625387594

Ganguly M (ed) (2015) "Whoever raises their head suffers the most": workers' rights in Bangladesh's garment factories. Human Rights Watch, New York

Garvin DA (1993) Building a learning organization. Harv Bus Rev 71:78–91

Gereffi G (1994) The organization of buyer-driven global commodity chains: how US retailers shape overseas production networks. Commod Chains Glob Capital 95(3). dukespace.lib.duke.edu/dspace/bitstream/handle/10161/11457/1994_Gereffi_RGC.pdf?sequence=1

Gereffi G (1997) The reorganization of production on a world scale: states, markets and networks in the apparel and electronics commodity chains. In: Regionalization and labour market interdependence in East and Southeast Asia. Palgrave Macmillan, London, pp 43–91

Gereffi G (1999) International trade and industrial upgrading in the apparel commodity chain. J Int Econ 48:37–70. https://doi.org/10.1016/S0022-1996(98)00075-0

Gibbon P (2001) Upgrading primary production: a global commodity chain approach. World Dev 29:345–363. https://doi.org/10.1016/S0305-750X(00)00093-0

Haider MZ (2007) Competitiveness of the Bangladesh ready-made garment industry in major international markets. Asia-Pac Trade Invest Rev 3:3–27

Hamilton SF, Zilberman D (2006) Green markets, eco-certification, and equilibrium fraud. J Environ Econ Manag 52:627–644. https://doi.org/10.1016/j.jeem.2006.05.002

Hannan MT, Pólos L, Carroll GR (2003) Cascading organizational change. Organ Sci 14:463–482. https://doi.org/10.1287/orsc.14.5.463.16763

Harrison J (2008) The Oxford handbook of international environmental law. Edinb Law Rev 12:156–158. https://doi.org/10.3366/E1364980908220226

Horbach J (2008) Determinants of environmental innovation—new evidence from German panel data sources. Res Policy 37:163–173. https://doi.org/10.1016/j.respol.2007.08.006

Houe R, Grabot B (2009) Assessing the compliance of a product with an eco-label: from standards to constraints. Int J Prod Econ 121:21–38. https://doi.org/10.1016/j.ijpe.2008.03.014

Human Rights Watch (2014) Bangladesh: protect garment workers' rights. Human Rights Watch. https://www.hrw.org/news/2014/02/06/bangladesh-protect-garment-workers-rights. Accessed 27 Sep 2018

Human Rights Watch (2015) Bangladesh: 2 years after Rana plaza, workers denied rights. Human Rights Watch. https://www.hrw.org/news/2015/04/22/bangladesh-2-years-after-rana-plaza-workers-denied-rights. Accessed 27 Sep 2018

Islam MS (2008) Towards certification and ecolabelling: a compliance study of Bangladesh shrimp aquaculture. Dissertation, UNU-Fisheries Training Programme

Islam MA, McPhail K (2011) Regulating for corporate human rights abuses: the emergence of corporate reporting on the ILO's human rights standards within the global garment manufacturing and retail industry. Crit Perspect Account 22:790–810. https://doi.org/10.1016/j.cpa.2011.07.003

Kotter JP (1995) Leading change: why transformation efforts fail. Harvard Business Review, Boston, pp 59–67

Kotter JP (2007) Leading change. Harvard Business Review Press, Boston

Lee D, Saez E (2012) Optimal minimum wage policy in competitive labor markets. J Public Econ 96:739–749. https://doi.org/10.1016/j.jpubeco.2012.06.001

Linton JD, Klassen R, Jayaraman V (2007) Sustainable supply chains: an introduction. J Oper Manag 25:1075–1082. https://doi.org/10.1016/j.jom.2007.01.012

Lips M, Tabeau AA, van Tongeren FW, et al (2003) Textile and clothing sector liberalization-Consequences for the Bangladesh economy. In: Global Trade Analysis Project (GTAP). Presented at the 6th Annual Conference on Global Economic Analysis, Purdue University, West Lafayette, The Hague, The Netherlands, 12–14 June 2003

Loureiro ML, Lotade J (2005) Do fair trade and eco-labels in coffee wake up the consumer conscience? Ecol Econ 53:129–138. https://doi.org/10.1016/j.ecolecon.2004.11.002

Mani M, Wheeler D (1998) In search of pollution havens? Dirty industry in the world economy, 1960 to 1995. J Environ Dev 7:215–247. https://doi.org/10.1177/107049659800700302

Mottaleb KA, Sonobe T (2011) An inquiry into the rapid growth of the garment industry in Bangladesh. Econ Dev Cult Change 60:67–89. https://doi.org/10.1086/661218

Oyewole P (2001) Social costs of environmental justice associated with the practice of green marketing. J Bus Ethics 29:239–251. https://doi.org/10.1023/A:1026592805470

Palmer D (2013) U.S. suspends trade benefits for Bangladesh over safety. Reuters. https://www.reuters.com/article/us-obama-trade-bangladesh/u-s-suspends-trade-benefits-for-bangladesh-over-safety-idUSBRE95Q15720130627. Accessed 27 Sep 2018

Piderit SK (2000) Rethinking resistance and recognizing ambivalence: a multidimensional view of attitudes toward an organizational change. Acad Manag Rev 25:783–794. https://doi.org/10.5465/amr.2000.3707722

Prybyla J (2014) Manufacturing miracles: paths of industrialization in Latin America and East Asia. Princeton University Press

Rehfeld K-M, Rennings K, Ziegler A (2007) Integrated product policy and environmental product innovations: an empirical analysis. Ecol Econ 61:91–100. https://doi.org/10.1016/j.ecolecon.2006.02.003

Rennings K, Ziegler A, Ankele K, Hoffmann E (2006) The influence of different characteristics of the EU environmental management and auditing scheme on technical environmental innovations and economic performance. Ecol Econ 57:45–59. https://doi.org/10.1016/j.ecolecon.2005.03.013

Rossi A (2013) Does economic upgrading lead to social upgrading in global production networks? Evidence from Morocco. World Dev 46:223–233. https://doi.org/10.1016/j.worlddev.2013.02.002

Selwyn B (2013) Social upgrading and labour in global production networks: a critique and an alternative conception. Compet Change 17:75–90. https://doi.org/10.1179/1024529412Z.00000000026

Senge PM (1990) The fifth discipline, the art and practice of the learning organization. Doubleday Canada, New York

Serrat O (2017) Building a learning organization. In: Knowledge solutions. Springer, Singapore, pp 57–67

Seuring S, Müller M (2008) From a literature review to a conceptual framework for sustainable supply chain management. J Clean Prod 16:1699–1710. https://doi.org/10.1016/j.jclepro.2008.04.020

Sikka P (2011) Accounting for human rights: the challenge of globalization and foreign investment agreements. Crit Perspect Account 22:811–827. https://doi.org/10.1016/j.cpa.2011.03.004

Sobhani FA, Amran A, Zainuddin Y (2009) Revisiting the practices of corporate social and environmental disclosure in Bangladesh. Corp Soc Responsib Environ Manag 16:167–183. https://doi.org/10.1002/csr.193

Solaiman M, Haque AT, Banik S (2011) Responding to climate change: a study on eco-labeling practices in consumer goods of Bangladesh. Eur J Bus Manag 3:45–52

Tula Connell (2014) Two years after fatal Tazreen fire, life worse for survivors. Solidarity Center. https://www.solidaritycenter.org/two-years-after-fatal-tazreen-fire-life-worse-for-survivors/. Accessed 27 Sep 2018

12 International Trade: Patterns and Prospects

Nazmus Sadat Khan

Abstract

Bangladesh has come a long way in terms of trade. The problems associated with import substitution policies that Bangladesh adopted in early years were soon very apparent. This prompted a change in trade policy and the government soon started to liberalize trade and promote exports. Over the years, several financial incentives were provided to change the business climate. This included the establishment of several export processing zones (EPZ) to attract Foreign Direct Investment (FDI), export subsidies, and privatization of several state-owned enterprises. The export composition has also changed from primary commodities to manufacturing goods. The share of industrial raw materials, capital machinery, and intermediate goods started to increase as a part of total imports and contributed significantly in the production of goods that were consumed domestically and were also exported in other countries. The change in policy helped Bangladesh to allocate resources more efficiently, increase GDP at a faster rate, and consequently reduce poverty. Creation of EPZ helped to attract FDI, increase foreign exchange earnings, create employment, and transfer new technologies and skills. Some significant progress has been made in terms of rationalization of tariffs and elimination of quantitative restrictions. However, the protection level is still very high compared to many neighboring countries and improvement in the business climate is needed in order to promote trade that can support robust growth. The aim of this article is to provide a brief description of current export and import patterns in Bangladesh and discuss few areas that can affect the future.

N. S. Khan (✉)
Economist, The World Bank, Dhaka, Bangladesh
e-mail: nkhan12@worldbank.org

Keywords

Exports · Imports · Automation · Export diversification · Regional cooperation

12.1 Trends in Exports and Imports

Exports and imports in Bangladesh have experienced significant growth in the last two decades. The average growth rate of exports and imports over the last 18 years from the fiscal year 2001 to 2018 were 11.2% and 12.4% respectively. This double-digit growth indicates the growing importance of international trade in Bangladesh's economy. However, the growth has not always been stable. As shown in Fig. 12.1a, growth rate became negative for both exports and imports in year 2002 and they both reached as high as 41% in year 2011. Share of exports and imports in GDP has also increased over the years.

In 2001, exports and imports were 6.5% and 9% of GDP respectively. These numbers increased to 15.2% and 20.9% in 2018.[1] However, as can be seen in Fig. 12.1a, there has been a decreasing trend in this ratio for both exports and imports since 2012. Though imports have picked up recently, share of exports in GDP is still in a declining path. In order to achieve export-led growth like many other Asian countries, it is important to detect the sectors with potential and take proper steps to promote exports before it is too late.[2]

[1] Source of all the data in this article related to export and import are Export Promotion Bureau and Bangladesh Bank respectively.

[2] See Shiraji and Manap (2007), Parida and Sahoo (2007), Jimenez and Razmi (2013), among others on export led growth in Asia.

R. Ahmed et al. (eds.), *Transforming Bangladesh*, World Regional Geography Book Series,
https://doi.org/10.1007/978-3-031-45093-8_12

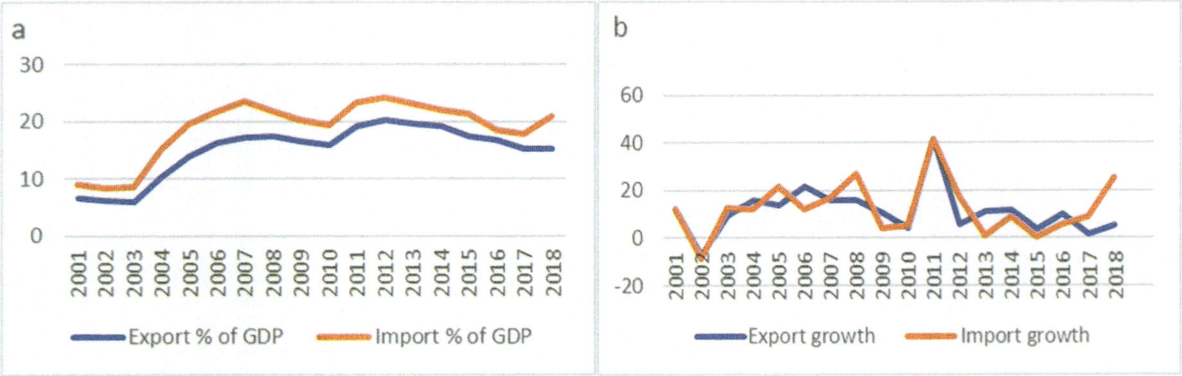

Fig. 12.1 (**a**) Export and import as % of GDP, and (**b**) export and import yearly growth (%)

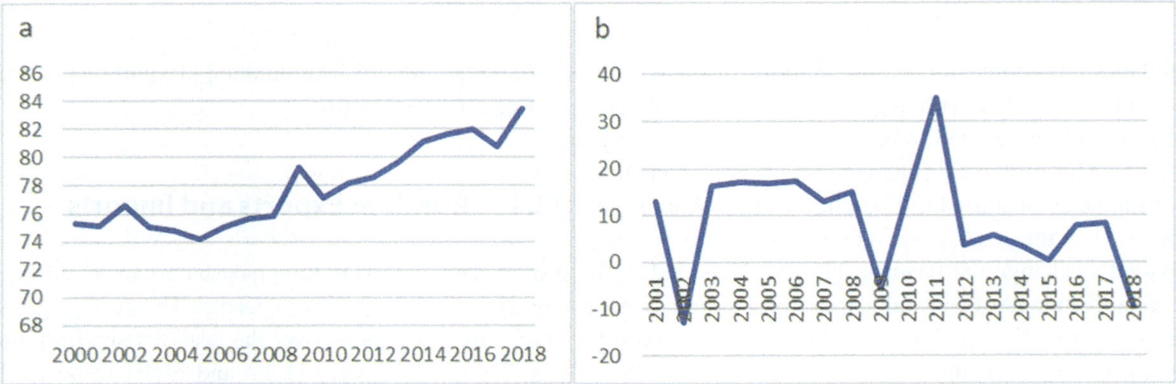

Fig. 12.2 (**a**) RMG as % of total exports, and (**b**) non-RMG growth (%)

12.2 What Do We Export?

Export from Bangladesh is largely driven by readymade garments (RMG) sector. In the fiscal year (FY 2017–18), RMG sector was responsible for 83.5% of total exports. It grew 8.8% from the previous fiscal year, exceeding the annual target of 7.1%.[3] Figure 12.2a shows the share of RMG sector in total exports since 2000 and it clearly depicts an upward trend. After a sudden drop in 2017, RMG exports picked up again in 2018. Since 1994, RMG sector has grown at a staggering 14.3% on average every year and has made Bangladesh the second largest exporter of RMG in the world after China. While these numbers are very imposing, in order to diversify risk, it is also essential to make sure the country diversifies its export basket (Sarker 2018).

Figure 12.2b shows the yearly growth of non-RMG sector products and it does not show any clear pattern. In fact, the average yearly growth rate of non-RMG products between 1994 and 2018 has been 7.6% only, which is almost half of the average yearly growth rate of the RMG sector for the same period. This growth also fluctuated considerably and went into negative territory in 2002, 2009, and more recently in 2018.

Figure 12.3 shows the percentage of non-RMG products in total exports between 2000 and 2018. Other than the subclass "other manufacturing goods" (which includes leather bags and purses, camera parts, melamine, tableware, and others), most of the other products either show a declining trend or contribute very less to overall exports. This is an alarming trend and the need to diversify export products cannot be overlooked. Promising sectors need to be identified and policies should be formulated to support these sectors. More on this is discussed in Sect. 12.3.

12.3 What Do We Import?

For Bangladesh, imports have been a crucial part of trade. After independence, there were very few good-quality industries and they were not capable of meeting the demands of the large population. As a result, almost all the necessary commodities were imported. It took some time for exports growth to come closer to the growth rate of imports.

Food items consisted of a large part of total imports for Bangladesh in its earlier years. Since then, thanks to the

[3]See World Bank (2018).

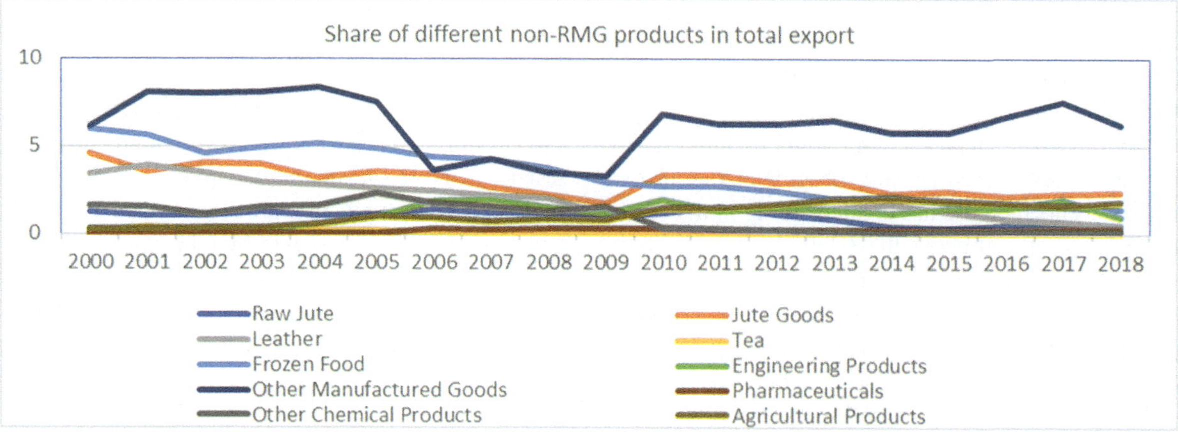

Fig. 12.3 Percentage of non-RMG products in total exports between 2000 and 2018

Fig. 12.4 (**a**) Import growth of food items, and (**b**) import growth of consumer, intermediate, and capital goods

"green revolution", food grain production has increased several times. From a chronical food deficit country, it almost reached self-sufficiency in rice production (Chowdhury 2009). Figure 12.4a shows the import growth of rice, wheat, and other food products since 2000. Rice and wheat imports faced negative growth in 2002, 2006, 2009, 2010, 2012, 2013, and 2016. This is a very promising sign. However, this does not mean there is room for complacency. Bangladesh is a country prone to natural disasters and cultivation of food items still depends largely on the weather. A bad harvest year can increase the necessity of food imports very quickly. This is also evident in Fig. 12.4a. A negative import growth is often followed by more than 100% increase of rice and wheat in the following years. For example, rice and wheat imports faced a 140% increase in 2018 due to flood-related production loss and the subsequent reduction of import duty on rice to meet the demand. Import duty has been increased since then as the situation normalized. Imports of other food items have been more stable, particularly in the last 5 years. These items include milk and cream, spices, oil seeds, edible oil, pulses, sugar, etc.

Imports are not necessarily bad for developing countries like Bangladesh. It can provide the consumers with more varieties and increase competition in the local market. Most importantly, they can promote growth in the long run if used in different productive sectors in an efficient manner (Herrerias and Orts 2011). The export sector in Bangladesh is dependent on imports of different intermediate goods (i.e., for the RMG sector this includes cotton, yarn, etc.). Also, building new infrastructure and industries require capital machinery and industrial raw materials. Some of the main consumer and intermediate goods that Bangladesh imports include crude petroleum, cement, chemicals, fertilizers, raw cotton, yarn, textile, staple fiber, plastics, etc. Figure 12.4b shows the yearly growth rate of consumer and intermediate goods (in grey) and capital and other goods (in purple). Except for 2002 and 2010, consumer and intermediate goods import growth has been positive since 2001. After maintaining a below 10% growth rate since 2012, it has started to pick up again and reached 18% in 2018. On the other hand, imports of capital goods have fluctuated since 2001. It was negative in 2004, 2008, and 2013 and the average growth rate over this period was 10%. However, capital goods imports have also increased more recently, and the growth rate crossed 30% in 2018. This increase has created a drag in the current account balance and affected the GDP growth rate negatively. The primary reason behind this sudden increase in capital goods is the country's involvement in several mega projects (i.e., Padma bridge, metro rail, Rooppur power plant, etc.). Though it is expected that these projects will yield benefits in the long run, the policymakers should make sure the abrupt rise in imports does not affect other parts of the economy too harshly in the short run.

Figure 12.5 shows the total imports of several products (in millions of USD) in the latest fiscal year (i.e., FY 2017–18). It gives a rough idea about the main import goods for Bangladesh. As can be seen in the figure, in terms of value, the contribution of food items is relatively smaller, while different capital goods/machinery and intermediate goods necessary for RMG production contribute the most in overall imports.

Going back to Fig. 12.1, it is clear that export and import growth in Bangladesh showed a very high correlation in the last 18 years. However, the two graphs have started to move in different directions in more recent years as import growth has surpassed export growth by a big margin. Since a good chunk of imports in Bangladesh is used for the production of goods that are being exported, policymakers should give careful attention to factors that are causing this divergence.

12.4 Looking to the Future

Future of trade is bright for Bangladesh. However, there are challenges and long-term vision is necessary to make the most out of trade. This section sheds light on few areas that are important for trade in Bangladesh in the future, particularly exports.

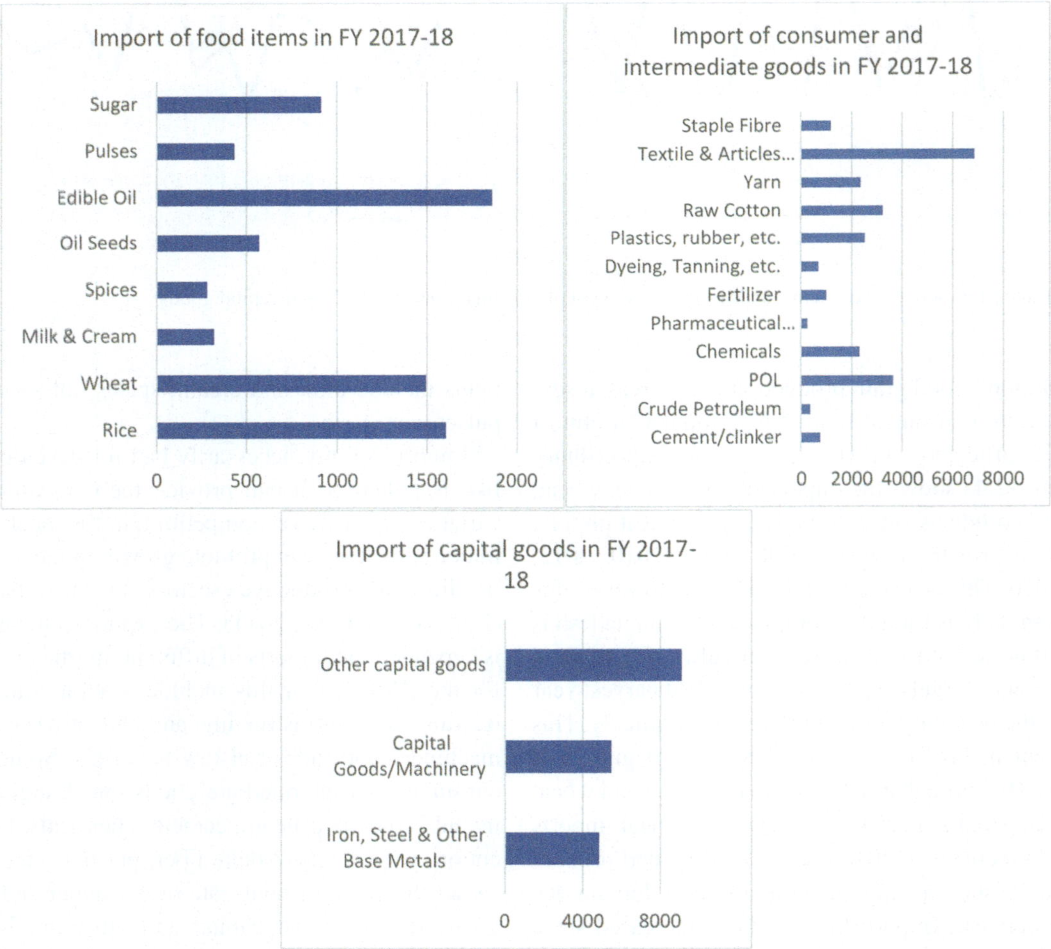

Fig. 12.5 Import of food items, consumer and intermediate goods, and capital goods in FY 2017–18

12.4.1 Export Diversification

In spite of tremendous growth in exports, diversification remains a challenge. As mentioned in Sect. 12.2, the RMG sector is responsible for more than 80% of total exports and its growth remains strong. In fact, in FY 2017–18, the growth of export from RMG was 8.5%, but the non-RMG sector growth declined by 9.5%. These are not good signs in terms of trade diversification, which Bangladesh badly needs. Such a high level of export concentration in one sector makes Bangladesh vulnerable to external shocks (Hossain 2020).

What are the sectors outside RMG that have the potential to become large? There are several sectors which showed promise in recent times. One of these is the pharmaceuticals industry. It has come a long way since its inception. The domestic demand of pharmaceuticals is estimated at USD 2 billion and currently the local companies meet almost all the demand. Bangladesh also exports to more than a hundred countries including the USA and many European countries. Output has grown a thousand times since 1982 and according to asset management firm LR Global, it has the potential to follow India's route to becoming a successful exporter internationally. Leather and leather goods (LLGs) have also shown promise and currently the second-largest export category for Bangladesh. The LLG market expanded several times in the last decade. Despite the challenges, the leather industry has huge potential particularly if one considers the availability of local raw materials and know-how of the supply management within the export industry (Khan 2014). Realizing its potential, Prime Minister Sheikh Hasina announced leather as the Product of the Year for 2017. The bicycle industry has also gained momentum and over the years and became one of the top exporters to Europe. Ship and shipbuilding industry, light engineering sector, cement, and ICT sector all have the potential to grow much larger.

However, many of these sectors flourished taking advantage of the preferential treatments that Bangladesh received due to its LDC status and not by competing in a level playing field. Particularly duty-free access to the European market under the Generalized System of Preferences status of GSP+ has been a major factor for growth for many industries including RMG. Drug Control Ordinance of 1982 and the relaxation of the WTO's agreement on Trade-Related Aspects of Intellectual Property Rights (TRIPS) have paved the way for Bangladeshi pharmaceuticals export to rest of the world. The bike export flourished due to anti-dumping duty on Chinese bicycles by the EU countries. The EU trade rules allow Bangladesh and Cambodia to export bicycles, parts and accessories to the EU's 28 member states without the regular 14% duty on complete bicycles and 4.7% duty on imported parts and accessories.

As Bangladesh prepares to graduate from the LDC category, the challenge will be to adjust in the new era of increased competition. Negotiations will be important. New countries should be explored where there are potential to increase exports and possibility of exporting new products should also be explored. Diversification should also be sought within the products. For example, for RMG products there are 105 HS codes in knitwear and 111 in woven products. For Bangladesh, top four knitwear products and top four woven products accounted for 68.9% and 70% of exports in these two categories respectively (World Bank 2018). This shows that there are opportunities for RMG exporters to diversify their products.

While Bangladesh might still have to rely on labor-intensive products for export earnings for now, long-term plans should be incorporated regarding how to move from labor-intensive products to capital-intensive products, keeping in mind the continuation of wage increase in the future. Improvement in innovation abilities, productivity, and skills will be crucial. Finally, to promote export diversification, the anti-export bias currently present in the tariff regime needs to be changed. Otherwise, it will be difficult for non-RMG sector to improve their export performance.

12.4.2 Automation

The economic success of the Asian tigers reveals a common pattern of development. After the Second World War, countries like Hong Kong, Korea, Taiwan, and Singapore started exporting labor-intensive products to rest of the world. As they became richer they specialized in much sophisticated capital-intensive industries like automotive and electronics. Later their places were taken by countries like China, Malaysia, Thailand, and Indonesia. As these countries are now trying to build capital-intensive industries and wage level is rising, the next tier of countries like Bangladesh, Cambodia, and Vietnam are now taking advantage of their low wage to export labor-intensive products to rest of the world. But will they be able to follow the same development pattern to reach the stage where Singapore and Korea are now?

Robots were never really thought of as an alternative to human in making labor-intensive products like clothes. Stitching takes several stages and long and delicate skills are required to complete a product. Also, entrepreneurs were not interested in designing automated systems for such low-margin industries. But things have started to change. More than 80% of garments industry workers in Southeast Asia might lose their jobs due to automation and ILO estimates this change might start occurring only 2 years from now (Chang et al. 2016). USA-based startup named "SoftWear Automation" has already started selling their invention "Sewbot" which can pick up flexible fabric and pinpoint where to stitch and cut. At present, it can produce products like pillows, towels, and mattress covers which require 10 steps or fewer, but they are working on upgrading the machine to produce t-shirts and jeans. Revenue of "SoftWear Automation" rose 1000% last year and they are likely to repeat this performance this year. It will not be surprising if the number of companies like SoftWear Automation increases in upcoming years.

This can be a worrying news for countries like Bangladesh where more than 80% export earnings come from the RMG sector. Not only RMG, but all sectors that are dependent on cheap labor doing repetitive jobs are also in danger of falling victim to automation. The presence of automation is yet to be felt in Bangladesh's export industry, but this should not mean the threats are not there. It is high time for Bangladeshi policymakers to think differently before the machines break the ladder they are wishing to climb in order to reach the higher stage of development.

12.4.3 Regional Cooperation

Trades are usually common among neighboring countries, mostly due to geographic reasons. That is why countries tend to find themselves in a win-win situation when they trade with countries with geographic borders rather than with countries situated in the other parts of the world where the costs of transportation and time required for shipments are higher. However, this has not been the case for South Asian countries. In spite of being the fastest growing region in the world, inter-regional trade among the South Asian countries is one of the lowest in the world. Most of the South Asian countries have their strongest trade ties not with one of their neighbors, but with countries in a different continent. It clearly suggests that countries are missing out on making the full use of several win-win situations through trade, which can further improve their growth prospects.

To improve inter-regional trade among the South Asian countries, South Asian Free Trade Agreement (SAFTA) was established in 2006. The goal of SAFTA was to reduce tariff among its participant countries gradually and foster trade and economic growth. However, SAFTA failed to achieve its goal. Though some of the tariffs had been reduced under SAFTA, countries kept on imposing para tariffs and non-tariff barriers to trade. Additionally, SAFTA did not include the so-called "sensitive" goods, and currently more than one-third of the goods produced in the region fall in this category. Intraregional trade in services has been reduced to visa restrictions and many other bureaucratic formalities. Areas such as education, tourism, and medical services have been affected negatively in this regard in spite of having great potential. All these made SAFTA one of the worst performing FTA in the world and it failed to learn from other successful FTAs in the EU, ASEAN, and North America.

Where does Bangladesh stand in this and how can they improve? Bangladesh's trade with South Asia is $7.6 billion, which is only about 9% of its global trade. So, there is a huge potential to improve in this area. Taneja (2018) conducted study on several products produced in Bangladesh, India, and Nepal and found out exports of at least three products (i.e., processed foods, readymade garments, and jute bags) from Bangladesh to India will make both countries better off if the countries decide to reduce the non-tariff measures and regulations. Nearly 40% of Bangladesh's import from the region fall under the sensitive list products. Bangladesh needs to negotiate with other South Asian countries, particularly India, in order to reduce the number of products that are currently considered as "sensitive". Proper implementation of BIMSTEC free trade area (FTA) agreement, another FTA Bangladesh is involved in, can lead to further increase in trade with neighboring countries. Bay of Bengal Initiative for Multi-Sectoral Technical and Economic Cooperation (BIMSTEC) was founded in 1997 and currently its members consist of Bangladesh, Bhutan, India, Myanmar, Nepal, Sri Lanka, and Thailand. If properly negotiated, Bangladesh can benefit from this FTA particularly because of central location. According to some experts, the BIMSTEC FTA has the potential to generate intraregional trade worth more than USD 50 billion. It can also create new opportunities for Bangladesh to strengthen trade ties with Myanmar. The relationship between the two countries has been going south for several years now. However, Bangladesh should explore more areas where they have a comparative advantage, particularly non-RMG sectors like pharmaceuticals where Myanmar is the number one importer of Bangladeshi products.

12.5 Conclusion

Bangladesh is a young and growing nation with huge potential. Trade can play a much more important role in this growth process. In fact, in FY 2017–18, contribution of net exports to growth was −3.6%.[4] This was largely due to increased growth in imports and relatively moderate growth in exports. While imports are not bad if they are used in productive sectors, continuation of this trend can create problems in the future. On the one hand, Bangladesh should make sure that imports are used wisely and they are contributing to society in the long run, on the other hand, steps should be taken to promote exports.

Can Bangladesh follow the leads of the Asian tigers to become a developed country through export-led growth? At the moment, it's difficult to predict. It is very important to

[4]Contribution of consumption and investment was 7.8% and 3.5% respectively.

have a vision and the ability to foresee the future. There are three things that the policymakers should have in mind. Firstly, the policymakers should understand the impact of potential geo-political changes and their effects. Whether it's a trade war between the top economies or Brexit, there will always be winners and losers. Based on the strengths of Bangladesh economy, it is important to take actions and react to these changes in an efficient manner. Going further, negotiation and diplomacy will be very important. This can be between governments to extend or start an agreement which can make export easier or between the businesses between Bangladesh and other countries regarding expansion of the product base and geographic locations where trade can be conducted. Success in these negotiations will form the backbone of future trading patterns. Secondly, the diversification of exports is critical. It will never be possible to flourish based on RMG export for decades after decades. Wages will ultimately rise, new competitors will emerge and the patterns of demand from importing countries may change. Whatever may be the reason, the policymakers should identify the potential sectors where there are opportunities and take proper policy measures now. Otherwise, in this highly competitive world, it might become too late to reap the benefits of globalization. Finally, there is no alternative to getting adjusted to the new technologies. As Bangladesh climbs up the income level, it will be necessary to invest more in research and development and invent new areas of comparative advantage and thus boost exports. Becoming a developed country will need a lot of other factors outside of trade to work effectively, but as history suggests, trade will always play an important part in this process.

Disclaimer The views and interpretations are those of the authors and do not represent those of the World Bank and its Executive Directors, nor the countries that they represent.

References

Chang J, Rynhart G, Huynh P (2016) The future of jobs at risk of automation. ILO Regional Office for Asia and the Pacific working paper no. 9

Chowdhury MAT (2009) Sustainability of accelerated rice production in Bangladesh: technological issues and the environment. Bangladesh J Agric Res 34(3):523–529

Herrerias MJ, Orts V (2011) Imports and growths in China. Econ Model 28(6):2811–2819

Hossain M (2020) Determinants of capital flows to least developed countries: lessons for Bangladesh. In: Hossain M (ed) Bangladesh's macroeconomic policy. Palgrave Macmillan, Singapore

Jimenez GH, Razmi A (2013) Can Asia sustain an export-led growth strategy in the aftermath of the global crisis? Exploring a neglected aspect. J Asian Econ 29:45–61

Khan W (2014) Leather industry in Bangladesh: opportunities and challenges. Am J Trade Policy 1(3):119–126

Parida PC, Sahoo P (2007) Export-led growth in South Asia: a panel cointegration analysis. Int Econ J 21(2):155–175

Sarker R (2018) Trade expansion, international competitiveness and the pursuit of export diversification in Bangladesh. Bangladesh Dev Stud 41(2):1–25

Shiraji NS, Manap TA (2007) Export led growth hypothesis: further econometric evidence from South Asia. Dev Econ 43(2):472–488

Taneja N (2018) A granular approach to addressing non-tariff barriers. A glass half full: the promise of regional trade in South Asia. Sanjay Kathuria (ed). World Bank Group

World Bank (2018, Oct) Bangladesh Development Update (BDU)

Part III

Environmental Transformations

13 Bangladesh Under Possible Climate Change Threat

Craig A. Meisner and Md. Yusuf Ali

Abstract

Food security in Bangladesh depends on agricultural growth and development, yet this sector faces a number of obstacles that prevent development and cause growth rates to stagnate. Some of these issues include: the gradual loss of arable land; declining soil fertility and salinity; inadequate investment in agricultural research and training; inadequate credit support for farmers; an unfavorable land tenure system; an exploitative market price for farmers that resulted in post-harvest crop loss; a large output gap; recent emigration and labor shortages in rural areas; and lack of capital. The primary challenge in Bangladesh is to assist farmers in switching from low-yield rice cultivation to higher-yielding, diversification, improved mechanization, water conservation, and better irrigation efficiency on the farm. Mass exports of potential agricultural products abroad are prevented by suboptimal adoption of good agricultural practices and flawed transportation/conservation systems. Bangladesh is the nation most impacted by climate change, according to the Global Climate Risk Index. Sea-level rise and saltwater intrusion, an increase in mean temperature (1.7 °C by 2050), greater precipitation variability, and an increase in the frequency and severity of extreme weather events are all likely to have a negative influence on the nation both now and in the next decades. Each of these will significantly affect the nation's agricultural output, and this chapter will go into great length on each of them. Many farmer-developed climate smart agricultural techniques are mentioned and described.

C. A. Meisner (✉)
Retired from IFPRI, CIMMYT FAO, Chapel Hill, NC, USA

Md. Y. Ali
Bangladesh Agricultural Research Institute, Rajshahi, Bangladesh

Keywords

Bangladesh · Climate smart agriculture · Climate change

13.1 Bangladesh Under Possible Climate Change Threat

Bangladesh's economy is heavily reliant on agriculture, with 14.7% of its GDP coming from this sector. The major industry in terms of employment in the nation is agriculture. According to the World Bank (2016b), 87% of rural residents make at least some of their money from agricultural pursuits. Since the 1980s, Bangladesh's population has doubled, and as of 2021, there are roughly 170 million people living there. This population growth has put significant pressure on arable land, which has decreased from 0.11 ha/capita in 1980 to 0.05 ha/capita in 2014 due to a high and growing population density (1100 people/km^2), increasing urbanization, and infrastructure development for industrialization (World Bank 2016a). With an average area of less than 1 hectare and 0.04 hectares of land per person, 99% of Bangladeshi farms are small and dispersed (BBS 2017).

13.1.1 Land Use

Approximately 9.1 million hectares, or 70% of the nation's land, is used for agriculture in Bangladesh. 6.5% of that area is used for permanent crops, while about 60% is thought to be arable. Approximately 11% of Bangladesh's land is forested, compared to 5% for permanent pastures, which are likely to decline as crop output increases (FAO 2015a, b). During the monsoon (July–October), water covers the remaining other lands. In the meantime, intensification has greatly grown, with double cropping occurring in the majority of regions, resulting in an average cropping intensity of 194% across the entire nation (BBS 2017). High-yielding

R. Ahmed et al. (eds.), *Transforming Bangladesh*, World Regional Geography Book Series,
https://doi.org/10.1007/978-3-031-45093-8_13

cultivars of rice, wheat, potatoes, sunflower, vegetables, mustard, maize, and mungbeans can be produced on some higher elevational terrain that is less subject to monsoon flooding (BBS 2017). Bangladeshi farmers currently cultivate more than 200 major and minor crops throughout the year for both family consumption and financial gain.

13.1.2 Agricultural Productive Systems

Bangladesh is a remarkably diversified nation in terms of geography, soil, and climate despite its modest size (147,570 km^2). While many locations have a shortage during the dry winter months, water resources are abundant during the wet season. The nation's nutrient-rich alluvial soils are extremely fertile, enabling the year-round cultivation of a wide range of food and crops. Bangladesh's agriculture historically focused on subsistence farming. The production of higher-value crops, livestock, and aquaculture has, nevertheless, transitioned toward more commercial production over the past 10 years. Most farms raise animals like cattle, goats, ducks, and chickens in addition to crops and vegetables (including those grown in backyard gardens), trees for fuel, fruit, and timber, and crops and vegetables. Numerous farmers in Bangladesh are also engaged in commercial shrimp farming in salt-water areas of the southwest districts and aquaculture production in ponds (Razzaque and Hossahin 2007).

The country is divided into 30 agroecological zones (AEZs) depending on the topography, floodplain, and soil type. 535 agroecological units and 88 agroecological subregions make up the remaining AEZ. The majority crop in the nation is rice (80% of all farmed land). Bangladesh consumes 200 kg of rice annually (about 417–454 g per person per day), making it the second-largest per capita rice eater in the world. In the nation, there are three primary types of rice systems: Aman (dependent on the tropical monsoon rains, which typically fall between June and October); Boro (during the winter dry season, which is completely irrigated since it is dry and does not receive any rain); and Aus (in spring, when it starts to rain as pre-monsoon rain showers in mid-April). 70% of Bangladesh's land is used to grow Aman monsoon summer rice (FAO 2015a, b). While using predominantly high-yielding varieties (HYVs) that are planted between December and May and take up around 60% of the acreage, boro rice is 100% watered during the dry season. Because it is primarily grown in the dry season just before spring storms, floods, or hurricanes, boro is sometimes thought of as being risk-free. As a result, it produces the most rice, 8 tons/ha, more than twice as much as monsoon rice. Only 20% of the space used for Aman rice is taken up by the third type of rice, aus, a direct-sown and rain-fed variety. Due to the expansion of conventional types, heavy weed infestation, and increased risk of storms and hurricanes this season, out-of-rice yields in Bangladesh are often the lowest. According to Shelley et al. (2016), the production of Aus, Aman, and Boro rice made up 7%, 38%, and 55% of the nation's total rice production in 2014, respectively. Instead of an increase in acreage, improved agricultural productivity in recent years has been mostly linked to increased rice yields from HYVs and hybrids.

13.1.3 Agriculture and Climate Change

13.1.3.1 Potential Economic Impacts of Climate Change

Food security in Bangladesh depends on agricultural growth and development, but this sector faces a number of obstacles that prevent expansion and cause it to stagnate. Some of these issues include the gradual loss of arable land, declining soil fertility, and salinity; inadequate investment in agricultural research and training; inadequate credit support for farmers; an unfavorable land tenure system; recent out-migration; labor shortages in rural areas; the need to deal with the growing effects of climate change; and the requirement to deal with these issues.

Supporting farmers to migrate away from low-yield rice cultivation and toward higher yields, diversification, improved mechanization, water conservation, and improved irrigation efficiency on farms is Bangladesh's main problem. In South Asia, Bangladesh currently has the highest level of tillage mechanization, with two-wheel tractors imported from China handling more than 95% of the tillage. Promoting even more appropriate mechanization in Bangladesh must be viewed as a potential strategy for rehabilitation, mitigation, and adaptation to address the dwindling time between cropping systems and climate-vulnerable ones as well as a way to replace the country's limited rural labor force with machinery for reasons of profitability and efficiency, assist farmers with transportation and livelihoods, as well as the speedy planting and harvesting of crops (Mandal et al. 2017).

In Bangladesh, crop yield has been required to rise in order to maintain food production due to declining land availability. In an effort to accomplish this, the government subsidized the use of urea fertilizers in the nation, which led to a dramatic increase in fertilizer use from 0.36 kg/ha in 1995 to over 298 kg/ha in 2007 and a nearly tripling of crop yields per hectare over the same period (Rana and Titumir 2014; FAO 2012). The habit of growing rice primarily in monoculture, along with an unbalanced use of chemical fertilizers and poor irrigation management, led to a reduction in soil fertility and soil organic matter even though this helped the country achieve rice self-sufficiency (Rana and Titumir 2014; Shelley et al. 2016).

The growth rates of grain crops have slowed recently, and rice yields are still below those seen in other South Asian nations. Future imports of wheat and corn into Bangladesh are anticipated to stay steady in order to satisfy domestic demand. In 2018, Bangladesh was the world's fourth-largest importer of wheat. Historically, many farming households were at danger of malnutrition due to a lack of diversification due to the predominate concentration on rice production (which accounts for 75% of the country's agricultural output) (FAO 2012). The government decided on a program to diversify crops in 2008 to combat this. Farmers are currently seeking to diversify their production more and more, concentrating in particular on high-value crops like flowers, fruit, vegetables, corn, oilseeds, or new potatoes. Creating new markets and including additional cattle in agricultural systems.

For instance, farmers in the Northwest forego growing monsoon rice in favor of planting early potatoes, which sell for much more than they would during the normal potato season. Providing services for animal health, as well as pertinent livestock research. Almost 50% of all animal deaths in the nation are caused by illnesses. The performance and size of subsectors are further constrained by the scarcity of permanent grazing pasture, ineffective livestock marketing techniques, and lack of access to improved livestock breeds (Ali and Hossain 2016). Cattle fattening became less lucrative as a result of the smuggling of local cow species from India, which contributed to some of this. Since India banned the export and smuggling of cattle in 2015, improved breed cattle fattening has gained popularity and economic viability in rural areas. Along with several local farmers who keep a few cattle in their homestead region for the purpose of selling them during the annual festival season (Edid-ul-Azha), there are three big beef firms that advertise better meat quality in the nation and even export beef.

Bangladesh is the nation most impacted by climate change, according to the Global Climate Risk Index (Harmeling 2011). Sea-level rise and saltwater intrusion, an increase in mean temperature (1.7 °C by 2050), variations in precipitation, and an increase in the frequency and intensity of extreme weather events are all anticipated to have an effect on the nation both now and in the next decades (CIAT 2017). The agricultural output of the nation will be significantly impacted by each of these variables. Bangladesh, an alluvial delta plain, features five major river systems, which have seen many location changes throughout the ages due to siltation and currents. Because of this, the nation is extremely vulnerable to inland water salinization and sea-level rise. By 2030, salinization is anticipated to advance 8 km further north in the nation, substantially diminishing the amount of land suitable for agriculture. Rather than sea-level rise brought on by climate change, transboundary flow factors will provide a greater infusion. Brine shrimp production has expanded over the past 20 years due to its higher profitability compared to other crops like rice, even if the injection of brine into rivers and canals continues to be a challenge for crop production (BRAC 2014). Only 4% of Bangladesh's annual rainfall falls during the winter months of November to February, with the majority of rain falling during the summer months between June and October. For instance, the early monsoon rains of April 2017 destroyed tender boro paddies in northeastern Haor (vast expanses of low-pressure depression). Climate change is predicted to cause an overall 9–12% increase in rainfall in Bangladesh by the year 2050 (CIAT 2017).

In South Asia, especially Bangladesh, large losses in rice harvests can occur if typical monsoon patterns are disrupted since summer monsoon rice yields depend on consistent, predictable rainfall. For instance, rain-fed monsoon rice, which accounts for approximately 38% of Bangladesh's total rice production, is extremely susceptible to changes in water supply brought on by variations in the timing of the seasonal monsoon (Mahmood et al. 2004). On the one hand, the early monsoon can result in flood damage if rice seedlings are inundated in the water during the early stages of growth, especially if farmers are not employing flood-tolerant varieties. A postponed monsoon, on the other hand, can cause water stress. The CERES rice model's findings demonstrate that excessive water stress throughout the flowering and ripening stages can cause yield losses in rice of up to 70% (Iqbal and Siddique 2014). Boro or winter rice cultivation may benefit from an increase in atmospheric carbon dioxide concentration, but these benefits will likely be offset by reduced yields during the winter harvest season and by increased temperatures during the flowering season. Variations in rainfall are less likely to influence the boro rice crop because this winter's crop is entirely irrigated. Floods and cyclones are two examples of extreme weather that Bangladesh is predicted to experience more frequently and intensely (Collins et al. 2013; Ramírez-Villegas and Jarvis 2008; Ramírez-Villegas and Thornton 2015). The amount of rice lost as a result of the cyclones in 2007 and 2009 was estimated to be over two million tons, or enough to feed ten million people. Bangladesh's southern, southwest, and southeast coastal regions are becoming more and more susceptible to powerful tropical storms and the ensuing saltwater intrusion.

13.1.4 Development of Climate Smart Agriculture (CSA) Technologies and Practices

Agricultural sector expansion, economic growth, and climate change mitigation are all possible with CSA technologies and practices. If a practice advances food security and at least one other CSA objective (adaptation and/or mitigation), it is said to be CSA. The term "CSA" refers to thousands of methods and technologies used all over the world.

In response to increased flooding and hurricanes, farmers on the southern coastal plains have long implemented many of the CSA methods outlined for Bangladesh. For instance, traditional gher farming in an aquaculture pond in non-saline wetlands with high dikes has evolved into a more complex method for growing vegetables while still enabling the production of shrimp, fish, and shrimp. On the trellises that span the pond, climbing vines are also frequently grown. The ancient method of stacking water hyacinths with soil in floating food gardens has now also been adopted at ghers in tidal flooded areas as a climate protection measure. Kangkong (water spinach) has traditionally been grown close to ponds, and productivity can be increased with new kinds. Farmers in Bangladesh have historically used a variety of ridge and furrow techniques in low-lying, swampy areas. The Sorjan method, for instance, is a variation on pyramidal farming that consists of high beds (30–90 cm) for the growing of vegetables and crops, alternated with furrows or ditches that are home to plants that can withstand submersion or are used to raise fish. In Bangladesh, the practice of raising small native fish in these canals is becoming more widespread, providing even small pond owners with access to wholesome food and higher income. Additionally, concrete paddy field fish rings or paddy field fish rings that shield fish from harm when paddy fields dry up have been put into place (Hossain et al. 2015). Even though these customs have been around for a while, they are currently gaining popularity quickly.

The problem of saltwater intrusion into Bangladesh's tidal rivers has gotten much worse during the last 40 years. Changes in sea level are quite likely to make the issue worse. Therefore, it's crucial to use high-yielding, salt- and submersion-tolerant varieties. The lengthy time between seed release and use in Bangladesh limits access to enhanced seeds (it takes roughly 15 years for producers to easily obtain released breeder seed). As a result, there has been little nationwide acceptance of rice types that are salt- or submersion-tolerant. The availability of flood-tolerant rice varieties, which first appeared on the market 7 years ago, has dramatically increased in recent years in response to losses brought on by changes in monsoon seasonality, especially for farmers in regions of the Northwest vulnerable by flash floods. After the monsoon, Bangladesh's winter is dry and mostly devoid of precipitation. As a result, any salt present in the soil evaporates to the surface, turning it salinous. To address this saline issue, bamboo and plastic were used to vegetable tower potted vegetables. Farmers in Bangladesh's flooded floodplains are not the only ones benefiting from CSA practices. Cattle were once used as draft animals and were eventually sold once they could no longer pull the plow. Currently, raising calves, goats, sheep, and buffalo for 3–6 months while giving higher quality feed, such as urea (a nitrogen molecule) and molasses feeding, has become a significant industry for young people. Others use intercropping and low and minimal tillage conservation agriculture techniques for grain crops.

The limited access to credit, the unfavorable ratios of advisory staff to farmers for the dissemination of new technologies and practices, the limited adoption of novel financing mechanisms, and the lack of safety net protections are all obstacles to the adoption of these and other CSA practices by small farmers in Bangladesh. For instance, Oxfam and a private insurer collaborated to construct index-based crop insurance, but the GoB or private businesses still find it challenging to scale up such interventions due to the absence of a clear business model.

13.2 Production System: Transplanted Monsoon Rice (T.aman)

The majority of the arable land is planted with transplanted aman rice, which is grown from July to December. It is a relatively inexpensive rice because it is primarily farmed with rainwater. Rice from T.aman is crucial for southern Bangladesh, where salt during the dry months is an issue. However, the river/canal water and soil turn non-saline with the arrival of the monsoon rains. As a result, the climate is favorable for rice farming. The CSA practice of using short-duration, high-yielding T.aman rice is appropriate for small- and medium-sized farmers, according to Table 13.1, where its climate smartness score is 1.94. Since the approach makes it possible to obtain early-season rice when farmers are experiencing a crisis of cereal food security, the impact on production is obvious. Traditional local T.aman rice is harvested in December, which is around 1 month later than short-duration varieties. In order to increase yields, farmers can cultivate high-value winter crops after the rainy season (also known as the rabi season) with the use of this CSA approach. The CSA technology is useful in adaption cases to prevent biotic stress and late-season climatic shocks like cyclones or tidal surges. The lifespan of this T.aman rice field is roughly 1 month shorter, which helps to prevent late-season dryness and increases water efficiency. Through ensuring healthy canopy and root growth, this method also contributes to a modest reduction in GHG emissions.

During the late monsoon, when heavy rains, upward receding water, and extra tidal water (full moon and new moon period) coincide at the same time, southern Bangladesh frequently experiences water congestion. This causes a significant loss in rice production since the cultivable area, which is mostly used for T.aman rice, submerges for several days to weeks. The above-water submersion shock can be prevented by implementing the recently adopted CSA prac-

Table 13.1 CSA practices of different rice crops for southern Bangladesh

Production system/Crop name (% area)	CSA practice	Applicable farm category	Climate smartness score[a]	Impact on		
				Productivity	Adaptation	Mitigation
1. T.aman rice (70)	1. Short duration high yielding (HY) varieties for facilitating timely planting of rabi/ winter crops	Small and medium	1.94	Increased and early season rice yield. Ensured winter crops planting	Resilience to biotic stress and climate shocks. Enhance water use efficiency	Moderate reduction in GHG emission
	2. Submergence-tolerant HY varieties	Small and medium	2.03	Increased and economic yield	Can endure tidal surge/ flood	Carbon sink increased
2. Boro rice (61)	1. Saline-tolerant HY varieties	Small and medium	2.60	Significant yield increase	Can endure soil and water salinity	Carbon sink increased
	2. Application of USG (urea super granular) as the N fertilizer for saving N and reducing N2O emission	Small and medium	1.71	Slight increased yield	Less N-fertilizer needs, sustained production	Reduced GHG emission
	3. Use of Axial Flow pump for irrigation to reduce irrigation cost by 30–40% in tidally flooded non-saline areas	Small and medium	1.80	Economic and increased yield	Better quality water, sustained yield	Reduced GHG emission
3. Aus rice (14)	1. Direct dry seeding for saving time, energy, and water	Small and medium	2.23	Slight increased yield	Reduces soil degradation, erosion	Reduced GHG emission
	2. Drought-tolerant, short-duration HY varieties	Small and medium	1.35	Increased yield	Tolerant to drought	C-sink increased
	3. Lodging tolerance taller varieties	Small and medium	2.29	Increased yield	Can cope with sudden water rise	C-sink increased

Source: Ali and Hossain (2019)
HY high yielding, *GHG* green house gas
[a]The average climate smartness score is calculated based on the practice's individual scores on eight climate smartness dimensions that relate to the CSA pillars: yield (productivity); income, water, soil, risks (adaptation); energy, carbon and nitrogen (mitigation) (CIAT 2017)

tice of choosing submergence tolerant high-yielding varieties. This method yields a score of 2.03 for climate smartness (Table 13.1), with higher yield and profitable return. The cultivation of these submergence-tolerant rice cultivars of crop ensures grain yield, root and canopy growth, increasing C-sink.

13.3 Production System: Boro Winter Rice

Despite having a lesser area (61%) than T.aman rice, boro rice, also known as winter rice, is the major type of rice grown in Bangladesh today. The process of turning photosynthesis into grain takes longer and requires more insolation in boro rice. However, during the winter or the dry season, salinity in the water and soil frequently has an impact on southern Bangladesh. So it's crucial to introduce CSA-Praxis salt-tolerant cultivars to guarantee a high and cost-effective yield. The practice has a moderate climate intelligence score of 2.60 (Table 13.1) with a significant increase in yield. The C sink increases together with grain yield, crown development, and root expansion. By lowering the volatilization of urea fertilizer in the soil after application (by up to 30%), the application of USG (Urea-Super Granule) in Boro rice is a CSA practice that lowers the needs of plants for nitrogen fertilizer. It applies to small- and medium-sized farmers and has a moderate climate smartness score of 1.71. The main advantage is the large reduction in nitrous oxide emission, which is typically released by the use of prilled urea fertilizer and is increased only little by the use of USG (Table 13.1). Central Bangladesh might save 30–40% on irrigation expenses (Table 13.1). Because AFP lowers the cost of irrigation and irrigation from surface sources uses superior quality water, larger economic returns are guaranteed. A centrifugal diesel pump is typically used for irrigation, which burns 30–40% more diesel than an AFP and results in higher greenhouse gas emissions. Thus, the usage of AFP has greatly decreased greenhouse gas emissions. The procedure has a 1.80 climate smartness rating and is relevant to small- and medium-sized owners.

13.4 Production Systems: Aus Rice

From March to July, 14% of the acreage is used to grow out or spring rice (Table 13.1). Droughts, storms, and other severe weather frequently affect it. In the early summer, it helps to secure the availability of grains for nourishment.

The CSA-Praxis-Direkt-Drocksaat aids in stock building at the ideal time, securing the yield. Similar to no-till farming, less disturbance of the soil results in less erosion and degradation of the soil. Additionally, it lowers greenhouse gas emissions because tilling the soil uses less energy. His rating for climate intelligence is 2.23. Short-term, high-yielding off-rice cultivars that are tolerant of drought may be used by the CSA to help prevent drought and ensure grain output. The C sink also rises because the culture's expansion is assured (Table 13.1). His Climate Smartness Score is 1.35. The CSA practice of housing tolerant, bigger rice varieties is suitable for the tidally inundated southern Bangladesh. Sometimes an impending rice crop is harmed by excessive tides or tide surges. Taller strains that can withstand submersion can therefore avoid this abrupt risk. More carbon sinks into the plant's canopy and subsurface root system as growth and yield are ensured. His high climate smartness rating of 2.29 (Table 13.1) is impressive.

13.5 Production System: Jute

In Bangladesh, jute is a significant fiber that covers 9% of the land in most communities. Summer through monsoon season (March to August) is when it grows. Jute is a crop with deep roots and luscious leaves that is considered essential for crop rotation. These leaves falling to the ground improve soil fertility for the following crop. It takes a lot of work to remove the fiber from the stick, and toasting it in pond or canal water significantly reduces the quality of the water. When there is not enough water in the canals during drought years, the rotting of the jute fibers might become an issue. Only a small amount of water that has been stored is needed to roast the jute sticks when belt roasting. In order to get good-grade jute fibers, belt toasting of jute fibers is a viable CSA. Although this CSA only has a 0.71 climate intelligence score (Table 13.2), it helps to lower greenhouse gas emissions. Salinity during the dry season is a challenge for healthy plant growth in southern Bangladesh. A good substitute for farming in salt-stressed locations is the CSA option, which uses a salt-tolerant jute strain. In saline environments, it is advantageous to improve and profit from jute production in addition to better carbon sequestration. He only has a 0.60 Climate Smartness Score (Table 13.2).

13.6 Production System: Spice Crops

No dish or curry in Bangladeshi cuisine could be made without chili, whether it was made with fish, vegetables, meat, or pulses. Approximately 7% of the agricultural area is devoted to growing spices. Its growth and economic output are hampered by both salinity and drought. Chili is a saline- and drought-tolerant spice crop that is used in CSA practices to ensure considerable output increases and a decrease in climate risk. C-sink is raised and GHG emissions are decreased with its cultivation. It has a 1.61 climate smartness score (Table 13.2).

Table 13.2 CSA practices of jute, spices, and oil seed crops for southern Bangladesh

Production system/Crop name (% area)	CSA practice	Applicable farm category	Climate smartness score	Impact on		
				Productivity	Adaptation	Mitigation
4. Jute (9)	1. Ribbon retting for jute fiber management (drought management and avoids water pollution)	Small and medium	0.71	Better quality fiber and market price	Suitable for less water area and reduced water pollution	Reduction of GHG-emission
	2. Saline-tolerant jute varieties	Small and medium	0.60	Increased and economic yield	Yield stability under saline conditions	Carbon sink increased
5.Spices-chilli (7)	1. Saline and drought-tolerant Chili cultivation in rabi and kharif I season	Small	1.61	Significant yield increase	Reduces climatic risk, so sustainable yield	Carbon sink increased and reduced GHG emission
6. Oilseed-Mustard (5)	1. Shorter duration and HY mustard	Small	2.90	Increased yield and income	Use of soil moisture, and avoid late rainfall	More C-sequestration and reduction of GHG
	2. Saline-tolerant HY mustard	Small	2.21	Increased production stability and income	Cope with climatic risk	Better C-sequestration

Source: Ali and Hossain (2019)

13.7 Production System: Oilseed Crop

The preparation of foods in Bangladesh uses a lot of vegetable oil, especially mustard oil, which also has some therapeutic benefits. Five percent of the land is used to grow mustard oilseed. Bangladesh now imports the majority of its other vegetable oils from other nations. At the end of the growing season, when the crop is planted later than usual due to a rise in temperature, mustard is frequently attacked by pests. For the next crop of boro rice to be grown on schedule, a short-season mustard crop is also required. In order to avoid climatic threats including early monsoon rain and pests, the CSA practice of adopting short-duration high-yielding (HY) mustard production is crucial. Additionally, it aids in lowering GHG emissions and increasing C-sink. The practice has a 2.90 climate smartness score (Table 13.2).

The salt-tolerant HY mustard variety is a prospective CSA strategy to obtain a greater and more cost-effective output in the salt-heavy southern region of Bangladesh. It can handle enhanced carbon sequestration as well as soil salinity in the winter and summer. It also has a 2.21 climate smartness rating (Table 13.2). See the link or CIAT reference below for a more thorough discussion of crops other than those included in Tables 13.1 and 13.2.

13.8 Other Potential Effects of Climate Change Besides Agriculture

Climate change is having a mixed effect on agricultural production in Bangladesh, according to an analysis using the International Model for Policy Analysis of Agricultural Commodities and Trade (IMPACT) (Robinson et al. 2015). For some crops, productivity gains and yield increases may result from climate change, while harvests may decline for others. In general, it is anticipated that climate change would have some impact on the majority of production systems in Bangladesh. The effects vary depending on the production method, with wheat, canola, and legumes/beans being the most impacted. By 2050, the model predicts overall production losses for grains, vegetables, jute, and wheat and yield gains for milk and meat, despite significant country-to-country heterogeneity. After then, the yields of wheat, rapeseed, and beans/legumes all decreased by 8.8%, 6.4%, and 6.3%, respectively. When compared to the No Climate Change (NoCC) value in 2050, the yields of rice, vegetables (as a whole), and other crops (including jute) are down 5.3%, 5.7%, and 3.3%, respectively. Both the CC and NoCC scenarios anticipate a significant increase from 2020 levels of about 52%; however, the growth under the CC scenario is marginally bigger by 0.2 percentage points (pp). Pig, sheep, and lamb are just a few examples of the numerous animal species that are less adaptable and are predicted to suffer from climate change.

The model also predicts that Bangladesh may depend more on imports of wheat, beans, corn, other crops (such as jute), vegetables (as a whole), and legumes as net agricultural trade shifts. Other crops, including jute, are predicted to have a 0.09 percentage point decrease in import dependency under CC conditions compared to NoCC, while corn, beans/legumes, vegetables (as a group), and wheat are predicted to have increases in import dependency of 4.5 percentage points, 3.3 percentage points, 0.16 percentage points, and 0.64 percentage points, respectively. The modeled results also indicate that exports of beef meat will rise in both the CC and NoCC scenarios, albeit the difference between the two is minimal (+0.07 pp). The model results most importantly indicate that Bangladesh may switch from being a net importer to an exporter of rice under both CC and NoCC. Under CC, this change will probably be 34 pp more noticeable than it would be under the NoCC situation. Demand shifts are ultimately influenced by relative commodity prices in both domestic and international markets. This makes it crucial to look at changes in supply-side drivers.

13.9 Conclusions

Bangladesh's existing policies and international agreements show that adaptation and mitigation in the agricultural sector continue to be top concerns. Such plans must have efficient delivery methods and a clear path for obtaining the extra funding required to carry out the vision. The GOB's annual budgets must be the starting point for these funding mechanisms, with a focus on research and monitoring for climate changes in their agriculture. Given that each hot spot has different limits, the road map should involve tackling the issues in each of the six hot spot locations separately (Bangladesh Delta Plan 2100). Bangladeshis don't have a diverse diet because 70% of their calories come from rice. Bangladesh spends more than 12% of its annual national budget on importing high-value, nutrient-dense crops such as pulses, edible oil, spices, fruits, soybean seed, and maize (BBS 2021). Due to the fact that rice crops take up land, it is impossible to expand their areas given the current per-unit yield of rice. Thus, at least 20% of rice acreage might be released for the aforementioned high-value crops in order to raise farmers' incomes and achieve dietary diversity in order to decrease malnutrition and food insecurity. This could be accomplished by greatly boosting per unit rice output and by decreasing per capita rice consumption. A national information, communications, and technology (ICT) plan was created by the government in 1997. It has evolved into a key force behind reforms, employment, sector growth, and better governance thanks to its ICT development process. Through those mobile phones and apps, the ICT, together with its highly competitive mobile phone and G3/G4 internet coverage, has the ability to encourage a higher adoption of CSA practices by farmers. Ninety-two million people currently have access to the internet and 156 million mobile phones are covered. A possible location to start in this regard is by strengthening village-level information centers as a part of the Department of Agricultural Extensions' Agricultural Information Service in the Union Parishad complex.

Boosting climate information services and making them easily accessible to farmers will considerably boost their capacity to adapt to change in addition to the ICT programs already stated. For instance, salt intrusion into the canals in southern Bangladesh may make it impossible to use them as irrigation for personal or commercial gardening. Farmers would be able to plan for suitable response tactics by being aware of where and when this salinity will occur through community-based data collecting, which would involve measuring the canal water with straightforward salinity meters. Given that the usage of briquetted urea (which can reduce urea consumption by 30%) is widely known, advising services on the technical application of more knowledge-intensive technologies like briquetted urea may allow farmers to significantly reduce agricultural emissions. However, it is crucial to improve and broaden the use of urea supergranular applicators in order for the technique to proliferate.

The creation of a clear, long-term vision for the nation's agricultural development depends on improving partner coordination. A step towards attaining this goal is the transparent provision and sharing of information through multi-stakeholder platforms, collaborative CSA activities, and regular opportunities for knowledge and experience sharing with various stakeholders involved in research, policy, and implementation.

References

Ali Z, Hossain I (2016) Barriers to development of livestock sub-sector in Bangladesh. Bangladesh Institute of Development. Bangladesh Agriculture Research Institute. Annual research report, 2018. Available at: http://bids.org.bd/page/researches/?rid=36

Ali MY, Hossain ME (2019) Profiling climate smart agriculture for Southern Coastal Region of Bangladesh and its impact on productivity, adaptation and mitigation. EC Agric 5(9):530–544

Bangladesh Delta Plan 2100. Baseline studies, volume 4. Agriculture Food Secutity and Nutrition. General Economics Division. Ministry of Planning. Government of the Peoples Republic of Bangladesh, Dhaka

BBS (2017) Yearbook of agricultural statistics – 2016. Bangladesh Bureau of Statistics. Ministry of Planning. Government of the Peoples Republic of Bangladesh. www.bbs.gov.bd

BBS (2021) Statistical yearbook of Bangladesh 2020. Bangladesh Bureau of Statistics, Ministry of Planning. Government of the Peoples Republic of Bangladesh, Dhaka. www.bbs.gov.bd

BRAC (2014) Final report on Ganges coordination and change enabling project (G5). Bangladesh CPWF, Dhaka. Available at: https://cgspace.cgiar.org/handle/10568/56871

CIAT; World Bank (2017) Climate-smart agriculture in Bangladesh, CSA country profiles for Asia series. International Center for Tropical Agriculture (CIAT); World Bank, Washington, D.C., p 28

Collins M, Knutti R, Arblaster J, Dufresne JL, Fichefet T, Friedlingstein P, Gao X, Gutowski WJ, Johns T, Krinner G, Shongwe M, Tebaldi C, Weaver AJ, Wehner M (2013) Long term climate change: projections, commitments and irreversibility. In: Stocker TF, Qin D, Plattner GK, Tignor M, Allen SK, Boschung J, Nauels A, Xia Y, Bex V, Midgley PM (eds) Climate change. The physical science basis. Contribution of Working Group I to the Fifth Assessment Report of the Intergovernmental Panel on Climate Change. Cambridge University Press, Cambridge/New York, pp 1029–1036. https://doi.org/10.1017/CBO9781107415324.024

FAO (2012) Country fact sheet on food and agriculture policy trends. Food and Agriculture Organization of the United Nations (FAO), Rome

FAO (2015a) FAOSTAT. Food and Agriculture Organization of the United Nations (FAO), Rome. Available at: http://faostat3.fao.org/download/T/TP/E

FAO (2015b) Mapping exercise on waterlogging in Southwest of Bangladesh. Food and Agriculture Organization of the United Nations (FAO), Rome. Available at: http://fao.fscluster.org

Harmeling S (2011) Global climate risk index, Briefing paper. Germanwatch

Hossain E, Nurun Nabi SM, Kaminski A (2015) Fish ring microhabitats: resilience in rice field fisheries. Penang, Malaysia: WorldFish. Program brief: 28. Available at: https://cgspace.cgiar.org/bitstream/handle/10568/51376/Fish%20ring%20microhabitats.pdf?sequence=4

Iqbal K, Siddique A (2014) The impact of climate change on agricultural productivity: evidence from panel data of Bangladesh. Discussion paper 14–29. Available at: http://econpapers.repec.org/paper/uwawpaper/14-29.htm

Mahmood R, Legates DR, Meo M (2004) The role of soil water availability in potential rainfed rice productivity in Bangladesh: applications of the CERES Rice model. Appl Geogr 24:139–159

Mandal MA, Sattar B, Stephen D, Justice S (eds) (2017) Rural mechanization: a driver in agricultural development and rural development. Institute for Inclusive Finance and Development (InM), Agargaon, Dhaka

Ramírez-Villegas J, Jarvis A (2008) High-resolution statistically downscaled future climate surfaces. International Center for Tropical Agriculture (CIAT); CGIAR Research Program on Climate Change, Agriculture and Food Security (CCAFS), Cali, Colombia

Ramírez-Villegas J, Thornton PK (2015) Climate change impacts on African crop production. Working paper no. 119. CGIAR Research Program on Climate Change, Agriculture and Food Security (CCAFS), Copenhagen. Available at: http://hdl.handle.net/10568/66560

Rana EA, Titumir RAM (2014) Recent trends of growth in agriculture, industry and power – Bangladesh economic update. Unnayan Onneshan 5(3)., Dhaka, Bangladesh

Razzaque A, Hossahin MG (2007) Country report on the state of plant genetic resources for food and agriculture Bangladesh. Bangladesh Agricultural Research Council. Ministry of Agriculture. Government of Bangladesh

Robinson S, Mason-D'Croz D, Islam S, Sulser T, Gueneau A, Pitois G, Rosegrant MW (2015) The international model for policy analysis of agricultural commodities and trade (IMPACT): model description for version 3 (IFPRI discussion paper). International Food Policy Research Institute, Washington, D.C.. Retrieved from http://ebrary.ifpri.org/cdm/ref/collection/p15738coll2/id/129825

Shelley IJ, Takahashi-Nosaka M, Kano-Nakata M, Haque MS, Inukai Y (2016) Rice cultivation in Bangladesh: present scenarios, problems, and prospects. Journal of International Cooperation for Agricultural Development 14:20–29

World Bank (2016a) World development indicators, Bangladesh. World Bank, Washington, D.C.. Available at: http://data.worldbank.org/indicator/NY.GDP.PCAP.KD

World Bank (2016b) Dynamics of rural growth in Bangladesh. Sustaining poverty reduction. Washington, D.C., World Bank. Available at: http://documents.worldbank.org/curated/en/951091468198235153/pdf/103244-REPLACEMENT-PUBLIC-Dynamics-of-Rural-Growth-in-Bangladesh-Reformatted-conf-version-May-17.pdf

Role of Adaptation Governance of Bangladesh in Global Climate Change Discourse

14

Shahrin Mannan, Fazle Rabbi Sadeque Ahmed,
Saleemul Huq, and M. Feisal Rahman

Abstract

Despite being an LDC and one of the most vulnerable countries to climate change, Bangladesh has taken several exemplary initiatives to tackle climate change as well as enhancing its own resilience and thereby established itself as a global teacher of climate change adaptation. Using the lens of this dual identity, this paper aims to explore how domestic policies, strategies as well as efforts of Bangladesh have influenced and shaped global climate change discourse and negotiations. To cater to this objective, the role of different initiatives undertaken by the Bangladesh government has been demonstrated under six thematic pillars. The discussion has shed light on Bangladesh's pioneering role in community-based adaptation along with its anticipatory development planning. The contribution of Bangladeshi scientists, researchers, and experts in knowledge generation has further reinforced its position in the global arena. Setting up of national financial mechanism in the form of Bangladesh Climate Change Trust Fund (BCCTF), climate budgeting at ministerial level, and capacity-building initiatives at all levels have influenced different countries especially the LDCs. Despite having environmental sustainability and economic growth in contradiction as well as different political economic dimensions, Bangladesh has been able to influence/shape global climate change discourse in different aspects.

Keywords

Bangladesh · Capacity building · Climate change trust fund · LDC

S. Mannan (✉) · S. Huq · M. F. Rahman
International Centre for Climate Change and Development, Independent University, Bangladesh, Dhaka, Bangladesh
e-mail: saleemul.huq@icccad.org

F. R. S. Ahmed
Palli Karma-Sahayak Foundation (PKSF), PKSF Bhaban, Dhaka, Bangladesh

14.1 Introduction

Bangladesh is known for its extreme vulnerability to natural calamities and consequent loss of life and resources, but it has been able to make a reputation as one of the most proactive countries when it comes to disaster risk reduction and climate change adaptation (Bailat 2017). Despite having very limited resources, huge population, and exposure to climatic stressors, the case of Bangladesh having this dual role has become a tested model to follow on climate change adaptation. In terms of raising voice in the climate negotiations for the rights of the developing and Least Developed Countries (LDCs), Bangladesh has always been in the forefront. For outstanding leadership on the front line of climate change, Prime Minister Sheikh Hasina even received the United Nations Champion of the Earth award in 2015 (Bailat 2017).

Alice Baillat, a French research fellow has used the term "weak power" for Bangladesh which refers to the ability to share resources to exercise influence in climate negotiations. Developing countries often get less attention in global arenas due to lack of expertise, small delegate, and absence of clear mandate. Bangladesh using its weak power has played a significant role in forming different groups such as the Least Developed Countries (LDC) group, Cartagena Dialogue, and Climate Vulnerable Forum (CVF). With the use of such coalitions, collective moral power, and external support from NGOs, LDC group including Bangladesh has been very successful in shaping global climate discourse.

This paper aims to explore how domestic policies, strategies as well as efforts of Bangladesh have influenced and shaped global discourse and negotiations while advancing toward achieving climate change agenda. In order to demonstrate the role of Bangladesh, the discussion in this paper has been structured under six thematic pillars. The overarching goal is to highlight how Bangladesh being an extremely climate-vulnerable country has proved itself to be one of the champions and shaped the global climate change arena.

R. Ahmed et al. (eds.), *Transforming Bangladesh*, World Regional Geography Book Series,
https://doi.org/10.1007/978-3-031-45093-8_14

14.2 Role of Bangladesh in Influencing and Shaping the Global Climate Change Discourse

14.2.1 Global Leader in Experiential Knowledge on Adaptation

Being a vulnerable country, Bangladesh has been giving tremendous effort to move out of the problem space, and become a global pioneer in adaptation. With high level of expertise and experiential knowledge in handling hydro-climatic disasters, Bangladesh has successfully implemented a range of adaptation projects especially around the concept of community-based adaptation (CBA). Strong national policy framework, government support, and a vibrant NGO community have further enabled Bangladesh to test innovative adaptation solutions and thereby gain experiential knowledge on dealing with the impacts of climate change. Over 500 adaptation projects both infrastructure and soft measures have been implemented using national climate funds. Furthermore, thousands of adaptation projects/programs have been implemented/under implementation using national budgetary allocation under the window of Annual Development Plan (ADP).

Using a combination of traditional and new knowledge, Bangladesh has successfully developed several adaptation techniques which could be useful for similar contexts elsewhere. These technologies include but are not limited to floating gardens (Box 14.1), flood and saline-tolerant seeds, aqua-geoponics, and sand-bar cropping. In addition to these technologies, Bangladesh has made tremendous progress on disaster risk reduction. With world-renowned community-based early warning system more than 2100 cyclone centers have been built in order to protect people from cyclones and tidal surges. The newly built multi-purpose cyclone shelters can be used as government offices, schools, and health centers. During cyclone Sidr in 2007, about 1.5 million people were sheltered according to a government estimate and the death toll was reduced by more than 95% compared to previous devastating cyclones in 1970 and 1991. Being one of the leading countries on global research and knowledge on adaptation, Bangladesh is now in a position to share experiential knowledge with other LDCs as well as developed countries.

Box 14.1 Floating Gardening in Southern Floodplains of Bangladesh

Floating gardening, a traditional agricultural practice in southern wetlands in Bangladesh has been recognized globally as adaptation sensation for having the potentials of tackling climate variability and at the same time offering different economic, ecological, nutritional and agricultural benefits. It is a hydroponics system where a reasonable sized platform is constructed using aquatic plants where vegetable and other corps are cultivated and seedlings are raised in rainy season. The residue from the platform is further used to prepare beds on the soil for the following winter vegetable cultivation. Water hyacinth is the most used construction material for the floating platforms (Pyka et al. 2020).

The traditionally tested floating gardening technique has been found to be an effective climate change adaptation measure to tackle extended water logging. Since the floating platform rises as flood water rises it allows farmers to continue agriculture during flooding (Irfanullah et al. 2011).

The technique has recently been tested in Vietnam and was also recognized as an adaptation option by IPCC for creating new and diversified livelihoods.

14.2.2 Contribution of Research Community in Knowledge Creation and Negotiations

Apart from exercising weak power in leading global climate negotiations, Bangladeshi scientists, experts, academics, and research communities have played a significant role to shape global discourse on climate change before the formation of IPCC and UNFCCC. Different research institutes and NGOs from Bangladesh as focal points for South Asia took strong positions in Inter-governmental Negotiation Committees (INC) prior to convening the United Nations Conference on Environment and Development (UNCED) in Rio de Janeiro in 1992 (Earth Summit). Besides, many individual scholars and organizations conducted applied research on climate change vulnerabilities on the early days.

Bangladesh recognizing its position as vulnerability took initiatives to establish its leadership in scientific studies of climate adaptation. By far, Bangladesh has contributed the most to the IPCC reports of any developing country. With more recognition and increasing availability of funding, different public and private universities are showing interest on climate change by setting up separate centers for climate change research and collaborating global universities for knowledge sharing (The Asia Foundation 2012).

LDC group has played a remarkable role in making Paris Agreement as well as ensuring the interest of the poorest and the most vulnerable countries. In COP21 held in Paris on 2015, two of the most politically debated issues were: (i) whether to limit the long-term global temperature at 2 °C or change it to 1.5 °C; (ii) recognizing loss and damage in the agreement. Bangladesh, as part of the Least Developed Countries (LDC) Group and the Climate Vulnerable Forum (CVF), was very active in advocating for both issues and also during the negotiations made strong cases for both issues. Eventually, the agreement recognized 1.5 degree as an aspirational goal with the main goal being "well below 2 degrees" in the Paris Agreement on Climate Change. Also, the agreement recognized Loss and Damage as a separate pillar under the Paris agreement*. As a member of the LDC group, Bangladesh negotiators along with their other LDC counterparts took an active role during the negotiations and made strong cases for both issues. These were perceived as major achievements for the vulnerable developing countries including Bangladesh.

14.2.3 Climate Resilient Development Planning

Despite being in the early years of institutional restructuring with a wide range of environmental and economic challenges, Bangladesh has made remarkable progress in climate-resilient development planning. The extensive experience of adaptation has been reinforced by the anticipatory adaptation planning by Bangladesh in the last decade (Huq and Khan 2017). Since then, substantial planning related to climate change adaptation took place in Bangladesh. Figure 14.1 demonstrates the chronological development of the policy framework with the main thrusts of the policies.

In the seventh Conference of Parities (COP7) in Marrakech, 2001, the decision to prepare National Adaptation Programme of Action (NAPA) for urgent and emergency adaptation actions for every LDCs was adopted. Bangladesh was one of the first two LDCs which developed the National Adaptation Programme of Action (NAPA) in 2005 identifying 15 priority adaptation programs and activities (Huq and Khan 2017).

However, recognizing the insufficiency of NAPA in tackling the medium-term and long-term impacts of climate change (Tawsif et al. 2022) and to include the other important areas of climate change such as mitigation, knowledge management & capacity building and to follow the Bali Action Plan in 2007, as an extension of NAPA, Bangladesh developed a decade long climate change strategy, the Bangladesh Climate Change Strategy and Action Plan (BCCSAP). Under six thematic areas. BCCSAP aimed at facilitating medium and long-term adaptation measures under its 10-year long time frame (Rai et al. 2014). The plan expired in 2018 and Bangladesh's government is now updating the plan which is set to be released in 2020. With an integrated approach toward climate change and sustainable development, the revised BCCSAP aspires to build resilience with poverty reduction (Huq 2019a, b, c).

Apart from climate change-specific plans and policies, the government toward mainstreaming climate change has given specific emphasis on the issue in its short-term (ADP-annual development plan), medium-term (Five-year plans), and long-term development plans (2041 Vision Document and the Delta Plan 2100). The latest Bangladesh Delta Plan 2100 is considered as one of the major milestones of the Bangladesh government in climate-resilient development planning by considering long-term impacts of climate change.

While there is a considerable implementation gap, having the upper hand in adaptation planning, Bangladesh is clearly in a position to share its experiential knowledge on mainstreaming climate change.

14.2.4 Financing Mechanism to Support Climate Change Adaptation and Climate Budgeting

In order to have sustained solutions and build resilience, the Bangladesh government has set up national financial mechanisms to support climate change adaptation measures and institutionalized them. For example, to implement BCCSAP and promote "Climate Proof Development", two funds were established in 2010 – "Bangladesh Climate Change Trust Fund (BCCTF)" with own resources and "Bangladesh Climate Change Resilience Fund (BCCRF)" from donor funding operated by World Bank (Haque et al. 2018). Around 77% of all the climate finance projects of Bangladesh are funded through the domestic budget. While article 9 of the Paris Agreement talks about extending financial support to most climate change-affected countries by the developed countries, Bangladesh has set a unique example for the entire world by establishing such an exemplary national mechanism with own resources to fund climate change projects.

The Government also took an exemplary initiative to track its own investments on climate change by introducing the climate budget reporting. Climate budgeting ensured that the national budget takes into account and has provision for reducing climate change impacts. Bangladesh's climate-related investments in the recent years for 25 of the most relevant ministries out of the 55 ministries can be found in Fig. 14.2. It is around 1% of the budget allocation. It might be higher at the time of disaster with higher climatic impacts. Other countries that are following the suit could learn from Bangladesh's experience.

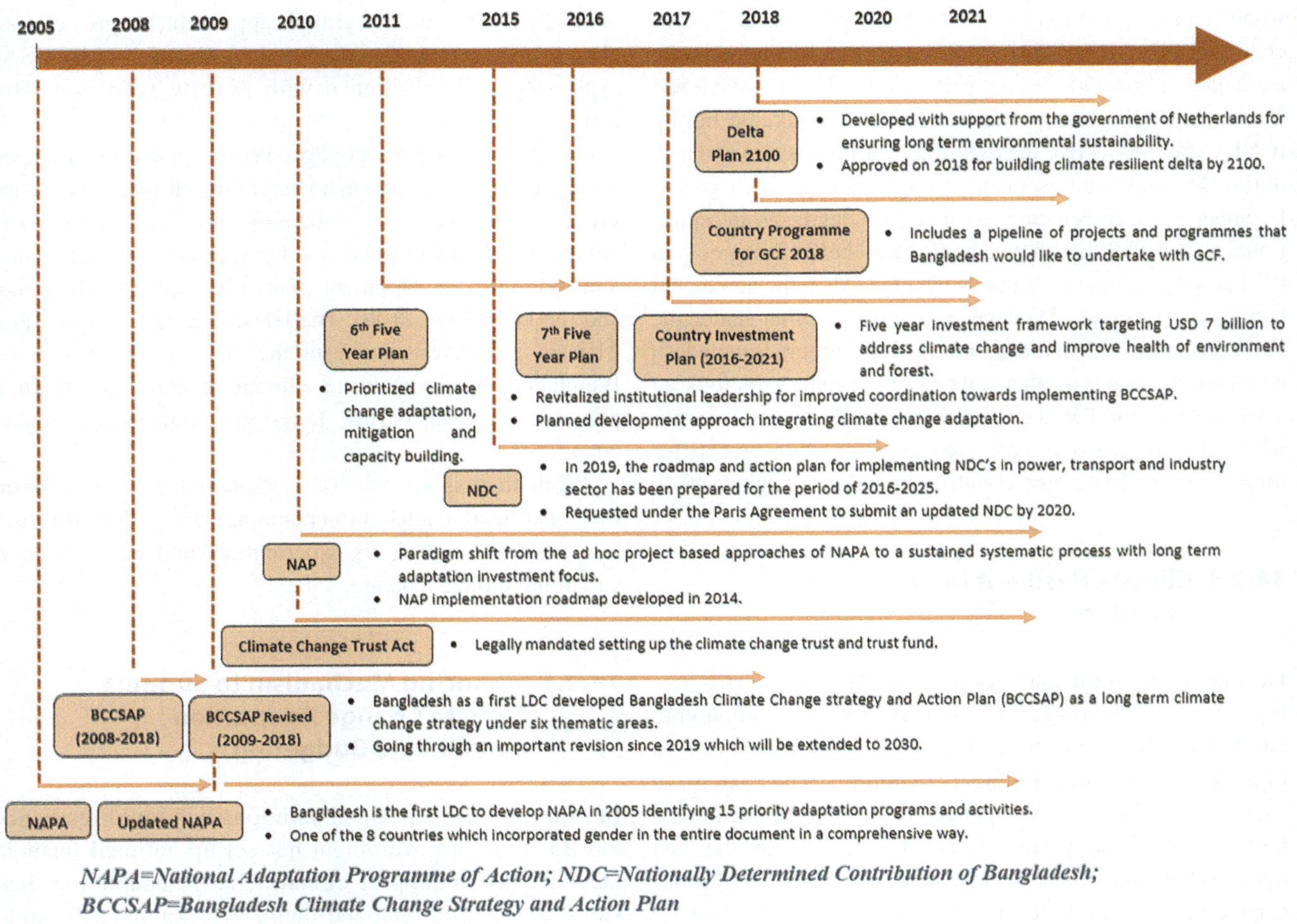

Fig. 14.1 Timeline for major climate change policy framework in Bangladesh, 2020–2021. NAPA National Adaptation Programme of Action; NDC Nationally Determined Contribution of Bangladesh; BCCSAP Bangladesh Climate Change Strategy and Action Plan

Fig. 14.2 Budget allocation in last 5 years for 20 ministries (in million USD). (Source: Finance Division, Ministry of Finance 2018)

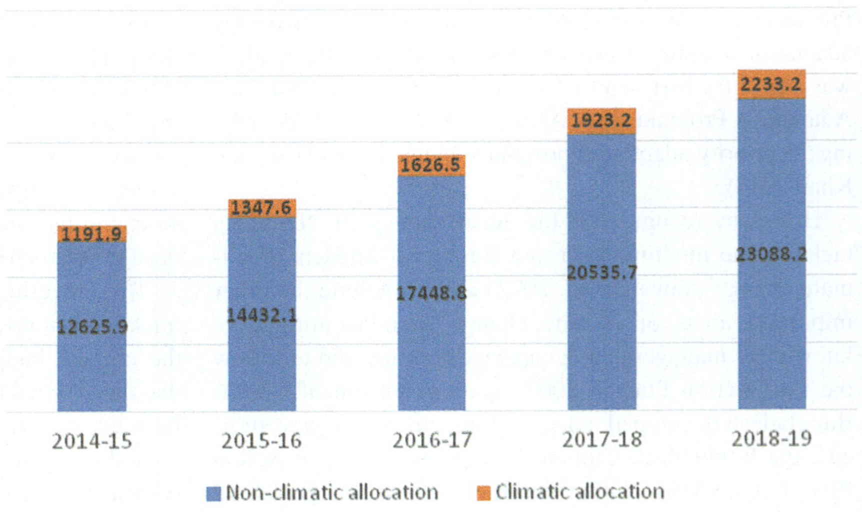

14.2.5 Building Long-term Capacity to Tackle Climate Change

Capacity building is a fundamental precondition to cut down emissions and adapt to climate change. But the prevailing paradigm of climate change capacity building has emerged from capacity building in development at large. The inefficiency and ineffectiveness of the ad-hoc project intervention-based capacity building has hardly been of any help for the most climate change-affected LDCs and Small Island countries.

To have a long-term sustainable capacity-building model, Bangladesh as a member of the LDC group engaged actively in the climate negotiations which resulted in important decisions in the Paris Agreement such as decision to establish Paris Committee on Capacity Building (PCCB) under Article 11, the Capacity Building Initiative for Transparency (CBIT) under Article 13, and promoting education, training and public awareness under Article 12 (Abeysinghe et al. 2016).

Nationally, to integrate climate change adaptation at all levels, regular training workshops, seminars, and policy dialogues are being implemented in Bangladesh to sensitize the government officials and policymakers on climate change. On the other hand, recognizing climate change as a long-term phenomenon, Bangladesh intends to invest more and more in making young boys and girls into climate change problem solvers. By building capacity at all levels, the country is pioneering in the application of the concept of whole-of-government as well as whole-of-society involvement not only in tackling climate change, but also for all 17 of the global Sustainable Development Goals (SDGs), where we aim to achieve those goals by 2030 (Huq 2020) (Box 14.2).

14.2.6 Approaches to Address Loss and Damage

Loss & Damage has become one of the most discussed issues in the global climate change discourse prompted by emerging evidences of Loss & Damage incurred in communities all over the world (Roberts and Huq 2015). This relatively new aspect stemmed from the collective global failure of both mitigating and adapting sufficiently to the adverse impacts of climate change resulting into losses and damages to people, infrastructure, and ecosystems. Being a highly politically sensitive issue, Loss & Damage has always been neglected by the main GHG emitters in developed countries and often portrayed as taboo words in international negotiations under the United Nations Framework Convention on Climate Change (UNFCCC). Despite all these efforts, developing countries led by Least Developed Countries (LDCs), including Bangladesh, Small Island Developing States (SIDS), and African countries, together have been able to achieve two important victories in recent years on getting this issue addressed under the UNFCCC. The first is the achievement of setting up the Warsaw International Mechanism (WIM) for Loss & Damage in 2013 during COP19 in Warsaw, Poland with an executive committee to develop a work plan. However, one of the shortcomings of this achievement was having WIM under the umbrella of adaptation rather than separate Loss and Damage (Huq 2018a, b, c).

Few years later in COP21, Paris, Loss & Damage was the flagship issue of LDCs including Bangladesh in Paris Agreement Negotiations. The three vulnerable groups demanded Loss & Damage to be included as a standalone article in the Paris Agreement and finally at the end of COP21, article 8 of Paris Agreement recognized Loss & Damage as a separate issue from adaptation that parties

Box 14.2 LDC University Consortium on Climate Change (LUCCC)

LUCCC is a network of 10 Universities in the LDCs that intends to exchange knowledge on climate change and build capacity through training and research. LUCCC's vision is to enhance capacity on climate change in all 47 LDCs through education, research and training to enable them to adapt effectively to the adverse impacts of climate change.The consortium was officially launched at the 11th International Conference on Community Based Adaptation (CBA11) to climate change in June 2017.

LUCCC was jointly initiated by the International Centre for Climate Change and Development (ICCCAD) based at Independent University, Bangladesh and the Makerere University, Uganda. Since its inception the platform has facilitated university-based exchange programme for climate change researchers in LDCs. Several consortium members are also conducting joint climate adaptation research projects.

LUCCC is now recognized as one of three long-term initiatives of the LDC Group on climate change- a network of 47 LDC countries who negotiate as a bloc at the intergovernmental negotiations under the UNFCCC sessions. By envisioning partnerships among the LDC Universities LUCCC has provided a necessary platform for these universities to share their respective expertise and create a pool of southern experiences and knowledge on adaptation.

should address (Abeysinghe et al. 2016). However, in the Paris decision 51 explicitly mentioned that "Article 8 of the Agreement does not involve or provide a basis for any liability or compensation".

Considering the importance and need to understand Loss & Damage for vulnerable countries, the government of Bangladesh initiated The Loss and Damage in Vulnerable Countries Initiative. For taking this debate into the next stage for the benefit of LDCs and other vulnerable countries, government of Bangladesh asked Climate and Development Knowledge Network (CDKN) to help building a common understanding around loss and damage and provide insight into what it entails for vulnerable countries. CDKN later formed a consortium of organizations including International Centre for Climate Change and Development (ICCCAD) from Bangladesh to execute the work. The strategic vision of this initiative was to proper a different angle that the decision-makers can use to perceive problems and identify solutions to climate change (CDKN n.d.).

Given the increasingly adverse impacts of human-induced climate change, government of Bangladesh has proactively developed Bangladesh Climate Change Strategy and Action Plan (BCCSAP) and then set up the Bangladesh Climate Change Trust Fund (BCCTF) in 2009 with our own money to implement it. BCCTF has contributions of USD 420 million from national exchequer to fund projects over the years. Furthermore, government is allocating around 1–2% of its GDP (currently it is more than 3 billion USD annually) for climate-linked investment. However, only two-thirds of the fund goes for implementing projects while one-third of it is kept in an interest-bearing account to be used during emergencies. Based on the decision of the inter-ministerial committee set up to decide the use of the reserve fund, Bangladesh government will pioneer in setting up a "National Loss & Damage Mechanism" using the reserve fund and develop a two-year long pilot program chalking out how to compensate the victims of climate change. If Bangladesh succeeds in using own money and develop a pilot mechanism in addressing Loss & Damage in a pragmatic way, it will set a unique example of tackling loss & damage and at the same time will be able to attract international funding from the global community (Huq 2018a, b, c).

14.3 Conclusion

Bangladesh having huge groundswell of experiential learning in adaptation coupled with anticipatory development planning has achieved the position of champion in the global climate change community. Despite being highly vulnerable to climate change, huge population, and lack of resources, the advanced position of Bangladesh as global leader in adaptation is mainly due to its strong political will, enabling environment, and capacity building at all levels. Considering the depth of the problem, the Bangladesh parliament has declared the current state of climate vulnerability as planetary emergency and thereby set a unique example for other governments on how to respond to such global issues. Though high investments in fossil fuel-based infrastructure and technologies pose challenge in the pathway; development of ambitious long-term strategy toward significant renewable energy power generation, energy efficiency, and afforestation will help align the future with the goals of the Paris Agreement and reap benefits for sustainable development.

References

Abeysinghe A, Craft B, Tenzing J (2016) The Paris Agreement and the LDCs: analysing COP21 outcomes from LDC positions. IIED, London

Bailat, A. (2017, Aug 11). Bangladesh as a 'weak power' climate leader? Dhaka Tribune

Climate and Development Knowledge Network. Retrieved from https://cdkn.org/project/work-programme-for-ldcs-on-loss-and-damage/?loclang=en_gb

Haque M, Pervin M, Sultana S, Huq S (2018) Towards establishing a national mechanism to address losses and damages: a case study from Bangladesh. In: Loss and damage from climate change, pp 451–473. https://doi.org/10.1007/978-3-319-72026-5_19

Huq S (2018a) Limiting global temperature rise to 1.5 degrees. The Daily Star, 18 July 2018

Huq S (2018b) Tackling loss and damage from climate change: an opportunity for Bangladesh to take the lead. Dhaka Tribune, 17 Mar 2018

Huq S (2018c) Making climate budgeting transparent. The Daily Star, 28 Nov 2018

Huq S (2019a) Bangladesh can lead the way. Dhaka Tribune, 17 July 2019

Huq S (2019b) How Bangladesh can improve access to climate funding. The Daily Star, 7 July 2019

Huq S (2019c) Changing the climate narrative. The Daily Star, 2 Jan 2019

Huq S (2020) What lies ahead for climate change. The Daily Star [Dhaka], 2 Jan 2020

Huq S, Khan M (2017) Planning for adaptation in Bangladesh: past, present and future. Retrieved from www.icccad.net/wp-content/uploads/2017/11/Policy-Brief_Planning

Irfanullah H, Azad M, Kamruzzaman M, Wahed M (2011) Floating gardening in Bangladesh: a means to rebuild lives after devastating flood. Indian J Tradit Knowl 10(1):31–38. Retrieved from https://www.researchgate.net/publication/236148093_Floating_Gardening_in_Bangladesh_A_means_to_rebuild_lives_after_devastating_flood

Pyka L, Al-Maruf A, Jenkins C, Shamsuzzoha M, Braun B (2020) Floating gardening in coastal Bangladesh: evidence of sustainable farming for food security under climate change. J Agric Food Environ 1(4):161–168. ISSN (Online Version): 2708–5694

Rai N, Huq S, Huq MJ (2014) Climate resilient planning in Bangladesh: a review of progress and early experiences of moving from planning to implementation. Dev Pract 24(4):527–543. https://doi.org/10.1080/09614524.2014.908822

Roberts E, Huq S (2015) Coming full circle: the history of loss and damage under the UNFCCC. Int J Glob Warm 8(2):141–157

Tawsif S, Alam MS, Al-Maruf A (2022) How households adapt to heat wave for livable habitat? A case of medium-sized city in Bangladesh. Curr Res Environ Sustain 4:100159

The Asia Foundation (2012) A situation analysis of climate change adaptation initiatives in Bangladesh. Retrieved from https://asiafoundation.org/resources/pdfs/SituationAnalysisofCCinitiatives.pdf

Part IV

Transformation Relating to People and Society

Livelihood of Coastal Community in Bangladesh: Options, Changes and Alternatives

15

Rabiul Islam

Abstract

This chapter explores the existing livelihood options, identifies the changes in livelihood and finds out alternative livelihood options in the coastal villages of Bangladesh. Using qualitative method, data was collected through Focus Group Discussions (FGDs) and Key Informant Interviews (KIIs) from the two coastal villages were severely affected by cyclone Sidr in 2007. The findings show that the coastal people mainly depend on fishing, agriculture and plantation of fruit and timber trees. These livelihood options have been facing many challenges due to over and illegal fishing, use of banned nets in fishing, saline intrusion in farming land, lack of saline-resistant crops and trees. Over time, the livelihoods of coastal people have been changed. In the changing context of environment, the coastal people have alternative livelihood choices. Saline-water-resistant rice and crop and plant varieties can be an important alternative to promote multiple uses of farming land. However, due to lack of saline-resistant crop varieties the people of other coastal areas of Bangladesh adopted shrimp cultivation. The study argues that coastal households cannot overcome the livelihood challenges alone due to the frequent attack of climate extreme events. This chapter suggests that various stakeholders, e.g., government agencies, NGOs, local government, law-enforcing agents and international organizations can play vital role to reduce livelihood vulnerabilities and introduce alternative livelihood options in the coastal communities of Bangladesh.

Keywords

Coastal villages · Livelihood options · Saline-resistant crop · Bangladesh

15.1 Background

Bangladesh is one of the country's most vulnerable to natural hazards in the world. Because of climate change the frequency of natural hazards, e.g., cyclones, storm-surges, floods, riverbank erosion, and salinity, has been increasing in the Bangladesh coast. These hazards create livelihood vulnerabilities of coastal people. Cyclones and other climate extreme events create disastrous situation for the large number of poor people – mainly dependent on climate-related livelihoods like agriculture and fishing (Masud-All-Kamal 2013). Coastal livelihoods are vulnerable for several reasons. Due to the increase of water salinity, traditional agriculture cropping is declining in coastal areas. Lack of saline-resistant crops and trees also threatened the farming-based livelihood. Fishing opportunities are decreasing due to the impact of cyclone and illegal human intervention, e.g., overfishing, use of banned nets in fishing and sea piracy. Because of these circumstances, the livelihoods of coastal people have changed over time. In the changing context of environment, the coastal people have alternative livelihood choices. Saline-water-resistant rice and crop and plant varieties can be an important alternative to promote multiple uses of farming land. However, due to lack of saline-resistant crop varieties the people of other coastal areas of Bangladesh adopted shrimp cultivation. Over the last three decades, shrimp culture has undergone a revolutionary development in coastal Bangladesh (Ahmed et al. 2018). Azam and Falk (2013) identify that due to salinity the transformation of traditional land use pattern has been changed from rice farming to shrimp culture in the southern coast of Bangladesh. Within 2 years of the devastating cyclone Sidr in 2007, when the

R. Islam (✉)
Department of Social Work, University of Rajshahi, Rajshahi, Bangladesh

Macquarie School of Social Sciences, Macquarie University, Sydney, NSW, Australia
e-mail: rabiul.islam@ru.ac.bd

R. Ahmed et al. (eds.), *Transforming Bangladesh*, World Regional Geography Book Series,
https://doi.org/10.1007/978-3-031-45093-8_15

coastal people of Bangladesh were trying to recover from damage and losses, another cyclone Aila hit the southern part of the country in 2009. These cyclones and storm surges washed away a huge number of houses, livestock, crops and other livelihood resources (Masud-All-Kamal 2013). To cope with the changing environment, coastal communities of Bangladesh have taken initiatives to secure existing livelihoods and seek alternative job options to survive. This chapter presents the existing livelihood options of coastal community in Bangladesh, explores the changes and challenges of livelihood and finally find out alternative livelihood options.

This chapter is written based on the experience and observation of the author during the doctoral study between 2011 and 2015. This study was conducted in the two coastal villages (South Charduani and Tafalbaria) of Patharghata Upazila, Barguana district, Bangladesh (Figs. 15.1 and 15.2) were severely affected by cyclone Sidr in 2007. Cyclone Sidr hit the south-west coast of Bangladesh on 15 November 2007. The Sidr was category 4 cyclone that caused enormous damage to housing, roads, bridges, and other infrastructure and cropping land as well. The number of deaths caused by Sidr was 3406, with 1001 missing, and over 55,000 people were physically injured (Economic Relations Division 2008). Using the qualitative method, data were collected through Focus Group Discussions (FGDs) and Key Informant Interviews (KIIs) from the two study villages. Author's observation was also a tool to collect additional data. Eight focus group discussions (FGDs) with the cyclone Sidr-affected villagers were conducted. Thirty-seven key informant interviews (KIIs) took place with local leaders, local government officials and village headmen to know the livelihood options, challenges and alternatives.

15.2 Livelihood Options and Recent Changes

15.2.1 Fishing

The livelihoods of coastal Bangladeshis are mostly based on fishing-related activities, for example, catching and drying fish, fish marketing, and fishing net and boat making. The recent change in the livelihood that the coastal people cannot

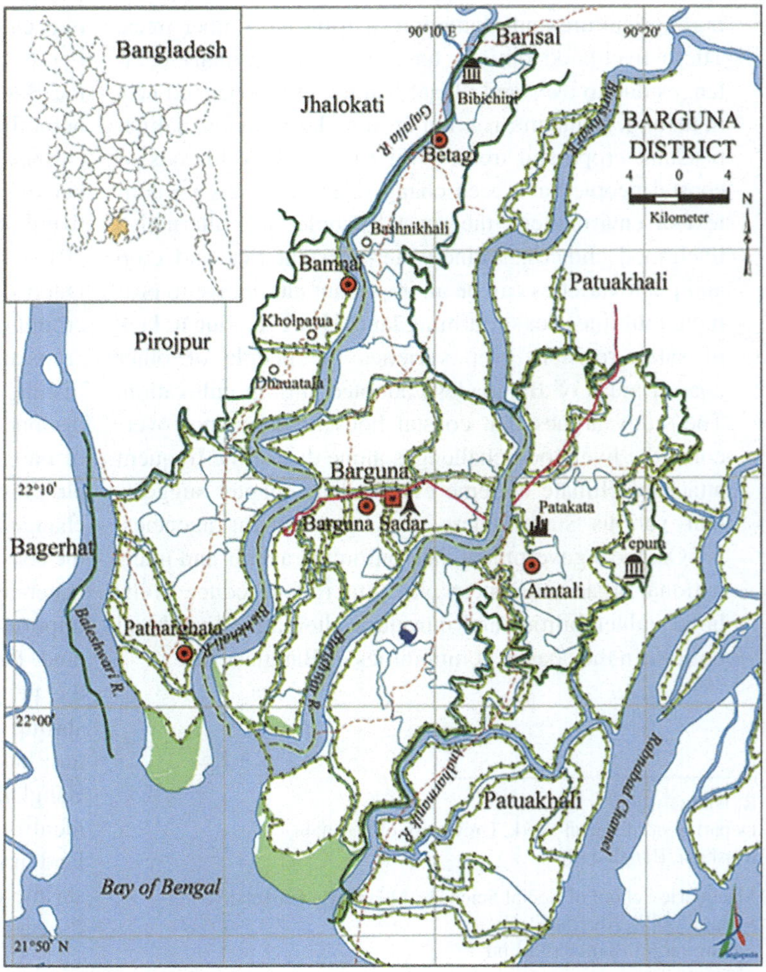

Fig. 15.1 Barguna district

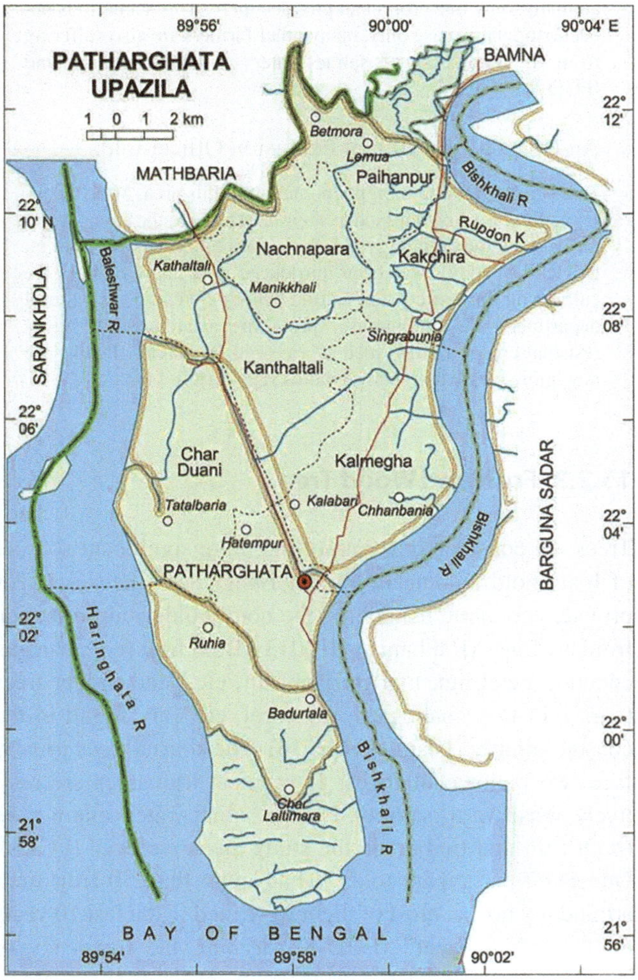

Fig. 15.2 Patharghata Upazila. Source: Banglapedia, 2015, https://en.banglapedia.org

catch fish throughout the year and fishing opportunities (specially catching *hilsa*[1] fish) are gradually declining (FGD-1,3, 5 and KII-8, 11). For example, coastal people can catch hilsa fish for only 5 months of a year, from mid-May to mid-October (Bengali month *Joystha to Aswin*). For another 5 months, from November to mid-March (Bengali month *Kartik* to *Falgun*), they are in fact involved with casual jobs, for example, mud cutting as a day labourer (at the rich peoples' houses of their own village or adjacent villages), catching small fish (such as *Bagdapona*, the fish spawn – juvenile size of shrimp) and *Kaun*[2] (species of catfish), and pulling *rikshaw* (three-wheeled passenger cart) in nearer city areas through short-term migration after disasters. The remaining 2 months from mid-March to mid-May (Bengali month *Chaitra to Baisakh*) are the neediest time for these people due to the incidence of unemployment (Islam 2015). During this time, seasonal migration occurred, as many people moved to nearer city areas to get casual jobs to overcome the hardship.

Coastal fishing in Bangladesh has been facing several challenges. *First,* catching fish during *ban time* poses a big threat to local livelihoods and to the future generations of hilsa fish (Ahmed and Troell 2010). The Ministry of Fisheries and Livestock of Bangladesh has initially imposed a 10-day ban (locally termed as *obarodh,* during full moon – *Purnima*) in October (*Aswin*) on catching, selling, marketing and importing hilsa, because during this time, adult fish migrate to the rivers from the deep sea for spawning. Recently the government has banned *hilsa* catching, selling, transporting, hoarding and marketing for 22 days from 9 to 30 October to protect mother *hilsa* for ensuring safe breeding (Bangladesh Songbad Songstha 2019). The ban time is beneficial for the long-term interests of fishers so that they can catch more fish in the future. However, some fishers continue to catch hilsa during ban time through providing bribes to coastguards and local police stations, which is a big threat to future generations of hilsa. *Second,* the use of banned nets (*current and badha*[3]) is also a major threat for the livelihoods of local fishers. A local fisherman reported, "Last season we weren't able to catch the expected quantity of fish, as the river was full of banned nets" (FGD-4). An Upazila fisheries officer reported that by using *badha* nets, some dishonest fishers regularly catch and kill hundreds of tonnes of juvenile fish, including younger hilsa. These small-sized fish are usually used as food for the cultivated fish in the private farms (*gher*) or in the ponds (KII-6). Another fisherman reported that if it is possible to stop Jatka[4] catching (the juvenile size of hilsa) during October and November (mid-*Aswin* to end of *Kartik*), they could catch hilsa fish the whole year round (FGD-8). Ahmed and Troell (2010) found that due to lack of alternative livelihoods, poor people sometimes engaged in larvae fishing, which is one of the major challenges during ban time. An Upazila Fisheries Officer said, "We usually run

[1] *Hilsa* is the national fish of Bangladesh. It is the most popular and tasty fish locally known as *Ilish.* Hilsa is found in the Padma-Meghna-Jamuna delta of the Bay of Bengal. Catching hilsa fish is a major livelihood option in coastal Bangladesh.

[2] *Kaun* is one of the species of catfish. Fishers need special skills to catch them, as they live in the soil holes under the water. Fishers must dive into the deep water (10–15 *hat* i.e. 15–20 feet) and set the nets outside their holes (FGDs).

[3] There are different fishing nets found in the coastal areas, for example (a) *Badha jal*—height 40 *hat* i.e. 60 feet, are used for catching small fish; (b) *Current jal*—height 30 *hat* i.e. 45 feet, are used for catching *Jatka* fish; (c) *Chandi jal*—height 20 *hat* i.e. 30 feet, are used for catching *hilsa* fish in the river/channel of sea; (d) *Lakko jal*—height 30 *hat* i.e. 45 feet, are used for catching medium and large fish; (e) *Lasher jal*—height 90 *hat* i.e. 135 feet, are used also for catching hilsa fish in the deep sea; (f) *Char jal*—height 15 *hat* i.e. 23 feet, are used for catching small fish. Among those, the *Badha* and Current nets are most harmful, as they kill small fish (FGDs, KIIs).

[4] *Jatka* is the younger size (smaller than an inch) *hilsa* fish. Catching this is prohibited. However, some fishers catch it illegally by providing bribes to the local law enforcing agents.

mobile courts, seize banned nets, and burn them. However, due to lack of manpower, we can't do it regularly" (KII-6). Along with Bangladeshi fishers, illegal fishing by the fishers of Myanmar and India is also a major problem. Indian fishers often enter 50–60 km inside the Bangladesh sea areas and catch fish freely. The Indian fishers also disturb Bangladeshi fishers by damaging their fishing nets and boats, even beating them when they protest (Rahman 2013). *Third,* fishers are sometimes tortured by the forest robbers and pirates (even sometimes by forest officers and coastguards), which makes earning a livelihood difficult for the poorer coastal households. For example, a fisherman told that he was kidnapped by pirates and finally released from captivity through providing a hefty ransom (Tk. 225,000 – USD2,102, 1 USD = 107 Tk. as of May 2023), though some fishers were killed by the pirates (FGD-4). Along with these challenges, moneylenders' loans pressure is also there on the fishing communities. Fishing households usually take loans from moneylenders to buy boat and net for fishing in the sea. And sometimes, they fall in debt trap due to the poverty and frequency of disasters in their areas (FGDs, KIIs). From a resilience perspective, coastal peoples' livelihoods are vulnerable due to uncertainties of fishing, proneness to illegal fishing, and fear of sea pirates (as fishers are being abducted along with their trawlers, fish and nets, and released only after they pay extortion money). Such vulnerabilities of livelihood and lack of alternative livelihood options in the villages obstructed the recovery of coastal households after cyclone Sidr.

15.2.2 Farming

Farming is another major livelihood option in the coastal areas of Bangladesh. Due to salinity, soil fertility is relatively low in the coastal regions of Bangladesh, which limits the productivity of agricultural crops throughout the year, and in severe cases, total yield can be lost (Haque 2006; Mahmuduzzaman et al. 2014). This scenario is very threatening to the primary production system and affects the dominant crop (*Aman* rice) (Haque 2006). Because of saline water intrusion in the agricultural land, farming production is unstable in the coastal villages, as farmers cannot cultivate cropping land the whole year round. They can grow only one crop per annum (locally called *aksona* land) (Islam and Walkerden 2014). In the study villages, many uncultivated lands were observed during data collection which were not cultivated due to saline water intrusion and lack of alternative saline-resilient crop varieties. One farmer told,

> I have 20 decimal farming land where my father was used to cultivate rice and vegetables 20 years back. But, the recent years, intrusion of saline water has been alarmingly increased. Therefore, I can't cultivate that land in whole year round, annually, only one crop I can produce there. This scenario is not occurring in my case only, many other farmers are also suffering from the proneness of saline water in their cropping land (FGD-7).

An Upazila Agricultural Extension Officer told,

> Sea-salinity is a common phenomenon in this area. Many farmers can't cultivate their land as their needs due to the saline water intrusion in the cropping land. Some saline-resistant crops (e.g., BIRRI 47 paddy) have been introduced, and many farmers are cultivating this variety and getting benefits. We, the agricultural department are continuously providing awareness to saline-resistant crop varieties, such as, rice and vegetables. People are now more aware than before in this regard (KII-11).

15.2.3 Fruit and Wood Tree

Trees are considered as one of the more significant sources of household income in coastal Bangladesh, because trees provide economic benefits to the households and protection from cyclones (Rabbani et al. 2013). Both fruit (e.g., banana, coconut, betel nut, mango, jackfruit, etc.) and timber trees (e.g., rain tree, mahogany, raj koroi, etc.) are found in the coastal villages (Islam et al. 2013). The coastal trees usually face two major challenges. *First,* most fruit trees are relatively sensitive to salinity. Due to saline water, many trees (both fruit and timber) in the study areas are dead (FGDs). One FGD participant told, "I had more than 50 fruit trees around my house, most of them were died in the last 10 years due to saline water" (FGD-5). *Second,* the frequency of cyclones harms and destroys (through uprooting) many trees. Damage to trees badly affects households' economy, as trees have market value that can enable households to recover from a disaster. There are some native plants that can grow well in saline soil, for example, coconut, betel nut, mango, jackfruit, guava and palm. Among them, coconut and palm are strong saline-tolerant species in the coastal areas of Bangladesh (Islam et al. 2013). Therefore, many NGOs are creating awareness among coastal people and recommending to plant saline-tolerant palm trees for reducing climate change consequences at the household level (Mangroves for the Future 2013). An NGO official told, "We distributed many saline-tolerant trees (e.g., coconut, betel nut, acacia, palm, etc.) among our microcredit borrowers. They are now planting them around their homes" (KII-9).

15.3 Alternative Livelihoods

Due to soil salinity, illegal fishing, impact of cyclone and storm surges, the livelihoods of coastal people are unsecured. Therefore, the coastal people have alternative livelihood choices.

15.3.1 Saline-water-resistant Rice Varieties

Saline-water-resistant rice and crop varieties can be an important alternative to promote multiple uses of farming land, as due to saline water intrusion local farmers can grow only one crop per annum. In this circumstance, few alternatives are introduced. For example, a salt-resistant paddy – *BRRI Dhan-47* (Invented by the Bangladesh Rice Research Institute) – has been cropping in the coastal region and the farmers are achieving good yields (FGDs, KIIs). This should be popularized (through NGOs and Upazila Agriculture Offices), and many other alternatives need to be adopted.

15.3.2 Farming Alternative Crops

Farm-based households mainly depend on paddy cultivation. Rice cultivation is in risk in the coastal villages due to soil salinity. Therefore, it is necessary to promote alternative farming (e.g., sunflower, watermelon, potato, etc.) and utmost use of farming land in the study villages. As a potential crop, potato cultivation should be promoted, as the local sandy-loam soils are suitable for potato farming, and cold storage has already been built in the study areas to promote potato farming (KIIs, FGDs). Government and NGOs should provide loans, seeds and technical support to promote this alternative farming. There is also a need to introduce saline-resistant fishing, animal husbandry, poultry and trees to cope with the changing climate because many local breeds cannot be sustained in the adverse situation of climate change (Islam 2015). In many coastal areas of Bangladesh, people adopted shrimp cultivation due to lack of saline-resistant crop varieties.

15.3.3 Increase the Number of Fruit and Timber Trees

Alternative income options should be created in the coastal villages in order to increase household capacities and build community resilience against the effects of disaster. Massive plantation of fruit and wood trees can also be an important alternative. Households could increase their number of fruit and timber trees and be more likely to withstand disasters, as households with strong resources (both monetary and in kind) can recover quicker than households who are less asset-based. Knowledge and skills on disaster preparedness and alternative livelihoods should be promoted, so that coastal people can avoid depending on a single livelihood option (KIIs).

15.3.4 Creating New Job Options

Alongside fishing and farming, local people desired more emphasis to be placed on the creation of new jobs as alternative livelihood options – for example, fish processing factories, cottage industries, tailoring for women, shipyards, and development of new sea-beaches for tourism – through the direct intervention of NGOs and local government, however, this intervention was lacking. Akter and Mallick (2013) and Mallick et al. (2005) found that local government and NGOs generated employment and income opportunities in the coastal villages after cyclones, however, this was not considerably evident in the study villages of the current research. Very few casual job options from NGOs and local government, such as the reconstruction and maintenance work of roads under the "cash/food for work" project of the national government and international organizations, were found during data collection in the study villages (FGDs, KIIs).

15.4 Stakeholders' Role for Securing Livelihood

To reduce livelihood vulnerabilities and introduce alternative livelihood options various stakeholders (e.g., government, NGOs, local government, and international organizations) can play vital role.

15.4.1 Role of Government

The government should provide interest-free/low-interest loans for both the farming and fishing community so that households can discard moneylenders' loans and emerge from the debt trap. The government also needs to put measures in place to discourage male students dropping out of school early and to discourage the early marriage of female students, in order to build human resources in the coastal region. The government and NGOs could introduce vocational training (e.g., mobile, TV, refrigerator, fan, air conditioning and rice cooker servicing, water sanitary works, sewing, etc.) facilities for coastal people, which would provide additional job opportunities for local people within the country and abroad.

15.4.2 Role of Law-enforcing Agents

Existing livelihoods are unsecured in many ways. Therefore, fishing and farming opportunities need to be secured. *Jatka* fishing and use of banned nets (e.g., *current* and *badha* –

small mesh that kills lots of juvenile fish) should be strictly controlled through joint actions and transparency of coastguards, local law-enforcing agents, and political leaders. During *ban time*, the ice factory should be closed so that opportunist fishers could not freeze their catch (FGDs). Local government leaders, NGOs and law-enforcing agents must work together with the owners of ice factories. Government and local law-enforcing agents should strengthen the security measures to protect against sea pirates through providing additional manpower for the coastguard. Physical torture of fishers and illegal behaviour of forest officers should be stopped, as they sometimes take fish forcefully from the fishers (FGDs).

15.4.3 Role of NGOs

Livelihood support is also an important factor contributing to rehabilitation of the affected households in the longer-term recovery phase. Due to cyclone Sidr, coastal households lost their livelihood options, for example, crops, poultry, livestock, fishery, trees, boats, nets and fishing accessories (Paul and Routray 2011). Households received livelihood support mainly from external organizations such as NGOs and local government. Mills et al. (2011) found that external aid agencies (national and foreign NGOs) were the major providers of livelihood support to the Tsunami-affected coastal dwellers of Indonesia, which supports the current findings. NGOs provided nets and boats, livestock, poultry, cropping seeds, and microcredit to the affected households (Islam and Walkerden 2015).

15.4.4 Role of Local Government

The Union Parishad (a local government unit) also provided livelihood support through cropping seeds, nets, and repair costs for boats (FGDs, KIIs). Livelihood support from Union Parishad was less than that from NGOs, and very few households received this support from Union Parishad (FGDs, KIIs). A possible interpretation of this finding is that the NGOs' emergency management capacity was higher than that of the local government, because NGOs had higher donor support than local government. Another conclusion is that corruption within local government was higher than within NGOs during the distribution of livelihood support; therefore, whatever livelihood-related support they had available to distribute to the affected households was not distributed equally or fairly.

15.5 Conclusion

The main aims of this chapter are to explore the existing livelihood options, identify the changes in livelihood and find out alternative livelihood options in the coastal villages of Bangladesh. The findings show that the coastal people mainly depend on fishing, agriculture and plantation of fruit and timber trees. These livelihood options have been facing many challenges in the recent time. For example, the coastal people cannot catch fish throughout the year and fishing opportunities are gradually declining due to climate change, overfishing, illegal fishing during ban time, fishing of jatka, use of banned nets during fishing, torture of the forest robbers and pirates, etc. Due to soil salinity, agricultural crop production is declining in the coastal region of Bangladesh. Because of saline water intrusion in the agricultural land farmers cannot cultivate cropping land the whole year round. They can grow only one crop per annum. Trees are considered as one of the more significant sources of household income in coastal Bangladesh because trees provide economic benefits to the households. The coastal trees are facing challenges due to salinity many trees (both fruit and timber) in the study areas are dead. Moreover, the frequency of cyclones harms and destroys many trees.

In these circumstances, the coastal people have alternative livelihood choices. Saline-water-resistant rice and crop varieties can be an important alternative to promote multiple uses of farming land, as due to saline water intrusion local farmers can grow only one crop per annum. The coastal people thought that along with paddy cultivation sunflower, watermelon, potato, etc. can be an alternative farming practices in this region. Massive plantation of fruit and wood trees can also be an important alternative. Households could increase their number of fruit and timber trees and be more likely to withstand disasters. Alongside fishing and farming, local people desired to create new jobs as alternative livelihood options – for example, fish processing factories, cottage industries, tailoring for women, shipyards, and development of new sea beaches for tourism.

The study argues that in the changing context of environment, there have been a transformation in the livelihood of coastal people, therefore, they have alternative livelihood choices. Due to the lack of saline-resistant crop varieties the people of many coastal areas adopted shrimp cultivation. Over the last three decades, shrimp cultivation has quickly grown in coastal Bangladesh. The study suggests that various stakeholders, e.g., government agencies, NGOs, local government, law-enforcing agents and international organizations can play vital role to reduce livelihood vulnerabilities and introduce alternative livelihood options. The empirical findings of this study provide a new understanding of how coastal livelihoods are vulnerable and what changes are made in the livelihood patterns over the time, and what might be the alternative options for sustainable livelihoods. The current study has only examined the overall livelihood situation based on qualitative data. Quantitative data like survey could have given more robust findings on this issue. This research has thrown up few vital questions, i.e., how the alternative farming practices (e.g., sunflower, watermelon, and potato) working in the coastal areas of Bangladesh demand further research.

References

Ahmed N, Troell M (2010) Fishing for prawn larvae in Bangladesh: an important coastal livelihood causing negative effects on the environment. AMBIO J Hum Environ 39(1):20–29

Ahmed N, Thompson S, Glaser M (2018) Transforming organic prawn farming in Bangladesh: potentials and challenges. J Clean Prod 172(20):3806–3816

Akter S, Mallick B (2013) The poverty–vulnerability–resilience nexus: evidence from Bangladesh. Ecol Econ 96:114–124

Azam M, Falk GC (2013) Governance of climate induced migration in the coastal regions of Bangladesh: a new transformation required? In: Earth System Governance Tokyo Conference Papers

Bangladesh Songbad Songstha (2019) Ban on ilish catching, selling from Oct 9–30. Dhaka Tribune, 7 Oct 2019

Economic Relations Division (2008) Cyclone Sidr in Bangladesh: damage, loss and needs assessment for disaster recovery and reconstruction. Ministry of Finance, Government of the People's Republic of Bangladesh, Dhaka

Haque S (2006) Salinity problems and crop production in coastal regions of Bangladesh. Pak J Bot 38(5):1359–1365

Islam R (2015) Social capital and cyclones: how households' social networks contribute to disaster resilience and recovery in Bangladesh. Macquarie University, Sydney

Islam R, Walkerden G (2014) How bonding and bridging networks contribute to disaster resilience and recovery on the Bangladeshi coast. Int J Disaster Risk Reduct 10(2014):281–291

Islam R, Walkerden G (2015) How do links between households and NGOs promote disaster resilience and recovery?: a case study of linking social networks on the Bangladeshi coast. Nat Hazards 78(3):1707–1727

Islam SA, Miah MAQ, Habib MA (2013) Diversity of fruit and timber tree species in the coastal homesteads of southern Bangladesh. J Asiatic Soc Bangladesh Sci 39(1):83–94

Mahmuduzzaman M, Ahmed ZU, Nuruzzaman A et al (2014) Causes of salinity intrusion in coastal belt of Bangladesh. Int J Plant Res 4(4A):8–13

Mallick DL, Rahman A, Alam M et al (2005) Case study 3: Bangladesh floods in Bangladesh: a shift from disaster management towards disaster preparedness. IDS Bull 36(4):53–70

Mangroves for the Future (2013) Palm tree plantation in households and fodder grass experiment on shrimp farm dyke in Shyamnagar coastal area Available at: https://www.mangrovesforthefuture.org/grants/small-grant-facilities/bangladesh/palm-tree-plantation-in-households-and-fodder-grass-experiment-on-shrimp-farm-dyke-in-shyamnagar-coastal-area/. Accessed 8 Apr 2015

Masud-All-Kamal M (2013) Livelihood coping and recovery from disaster: the case of coastal Bangladesh. Curr Res J Soc Sci 5(1):35–44

Mills DJ, Adhuri DS, Phillips MJ et al (2011) Shocks, recovery trajectories and resilience among aquaculture-dependent households in post-tsunami Aceh, Indonesia. Local Environ 16(5):425–444

Paul SK, Routray JK (2011) Household response to cyclone and induced surge in coastal Bangladesh: coping strategies and explanatory variables. Nat Hazards 57(2):477–499

Rabbani G, Rahman A, Mainuddin K (2013) Salinity-induced loss and damage to farming households in coastal Bangladesh. Int J Glob Warm 5(4):400–415

Rahman H (2013) Jatka catching with banned nets, fish looting by intruders go on. The Daily Star, 13 Oct 2013

Environmental Governance of the Municipalities in Bangladesh: Policies, Actions, and Challenges

16

A. K. M. Mahmudul Haque, Tasnim Nazira Rida, and Abdullah Al-Maruf

Abstract

With the integration of Sustainable Development Goals into every sphere of governance in Bangladesh, the environmental issue has confronted every level of government: national, regional, and local. Municipalities, the local governments for large towns, not large enough to be designated City Corporations, are considered the centers for development in their geographic areas. Municipalities too, in their development planning, although not technically required to, should consider environmental issues because neither development nor the environment stands in a vacuum: both are interrelated as a part of greater wholes. Thus, "development planning" cannot be effective: only holistic planning, which balances development, environment and their mutual effects in their larger natural and social contexts, is realistic. The main objective of the study is to examine how they incorporated environmental considerations into their development planning, and identifying the obstacles and issues they encountered in doing so, in practice, considering two of the Municipalities of Bangladesh as case studies. A mixed method of data collection and analysis was employed, including qualitative and quantitative data and methods. The study revealed that municipality development governance, in practice, does not exist: it is only on paper. This makes such development in most municipalities unsustainable and even unsafe for inhabitants. Central government must persuade municipality councils and the stakeholders who live in those municipalities of the importance of environmental protection and how to play their role in it and then either centralize environmental governance, removing the role of municipalities, or devolve it to municipalities with full lawmaking and taxing power, without central government control. This would induce a holistic approach to development governance.

Keywords

Environmental governance · Environmental issue consideration · Municipality · Development activities · Environmental adversities

16.1 Introduction

In cities and towns in the developing world, centers of growth and development, the environmental problems created by such development are inevitable and obvious to all. Development is, at the same time, often creating improvements in living standards and livelihood opportunities and also making those improvements unsustainable. The effect can be creation, not of long-term development, but of "development bubbles" that advance the local economy quickly, "pop" due to negative environmental impact and then leave the local standard of living in a slow state of decline. The ponds of the city of Rajshahi, Bangladesh offer a memorable example: before development, the city was adorned with large ponds that provided clean water and even fish to the community. Development often meant filling the ponds with sewage to get permission to eliminate them as a public nuisance, then filling them with soil and building million-taka residential and commercial developments where they once were. Now Rajshahi's ponds are almost gone, the city has a water shortage and has had to go into intensive groundwater

A. K. M. M. Haque
Department of Political Science, University of Rajshahi, Rajshahi, Bangladesh
e-mail: akmmahmudul@ru.ac.bd

T. N. Rida
Department of Philosophy, University of Rajshahi, Rajshahi, Bangladesh

A. Al-Maruf (✉)
Department of Geography and Environmental Studies, University of Rajshahi, Rajshahi, Bangladesh
e-mail: ammaruf4@ru.ac.bd

R. Ahmed et al. (eds.), *Transforming Bangladesh*, World Regional Geography Book Series,
https://doi.org/10.1007/978-3-031-45093-8_16

pumping to get water, which removes the ground support for those large buildings, causing them to subside over time. After long enough, the net result could be no ponds, no water and crashing development which is unsafe to live or work in.

The Municipalities of Bangladesh are trying to break the shackles of historical poverty. To do that, they are making big plans and building big infrastructure. In addition, they are administering environmental regulations loosely, to help the private sector to invest and create jobs, opportunities, and a better standard of living (Haque 2019; Al-Maruf et al. 2022). Figure 16.1 illustrates the processes at work.

Urban development, in such processes, becomes showy, big, inspiring, rapid, productive, hazardous, and harmful. Considering the policies' impact on environment, according to GOST standards (state standard of Russia):

> All hazardous and harmful production factors are divided by the nature of their impact on the human body into four groups: physical, chemical, biological, and psycho- physiological. The physical factors include: high rate of dustiness and air pollution; air and surrounding surfaces temperature rise and drop; high or low barometric pressure; elevated level of ionizing radiation; elevated level of electromagnetic radiation, ultraviolet and infrared radiation. (Rastyapina and Korosteleva 2016)

An effective policy framework is important in developing an integrated environmental governance system. The frameworks will allow and encourage the participation of all stakeholders: government, industry, and public. To do this, they must ensure adequate and transparent dissemination of information, and constantly build the capacity for interactions of all sectors. Therefore, considering environmental issues during both the formation and implementation of policies can be addressed as environmental governance. Figure 16.1 clarifies the concept.

These may be naturally occurring or anthropogenic by origin. The chemical hazard factors include all substances generated during production or emanating from other transient sources of air pollution. These are the anthropogenic factors. The biohazards are pathogenic organisms (bacteria, viruses), metabolic byproducts, plants, and animals. The anthropogenic factors are considered to be man-made, even though they interact with the environment and are shaped by it too, because they originate, as *sine qua* non-causes, from anthropogenic events such as accidents at biotechnical factories, at waste treatment facilities and due to effluent, even treated. The psycho-physiological factors are those caused by deliberate human changes to the environment, which are driven by and drive development. All listed anthropogenic and natural factors influence the psycho-emotional state of the urban population (Rastyapina and Korosteleva 2016). Therefore, we can see that development is not just a matter of economics but an interaction among nature, humans, natural processes, and anthropogenic processes which produce economic, environmental, social, and psychological results. Policies and administrative practices that ignore this interconnectedness and consider development an isolated goal quickly lose control of the situation, fail in some sense, and produce unintended consequences. Thus, municipal development planning must consider the environment as a part of holistic and realistic planning.

Yet the reality in Bangladesh planning has hardly been holistic as described here. The environmental laws have not been upgraded to meet the challenges of an increasing pace of urbanization and industrialization. Environmental objectives have also not been spelled out clearly for the town governing bodies.

In the Constitution of Bangladesh, Articles 31 and 32 have relevance to environmental issues and more than 200 laws have been enacted to safeguard the environment of Bangladesh and/or provide environmental governance. Yet still environmental exploitation and protection go on (Akhtar

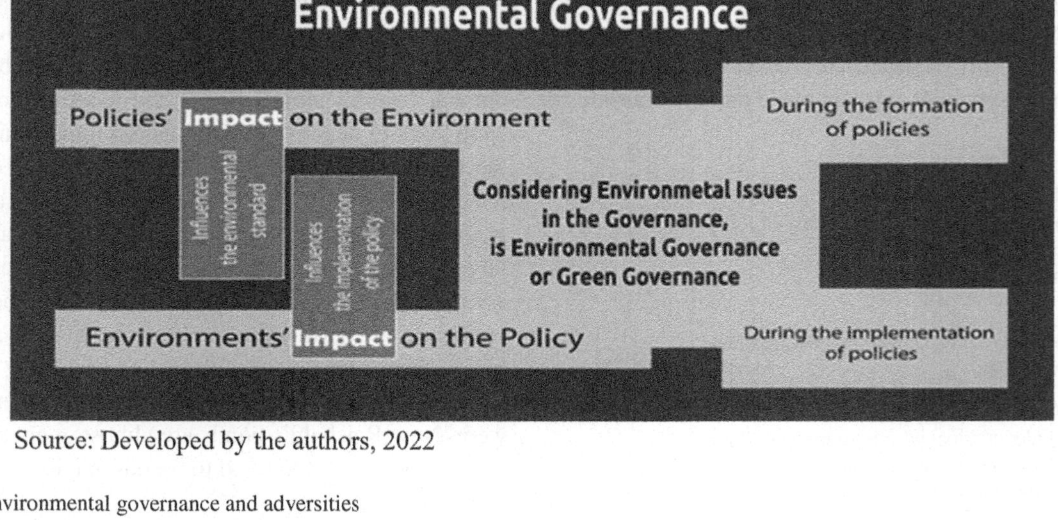

Source: Developed by the authors, 2022

Fig. 16.1 Environmental governance and adversities

2009). Bangladeshis are experts in making illegal conduct appear to be legal and buying, intimidating, or otherwise neutralizing anyone who objects. So there is a cognitive dissonance in what appears to be Bangladesh's leadership in environmental protection among the developing nations. The law and policy are in one world and reality is elsewhere: they often have little impact on, nor any relationship with, one another. Municipal environmental policy turns out to be one of the most glaring examples of this curious state of affairs. Making environmental planning not only holistic but realistic and a factor in how lives are being lived in the larger towns can be the first step toward doing so in the whole country.

This study is an evaluation of the Municipalities' environmental governance. Through this research, we seek to understand the policies and their impact on environmental issues in development in practice. If the effect is not consistent with the goals of the policies, we will propose to change both the policies and the impact as needed to remove that dissonance.

16.2 Materials and Methods

The research combines qualitative and quantitative approaches to assure data's validity and reliability and allow for objective and insightful analysis. Data have been collected from both primary and secondary sources. Primary sources include field studies, interviews with respondents, and records of the environmental management activities of Municipalities. Research reports and articles, official statistics, relevant books, thesis/dissertations, daily newspapers, and government rules and policies have been used as the secondary data sources. 255 stakeholders, officials, and public representatives of the Municipalities provided data by documents study and/or interviews. Questionnaires were administered to respondents, including open-ended structured and close-ended unstructured questions. The respondents were interviewed by the researchers and by trained interviewers. Observation of the actions of municipality staff in dealing with environmental policy compliance was conducted. Collected quantitative data have been analyzed using statistical techniques, whereas qualitative data have been analyzed through standard qualitative data analysis tools like thematic analysis, coding, and contextual analysis.

16.3 Results and Discussions

In collecting field data, the study used a five-point Likert scale to measure the attitudes and clarify the respondents' opinions about environmental governance in the selected Municipalities. Respondents were offered a choice of five pre-coded responses with the neutral point being neither agree nor disagree. It allowed the respondents to express the intensity of their agreement or disagreement with the statements in the questionnaire. Questions concerned the activities of the Municipalities in environmental matters and to challenges and successes in the Municipality's environmental activities.

16.3.1 Impact of Development Works on the Environment

Chart 16.1 illustrates the opinions of the stakeholders regarding the impacts of Municipality environmental policies on economic development. The choices given were: severe, troubling, neutral, tolerable, or safe.

It appears that multiple factors affect the environment as a result of Municipalities' development governance. For example, 38% of respondents thought that the Municipalities' inability to stop the dumping or placement of building materials anywhere and everywhere was a "severe" environmental problem. Such failure to properly place building materials in an enclosed space creates air pollution when wind blows the sand or brick chips, often blocks roads and pedestrian footpaths, and can contribute to auto and pedestrian accidents and injuries. Bitumen in such materials for road construction might cause severe health hazards (ANSES 2014). Half of the respondents informed that development works often take a long time, leaving unused building materials in place for a long time and increasing the negative impact of such materials on health and transport.

"*Smog is a mixture of pollution from coal, the primary source of energy for most homes and businesses in the region, as well as emissions from vehicles* (National Geographic n.d.)." When these particles are exposed to sunlight they turn into harmful airborne particles. As a consequence, many respiratory diseases occur, e.g., asthma. A report of Bangladesh Tribune says that over 123,000 people died in Bangladesh in 2017 due to indoor and outdoor air pollution. The Tribune article also reports that long-term exposure to outdoor and indoor air pollution contributed to nearly 5 million deaths from stroke, diabetes, heart attack, lung cancer, and chronic lung disease in 2017 in Bangladesh ("Report: Air Pollution Killed 123,000 People in Bangladesh in 2017" 2019).

On the other hand, about 60% of respondents from both study areas have expressed their opinion that urban flooding is also caused by unplanned development activities, which can lead to diverse environmental hazards. Urban flooding results when rainwater, especially the sudden, heavy rainfall of the monsoon season, simply builds up on the surface of the soil or road and spreads as far as it can (Tawsif et al. 2022). In Bangladesh, this urban flooding is mostly caused

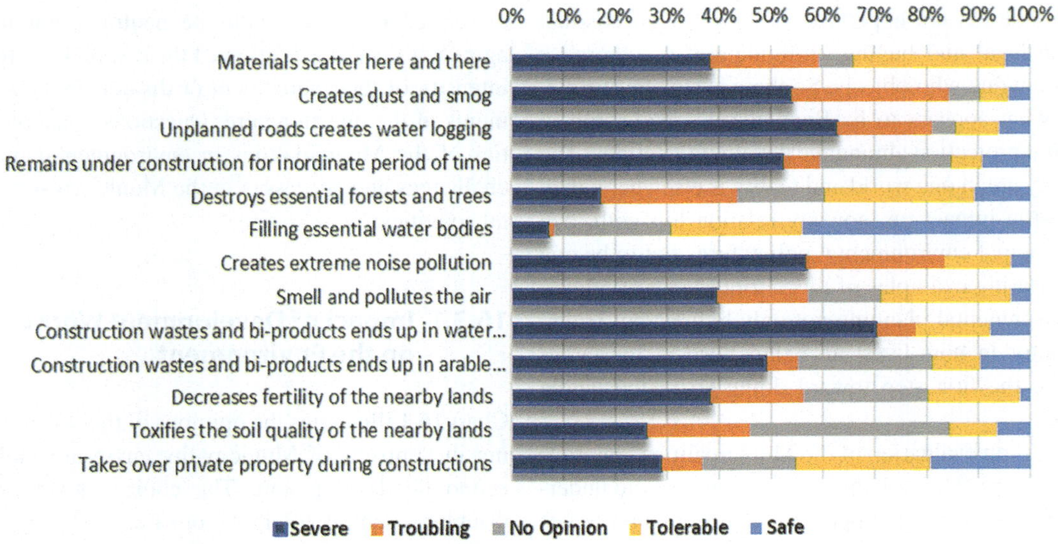

Chart 16.1 Impact of municipalities' development work on the environment (Overall). (Source: Field survey 2020)

by insufficient drainage (Al-Maruf 2017) as many of the sewerage systems were built by the British before 1947 and were never modernized or expanded to deal with larger and more dense population in their areas, due to lack of public sector funds under public ownership. . Unplanned urbanization has also exacerbated the problem. The effects of such sudden urban flooding include damage to property and other infrastructure, the spread of waterborne disease and disease vectors, distribution of sewage in roads, lanes and even in housing, disruption of business and employment due to people being trapped in their homes and lack of food and potable water, sometimes for extended periods (Anisha and Hossain 2014; Pyka et al. 2020).

Noise pollution is another environmental issue in the larger towns. 56.8% of interview and survey respondents reported that they face severe noise pollution and 26.8% responded that they have found noise pollution troubling. 90% of respondents to a survey in Dhaka said that noise pollution has severe impacts on the environment in that city, even interfering with speech communication and sleeplessness (Razzaque et al. 2010; Haque et al. 2021). That is how bad the towns can become in future if action is not taken now. The Australian Academy of Sciences reported:

> Exposure to prolonged or excessive noise has been shown to cause a range of health problems ranging from stress, poor concentration, and productivity losses in the workplace, and communication difficulties and fatigue from lack of sleep, to more serious issues such as cardiovascular disease, cognitive impairment, and tinnitus and hearing loss (Australian Academy of Science n.d.).

Hence, noise pollution is more than a mere inconvenience. Field observation showed that most of the sewerage drains of Naohata and Naogaon empty into or pass-through arable land. Consequently, these arable lands are becoming toxic over time, which must make the crops harvested from them toxic and damage their fertility. Koul and Taak report:

> 15% of the total arable land of the world has been extensively degraded due to the mixing of several toxic pollutants into the soil. Illegal dumping of the solid wastes, polluted water absorbed by the soil, use of chemical fertilizers and pesticides, dumping of minerals, oils, and radioactive wastes are the major causes of soil pollution. Soil fertility is severely affected due to the above-mentioned factors. Till date, soil pollution has spread in many developed and developing countries of the world (Koul and Taak 2018).

38.4% of respondents believe that sewerage is degrading the soil quality and reducing the fertility of arable land to a large extent. 20% of the respondents also believe that the soil in their municipality is becoming toxic due to contamination by sewerage drains.

16.3.2 Waste Management System and Its Impacts on the Environment

In accordance with the Local Government (Municipality) Act 2009, a Municipality of Bangladesh is empowered to take appropriate steps to effectively and sustain the solid waste in their jurisdiction. A Municipality shall make adequate arrangements for collecting and removing refuse from all public streets, public latrines, urinals, drains, and all buildings and land vested in the Municipality. All of the refuses removed and collected by a Municipality or under its control and supervision shall be the property of the Municipality. A Municipality must ensure the cleanliness of public latrines and urinals, and ensure that private toilets are cleaned and operated following basic hygiene standards. The Municipality must provide an adequate system of public sewerage and water drains in urban areas. All such drains must be constructed, maintained, cleared, and emptied with due regard to the health and convenience of the public.

A Municipality is responsible for establishing and maintaining public markets and ensuring proper maintenance, including waste management. It will also provide and keep at least one abattoir in each urban area. Streets and lanes must be maintained in a reasonably clean condition, not damaging to public health, including using water for this purpose and sweeping. A Municipality must include Standing Committees on Waste Management, Environmental Development, Disaster Management, etc. among the Standing Committees of its Council.

Table 16.1 shows the data from respondents to questionnaires, painting a picture of towns where rubbish and sewerage in public places is plentiful, polluting the air, the water, and the environment generally. The picture is disgusting. Whatever municipal environmental governance is in place, it is not working. This table has the most important data in this research. In waste disposal, what one sees and smells counts most: no system or scheme matters more than that.

Some data from both study areas are worth highlighting to understand the extent of the failure of municipal environmental governance: 100% have seen waste dumped into watercourses like drains, streams, and rivers; 88% of respondents have smelled stinking waste vehicles; 74% respondents say that wastes are not collected in a timely fashion; 72% see waste piles around town; 68.4% believe there is a lack of human resources to implement waste disposal; 50.8% say that there is a lack of waste transport vehicles for waste disposal. In addition, researcher observation found that hazardous and non-hazardous wastes are disposed of in the same way, in the same places, a health hazard for disposal staff and residents. These statistics, if the residents' perceptions are not distorted, suggest that the Municipalities are not trying and failing but are not even trying to provide the resources for basic services in waste disposal. Thus, faulty and inadequate waste disposal is a major driver of environmental pollution in Bangladesh and this falls squarely into the responsibility of the Municipalities. Solid wastes contaminate surface water because solid wastes are dumped near ponds, canals, and rivers. As a result, surface water is degraded to a large extent. Numerous biochemical components pollute the surface water and they impact fishing, and the collection and use of surface water by residents for daily needs (Ahmed and Hussain 2011). Since wastes are dumped into the rivers Padma in Rajshahi and Choto Jamuna in Naogaon, people who use these rivers for bathing, transport, fishing, and daily use are at risk of health consequences. Groundwater is no safer than surface water. Most of the deep tube wells which are situated near the waste pit or dumping zones have been abandoned by the people due to contamination of the water with waste (Haque 2020).

Another severe problem identified in both of the Municipalities is the dumping of rubbish into sewerage drains. All sorts of wastes, including plastics, are dumped in the drains. These tend to clog the drains and cause overflow of sewerage and rubbish mixtures into the streets and footpaths. Sometimes, the Municipality will stop resulting flooding by "clearing" the drains: merely digging rubbish out of the sewerage drains and piling it in the road, letting it slowly dissipate back. However, rag pickers will comb it for any rubbish that can be sold for recycling (without necessarily disclosing its collection point) at significant risk to health. Other sewerage pickers will extract human and animal waste from the Municipality's clearance pile and dry it in the sun, to sell as fire fodder to those who cannot afford to buy wood or liquefied gas. Sometimes the Municipality's clearance piles also dissipate into arable land, rivers, and streams. For good environmental management, this whole process, start-

Table 16.1 Waste management system and its impact on the environment

	Yes		No	
Waste management practices and its impact (Number of respondents: 250)	Freq.	(%)	Freq.	(%)
Wastes are not collected in time	185	74	65	26
Lack of workforce	171	68.4	79	31.6
Lack of transport facility	127	50.8	123	49.2
Lack of equipments	82	32.8	168	67.2
Conservancy workers don't use safety equipment	111	44.4	139	55.6
Garbage is piled randomly	180	72	70	28
Garbage truck stinks	220	88	30	12
Garbage pit causes high pollution	135	54	115	46
Garbage is thrown into the drain	240	96	10	4
Garbage is not separated in household	113	45.2	137	54.8
Garbage is not collected separately	245	98	5	2
Environment pollution due to waste mismanagement	126	50.4	124	49.6
Waste disposal into water body	250	100	0	0

Source: Field survey 2020

ing with dumping rubbish into sewerage drains, simply should not happen. In Western countries, this is prevented by putting sewerage drains under the street. Bangladesh cities often do this to some extent but Municipalities usually do not. Digging up roads to cover drains is very costly and disrupts transport and trade in the process of doing so.

So, in towns in Bangladesh, people throw rubbish into the sewerage drains. They think that this is more hygienic than throwing the rubbish into the street. The Municipality lacks people and vehicles to collect the rubbish but sometimes they do so. In cities, where the sewerage drains are under the road, if a sewer grating in the road is nearby, people will get the rubbish into the sewerage drain that way. If not, they will throw the rubbish into the street, the footpath, or any grassy land where it is not so visible. The "improvement" in the cities over the Municipalities, from burying the sewerage drains under the roads, is, frankly speaking, debatable. It is certainly one reason why the large cities of Bangladesh are notorious for looking like rubbish tips but the Municipalities look comparatively cleaner.

16.3.3 Defining the Problem of Municipal Environmental Governance

The existing environmental policies and regulations of Bangladesh prescribe the compliance of environmental issues by the local government bodies when planning and implementing infrastructure development projects such as roads, culverts, bridges, potable water supply, flood control, irrigation infrastructures, etc. (Asian Development Bank 2004). Therefore, these bodies are responsible for identifying and determining the impacts of such projects on the environment and natural resources and taking necessary steps to avoid or reduce the adverse environmental impacts of these projects (Ministry of Environment and Forest 1992a, b).

Yet, due to poor planning and implementation of development projects, without adequate assessment of potential impacts, the physical and biotic environment is significantly changed, causing environmental damage. Also, much of the environmental damage that occurs is not the result of specific projects but a product of problems in the central-local government structure and the culture of Bangladesh. Finally, the local governments often fail to implement the existing rules and regulations described above.

Table 16.2 demonstrates the opinion of the stakeholders who participated in the data collection during this research. 54% of respondents from both Municipalities think that political influence severely impedes local environmental governance, with an additional 30% saying that such influence "impedes [municipal environmental governance] a lot". 46% of respondents believe that other local influence "impedes [local environmental governance] a lot" and an additional 43% say local influence impedes it severely. However, a greater percentage of Naohata residents blamed local influence than did Naogaon residents. 49% of respondents thought that financial problems "severely" impeded municipal environmental governance and an additional 30% said that financial problems interfered "a lot", a total of 79%. A large number of respondents (79%) identified the knowledge gap of the stakeholders as a major problem which impedes the environmental governance process "severely" or "a lot".

As for issues which are not part of the problem of poor environmental governance, 88% of Naogaon respondents said that industrialization does not impede the environmental governance process. 95.6% of respondents do not think of geographical location as a contributing factor. More than three-fourths of the respondents do not think that the Government's financial policy is a barrier in this regard.

From informal discussion with the town dwellers and field observation, it is found that political influence, local influence, financial problems (Al-Maruf et al. 2021), knowledge gap, non-cooperation, and town planning are the biggest obstacles in both Municipalities. Town planning seemed to be problematic during the field observation of Naohata.

Table 16.2 Obstacles in the environmental issue consideration of municipalities

	0		1		2		3		4	
	Do not impede		Impede a little		Impede		Impede a lot		Impede severely	
Intensity of problems	Freq.	(%)	Freq.	(%)	Freq.	(%)	Freq.	(%)	Freq.	(%)
Political influence	16	6.4	21	8.4	3	1.2	75	30	135	54
Local influence	23	9.2	3	1.2	1	0.4	116	46.4	107	42.8
Financial	23	9.2	11	4.4	18	7.2	75	30	123	49.2
Knowledge gap	0	0	4	1.6	24	9.6	132	52.8	90	36
Non-cooperation	21	8.4	7	2.8	5	2	42	16.8	175	70
Geographical	239	95.6	2	0.8	5	2	1	0.4	3	1.2
Industrial	116	46.4	20	8	32	12.8	13	5.2	69	27.6
Town planning	8	3.2	8	3.2	1	0.4	18	7.2	215	86
Govt. financial policy	198	79.2	0	0	20	8	0	0	32	12.8

Source: Field survey 2020

No specific plans or indications from the authority were visible inside the residential area. Most residents do not have a proper drainage facility or waste management facility. In many cases, wastes are piled in holes by the house owners. Some of these are burned which causes further air pollution. The local residents also blame their local leaders for not building a proper drainage system in Naohata Town. According to the local residents, the local leaders proposed to build waste management infrastructure like sewerage drains and rubbish tips but they let important local people forestall them in their own personal interest.

16.3.4 Functionality of Environmental Committees of Municipality

When the Standing Committees of the Councils of both Municipalities were mentioned to respondents in their respective Municipalities – Finance and Establishment; Waste Management; Education, Health and Family Planning; Urban Planning and Development; Urban Infrastructure Development and Conservation; and Environmental Development – were mentioned to local residents in their Municipalities, no one knew anything about them. At worse, this suggests that the Committees are not doing anything. At best, it could mean that the Committees are working but have not created effective channels of communication with residents to build knowledge of and support for environmental governance, as well as that such governance is not very participatory.

Members and staff of the committees also gave the researcher mixed responses. Some of the respondents suggested that these committees are just on paper and neither exist nor function in any way. Others reported that these committees hold meetings occasionally. Whether the committees hold occasional meetings or not is not determinative: even if they do meet, this is not a description of a working municipal environmental system of governance.

More critical in assessing the municipal environmental governance, no consultation was ever done with the environmentally related committees before, making any Council decision or policy. No one has informed the committees of, or protested to them about, any environmental degradation. None of the committees of these Municipalities have ever made any environmental decision, or policy, or reviewed the operation of any Municipal action from an environmental point of view. If there are meetings, they are "low tea and chat" meetings having nothing to do with the statutory purposes of the committees.

The Municipalities' environmental governance processes are found to be moribund (see Tables 16.3 and 16.4). Given the seriousness of Bangladesh's environmental problems and the central role of the Municipalities under Bangladesh law and policy in solving them, this is a disaster.

Table 16.3 Functionality of the environmental issue related committee of municipality

Functional committees	Respondents familiar with (%)	Functionality
Finance and Establishment	0	Functional
Waste Management	0	Partly Functional
Education, Health, and Family Planning	0	Not-Functional
Urban Planning and Development	0	Not-Functional
Urban Infrastructure Development and Conservation	0	Not-Functional
Environmental Development	0	Not-Functional

Source: Field survey 2020

Table 16.4 Environmental issue consideration of municipality

Variables	Yes (%)	No (%)
Familiar with the environmental committees of Municipality	0	100
Familiar with the waste management committee or responsible	0	100
Familiar with the drainage and sanitation-related committee or responsible	0	100
Consulting with these committees	0	100
Committee's consideration of environment during plans and activities	0	100
Informing the environmental status to the authority	0	100

Source: Field survey 2020

16.4 Conclusions and Recommendations

The major significant conclusion of this study is that, at least in the two local governments used in the case study, municipal environmental governance is not working and barely exists. The laws and policies are sound in principle, although they are too piecemeal and do not represent the holistic and realistic environmental policy described at the outset of this article. However, they are dead letters in practice. Thus, there is no use in bothering to criticize them.

Yet the Municipalities of Bangladesh do desperately need an efficient and effective environmental governance system. They are drowning in rubbish, human and animal waste, and urban flooding. Respondents to this study see it, feel it, and smell it. As population and economic development continue to increase in the Municipalities, this can only worsen, although one cannot bear to think what that may look like, feel like, and smell like for residents.

One fact is clear: municipal environmental governance has not been and cannot be imposed from the top down. The last thing Bangladesh needs is another law on environmental governance that will be ignored and become a failure.

We have to get at the root causes of those kinds of failure and those roots run very deep: into the unsatisfactory distribution of power and revenue between the central and local governments, into Bangladeshi culture, into the failure of local government generally in Bangladesh because it did not develop from a thirst for democracy or devolution of the people but was always a concoction of administrators in the capital, first colonial administrators and later bureaucratic administrators. All that cannot be changed, if only because a lot of it is simply the country's history. One should not be too ambitious in planning to solve the collapse of municipal environmental governance. Indeed, whether that collapse can be solved at all is a very open question. Some real incremental improvement would be a nice change worth working for.

Let us start with a realistic hypothesis that, if municipal environmental governance cannot be imposed from the top down, it is worth trying to see if it might be built from the bottom up. The first part of the problem in trying to build a popular movement for municipal environmental governance that will induce social, political, economic, and cultural change in the Municipalities of Bangladesh is that Bangladesh is not Germany. German culture is obsessed with order: people obey rules, value cleanliness and predictability, derive value from clear and logical structure and organization, and respect the environment. Thus, German cities are clean and orderly from the behavior of the individual citizen to the structure of their governance. This is bolstered by planning legislation which needs little enforcement because everyone follows it and by expensive infrastructure to treat waste, dispose of rubbish, and make economic development environmentally sustainable. It must be conceded that some of this is rooted in Germany's economic success (or maybe that the economic success is rooted in the culture): rich people have the luxury of organizing things very well because they are not preoccupied with surviving, the environment be damned, and they have the resources needed to give the environment the best protection that modern science has to offer. Yet not all of it is rooted in wealth. Some of it is just how the people think.

When we say that Bangladesh is not Germany, we mean that Bangladeshis, like most South Asians, do not value order, at least for its own sake. Germans are logical people, many say coldly logical. South Asians are passionate and intuitive, to the extent that they feel somewhat more comfortable with the disorder than constricting order. Rules are something to be got around: by reinterpretation, ignoring, if necessary, by downright lying and bribery. While personal cleanliness is valued, Bangladeshis do not see any immediate benefit in a clean environment and would rather be free of any headache over throwing rubbish in the road or the sewerage drain than to worry about finding somewhere more environmentally conscious place to put it. Structure and organization is viewed instinctively with suspicion. Bangladeshis do not mind having hundreds of committees but when they actually join together, they would rather partake in lots of tea, sweets, *biryani* if available and storytelling than to write detailed plans. When one understands the culture, why municipal environmental governance is as it is in Bangladesh is patently obvious.

Can culture be changed? Optimists can say that Western people have been induced to stop smoking, stop beating their children and pets, respect women and minorities and give them a lot more equality of treatment, give workers some basic standards of fair treatment, eat a healthier diet and, yes, stop throwing rubbish in the street or in the sewers, in barely one generation. Pessimists will quickly snap back that none of the above has been achieved in Bangladesh, except among a narrow city elite. Yet, to avoid gaining nothing, municipal environmental governance advocates might just want to have a go anyway. One thing we can be sure of is that, if the environmental culture of Bangladeshis can be changed for the better, the governance mechanisms would probably become workable, if they do not completely fall into line naturally. A committee that truly cares about the environment could not conceive of failing to do some environmental action in between those sips of tea and telling of stories.

One thing we know about culture is that it cannot be imposed from the top down. That is a major reason why environmental consciousness, which involves cultural change, has not been imposed by central government environmental law and policy. That cultural change cannot be imposed from the top down is why colonialism failed. Few people know that the British made a real coordinated effort to stamp out female genital mutilation in Kenya in the 1920s and child marriage and dowry in India in the 1800s. All they got from these initiatives was a burgeoning independence movement that later became violent and uncontrollable. After long enough trying to "civilize" their colonies and getting nowhere, largely by changing their cultures, the British threw their hands up and went home. They were great at building railways, schools, hospitals, and sewerage and water lines because they had the money and the technology to do it: but they could not change the leopards' spots or the people's minds. So any thoughts that we can stop people throwing rubbish in the sewerage drains by bringing back the medieval penalty of "drawing and quartering" must be squelched instantly.

Education is one factor we know that helps to change culture. Propagandizing and marketing does so even better. Indoctrination is even better. All the changes in culture referred to in the west above, like personal habits (smoking, eating habits), relations in the family (gender, age), and the workplace (worker rights) began with education. So, building a Bangladeshi culture of environmental protection, which has been wholly neglected in Bangladesh's legislation on the environment since the 1990s, would take a strong step forward by education, persuasion, propagandizing, marketing, and, where possible, indoctrination. Some old-fashioned NGO community organizing, on environmental action at the community level, might be a good start. Then the structures might work on their own or, if new laws were needed, people might actually obey them because they believe in the importance of such laws. Laws in advance of the culture change will be fantasy laws, of no impact, like the ones we have on municipal environmental governance now.

Beyond cultural change, the whole structure of local government in Bangladesh needs to be reviewed. Local governments in Bangladesh, including Municipalities, have a bad history of being unable to collect much revenue and thus, in environmental governance, come to the table with nothing. They are wholly dependent on central government subsidies and are also tightly controlled by the central government's District Commissioners and other officers like Upazila Nirbahi Officers, who must approve almost everything they do. Central Governments have arbitrarily used these subsidies and controls to build up their local Parties and gut their local oppositions. These appendages of the central government are incapable of performing the functions of environmental governance allotted to them by local government and environmental legislation. Either environmental governance should be "nationalized" in the central government, which has the power and resources to do it, with no role for the municipalities, or the Municipalities should be empowered to levy and collect taxes and make laws without the permission of central government officers, which would give them the power and resources to do it. The present structure of putting all the resources and power in the central government and all the responsibility for environmental governance in the Municipalities is like a predictable plan for failure – and it works to do that.

In conclusion, two steps:

1. A massive public awareness-building campaign, from the village level up to and through the Municipal and up to the national level, including education, propagandizing, marketing, and indoctrination, should be conducted by community organizing, advertising, speeches by popular personalities, etc. to convince ordinary Bangladeshis of the importance and benefits of environmental protection and of playing their role in it, from putting rubbish in proper containers and obeying environmental laws up to holding politicians accountable by "voting green (not the Green Party but the green concept)" in local and national elections.
2. A reform of environmental responsibility based on complete centralization, taking the Municipalities out of the process, or complete devolution, giving Municipalities full taxing and lawmaking powers on environmental issues within their geographic jurisdictions, without central government control or subsidy; should set the preconditions for a workable environmental governance policy to be developed over time, in place of the nothing that exists, in reality, today.

Acknowledgment The research was funded jointly by the University Grants Commission and the Faculty of Social Science of the University of Rajshahi, Bangladesh. The grant number is: 1337/5/52/RU/Soc-02/2019-2020, dated: November 11, 2019. The authors would like to give special thanks to the Faculty of Social Science of the University of Rajshah for providing support in research grants. They are also thankful to the study participants, the research assistants, and the staff of the selected Municipalities of this research.

References

Ahmed A, Hussain SH (2011) Solid waste management system of Narayanganj city: an environmental study. Department of Geography and Environment, University of Dhaka, June (2011)

Akhtar R (2009) National environment policy of Bangladesh: a critical review. Unpublished MSS Dissertation. Institute of Governance Studies, BRAC University, Dhaka

Al-Maruf A (2017) Enhancing disaster resilience through human capital: prospects for adaptation to cyclones in coastal Bangladesh. Doctoral dissertation, Universität zu Köln

Al-Maruf A, Mira SA, Rida TN, Rahman MS, Sarker PK, Jenkins JC (2021) Piloting a weather-index-based crop insurance system in Bangladesh: understanding the challenges of financial instruments for tackling climate risks. Sustainability 13(15):8616

Al-Maruf A, Pervez AK, Sarker PK, Rahman MS, Ruiz-Menjivar J (2022) Exploring the factors of farmers' rural–urban migration decisions in Bangladesh. Agriculture 12(5):722

Anisha N, Hossain S (2014, December) A case study on water logging problems in an urban area of Bangladesh and probable analytical solutions. In: 2nd international conference on advances in civil engineering 2014 (ICACE-2014), CUET, Chittagong, Bangladesh

ANSES (2014) Exposure to bitumen. https://www.anses.fr/en/content/exposure-bitumen#:~:text

Asian Development Bank (2004) Country environmental analysis: Bangladesh. ADB, Manila

Australian Academy of Science (n.d.). Health effects of environmental noise pollution. https://www.science.org.au/curious/earth-environment/health-effects-environmental-noise-

Haque AKMM (2019) Local environmental governance in Bangladesh: assessing Paurashava's initiatives in achieving environmental sustainability. Unpublished Research, University Grants Commission of Bangladesh

Haque AKMM (2020) Field observation of Naogaon Garbage Pit. March 30, 2020

Haque AKM, Al-Maruf A, Rida TN (2021) Governance for sustainable development to combat the impact of climate change in a medium-sized city in Bangladesh. In: Bangladesh II: climate change impacts, mitigation and adaptation in developing countries. Springer, Cham, pp 357–373

Koul B, Taak P (2018) Soil pollution: causes and consequences. In: Biotechnological strategies for effective remediation of polluted soils. Springer, Singapore. https://doi.org/10.1007/978-981-13-2420-8_1

Ministry of Environment and Forest (1992a) The environment policy, 1992. Ministry of Environment and Forest, Government of the People's Republic of Bangladesh, Dhaka. Article, 3

Ministry of Environment and Forest (1992b) The environmental action plan. Ministry of Environment and Forest, Government of the People's Republic of Bangladesh, Dhaka. Article, 5

National Geographic (n.d.) Shanghai Towers above the Smog. https://education.nationalgeographic.org/resource/pollution

Pyka L, Al-Maruf A, Shamsuzzoha M, Jenkins C, Braun B (2020) Floating gardening in coastal Bangladesh: evidence of sustainable farming for food security under climate change. J Agric Food Environ 1(4):161–168. ISSN (Online Version): 2708-5694

Rastyapina OA, Korosteleva NV (eds) (2016) Urban safety development methods. Procedia Engineering, International Conference on Industrial Engineering, ICIE 2016: Vol. 150 (2016) 2042–2048. https://doi.org/10.1016/j.proeng.2016.07.292

Razzaque M, et al (2010, July) On the impacts of noise pollution in Dhaka. In: 17th international congress on sound and vibration 2010, Cairo, Egypt

Report: Air pollution killed 123,000 people in Bangladesh in 2017 (2019, April 4). Dhaka Tribune. https://archive.dhakatribune.com/bangladesh/nation/2019/04/04/report-air-pollution-killed-123-000-people-in-bangladesh-in-2017

Tawsif S, Alam MS, Al-Maruf A (2022) How households adapt to heat wave for livable habitat? A case of medium-sized city in Bangladesh. Curr Res Environ Sustain 4:100159

Gender Disparity in Drought-Prone Areas in Bangladesh

17

Md. Kamruzzaman and Ferdous Farhana Huq

Abstract

The north-western part of Bangladesh has been experiencing extreme drought conditions compared to other parts of the country. Water crisis and drought environments cause eternal sufferings to the women who usually tackle all household activities. This paper aims at examining the prevailing gender disparity in a drought-prone region of Bangladesh by examining the daily activities, mobility pattern, water management, and job opportunity parameters. Questionnaire survey, Participatory Rural Appraisal (PRA) Survey, Focus Group Discussion (FGD), and direct observation methods were applied to assess the drought impact on gender dimension in the study region. It is evident through the study that gender inequality prevails in the communities and the women are deprived in terms of social respect, recognition of work, wage, income opportunities, scope of socialization, leisure, and marginalization despite people do not aware of the issue.

Keywords

Water crisis · Mobility · FGD · Marginalization · Social respect · Gender

17.1 Introduction

Impact of climate change and associated problems are more common in least developed countries (IPCCC 1998). The poor are the most sufferers from the thrust of climate change impacts (NDMC 2006). Flood, earthquake, riverbank erosion, and cyclones are most common disasters in Bangladesh and occur recurrently. In the northern part of Bangladesh, droughts are prevalent. Manmade and climatic reasons are threatening issues to cause drought. Problem of groundwater depletion is the single most contributing reason to cause drought. Extreme climatic conditions and huge suction of groundwater for agriculture and domestic uses are contributing to worsen the drought situation with every passing day. Observing the rural poor inhabitants struggling to survive within the extreme drought events within this region provoke to conduct this study. The Barind tract region covering the whole north-western part of Bangladesh is selected to carry out the case study and to understand the incidents of gender disparity. The study uncovers inequality, marginalization, and disadvantaged situations of women within the drought environments. The objective of this paper is to examine the gender disparity in prevailing drought scenario within the household functional activities to manage water crisis, mobility pattern for water collection, job opportunities, wage and livelihood pattern comparing the opportunities for both men and women through a gender lens.

17.2 Study Area

This study was conducted in three villages, i.e., "Natunpara" "Khatirpur" and "Bandhupara" within Porsha Upazila of Naogaon District in Bangladesh. It is 240 km North-West from Dhaka, the Capital City of Bangladesh. A typical dry climate with comparatively high temperature prevails in the study area and belongs to dry humid zone. The average annual rainfall in this region is 1329 mm; while the rest of the country receives a rainfall of 4338 mm (Shahid et al. 2005). Temperature ranges from minimum of 8 °C in the winter to maximum of 44 °C in the summer. A gradual increase in daytime temperature and extreme environment was observed by both meteorological department and the local dwellers of the study area.

Md. Kamruzzaman
Department of Civil Engineering, Rajshahi University of Engineering & Technology (RUET), Rajshahi, Bangladesh

F. F. Huq (✉)
Department of Urban & Regional Planning, Rajshahi University of Engineering & Technology (RUET), Rajshahi, Bangladesh

R. Ahmed et al. (eds.), *Transforming Bangladesh*, World Regional Geography Book Series,
https://doi.org/10.1007/978-3-031-45093-8_17

The indigenous community of the study area has retained the traditional lifestyle. According to BBS (2011), there are about 175 households having 676 inhabitants out of which 48% are male and 52% are female. In case of gender dimension, literacy rate for male is about 53.10% and for female is about 60% (BBS 2011). Although statistical sources reveal mentionable education level in the study area but ground scenario is different. From the FGD, it is found that most of the people are illiterate and lack the opportunity of formal education. Majority of the community people are landless poor. The main occupation of the respondents is agricultural day labor, wage labor, livestock rearing, and driving three wheelers. Females are totally engaged in household works; majority of them help in agricultural works with the men.

Statistical record confirms that there presents a declining trend of annual fluctuation of groundwater table in the northwest region of Bangladesh over the last 30 years (1981–2011), which implies that groundwater use is not sustainable in the study area (IWM 2006). Over-exploitation of groundwater and decreasing annual rainfall contributed to this incident. Over-exploitation of groundwater for agricultural purpose has triggered severe water crisis. Therefore, many shallow tube wells fail to supply drinking water and accelerated the sufferings of the local residents.

17.3 Methodology

Selection of the area to be studied is guided by three general principles. The first principle is to ensure that the area represents a typical drought-prone area. The second is to consider the presence of transitional climate. Finally, it is important to ensure that the distinction persists between male and female victims from the sufferings due to water crisis. This enables us to better understand the gender disparity from drought scenario.

The study utilizes Participatory Rural Appraisal (PRA) Survey, Focus Group Discussion (FGD) and direct observation methods to intensely and effectively explore the drought impact on gender dimension in the study region. Daily activity schedule and mobility pattern tools from PRA surveys were used to generate the lifestyle and livelihood scenario of women and men. FGDs were conducted with villagers to understand the impact of drought on their livelihood and daily life of the people living in the study communities. Direct observation method was used to understand the social setting of the communities. The data analysis tools included content analysis and simple statistical methods to complement the qualitative data.

Out of 175 households; about 57 households were selected by random sampling method for the questionnaire survey. Ten FGDs from the communities were conducted. The average number of participants in the focus group was 8 to 10. Both male and female participants became the part of the FGD. PRA surveys were conducted with the help of these focus groups. Daily activity schedule and mobility mapping PRA tools were applied to understand the gender dynamics of the study area. Direct observation was the essential component of field survey throughout the field survey to get the clear view of social lifestyle of male and female of the region. Photographs were taken during direct observation method.

17.4 Study Findings

Drought situation in Barind tract has made visible distinction and impacts on gender aspects. The field survey addressed and scrutinized both men's and women's roles into household water management issues and has identified the inequality.

The traditional lifestyle of women in the study area certifies their critical role in serving all household activities. Engagement and responsibility of females in domestic activities make a noticeable difference compared to the males. From dawn to dusk, they need to do various kinds of activities which include cooking, dishwashing, cleaning, collecting water, rearing children, rearing domestic animals, gardening, and farming. They spent a significant amount of time and energy on household water collection, storing, and management. Therefore, women suffer the most during the water crisis.

Generally, domestic water issues are usually framed within the context of social rights, sanitation and hygiene, health, and other basic household needs; while issues of water relating irrigation are solely associated with production and economic profit. Such preconceived mentalities have resulted into allocating women with many household responsibilities and men with most powers and rights. As a result, during the water crisis, women become the most vulnerable group and they suffer from mental as well as physical trauma of managing water-related responsibilities (Bennett et al. 2005).

In this study, the impact of drought on gender has been attempted to explore by analyzing the disparity with respect of: (1) Daily activities; (2) Mobility pattern; (3) Social recognition and respect; (4) Household water management; and (5) Inequality in job sector.

17.4.1 Daily Activities

Drought events influence the daily activities and the lifestyle of the local people.

Daily activity schedule of men and women (Table 17.1) shows that women perform various household activities throughout the day whereas men's activities are limited to personal activities, working outside, and spending time with family members. Women's major works are praying, firewood collection, cooking, collecting water, dishwashing, laundering, taking care of children, and livestock rearing. On the other hand, men work outside for agriculture, farming, and fishing. Daily activities table does not show any common activity between men and women. According to the daily activity schedule both men and women wake up in the morning at 6:00 AM. Man works from 6.30 to 13:00 and from 15:00 to 18:00 in the fields whereas woman has daylong activities to do. In every hour they need to do different types of household activities. Woman takes break only after lunch for a while and again they start doing different household activities. Along with the household activities women also take care of the livestock. Man works outside home and does not contribute to household activities. Specially after coming back from work to home, man spares some relaxing time whereas woman keeps herself busy in preparing dinner. The leisure-time equality principle says that gender equity in a society requires equality in the division of leisure time. But the study clearly shows that women do not get the chance to enjoy leisure time whereas men enjoy the opportunity to do so after coming back from work. Another observation from the field survey is that the women spend time for collecting water from distant places to their houses. As a result, they do not get the opportunity to socialize. On the contrary, men after coming back from the work, visit market, gossip in tea stalls with other people, and socialize. Lack of socialization and leisure time hinders female's mental growth and satisfaction towards life. As women spend a significant time in collecting and storing water, drought has a great impact on their daily life activities and satisfaction toward life. During summer time they collect water from distant places in extreme hot weather which affects their physical and emotional welfare.

Table 17.1 Disparity in daily activities between female and male

No.	Time	Activities by female	Activities by male
1.	6:00	Rising from the bed	Rising from the bed, ablution, and prayer
2.	6:15	–	Take livestock in the field
3.	6:30	–	Go to work and take the previous days food as breakfast
4.	7:00	Ablution and prayer	–
5.	7:30	Sweeping and cleaning house	–
6.	7:45	Washing domestic animal	–
7.	8:00	Go to fetch water	–
	8:30	Start cooking (for breakfast and lunch)	–
8.	9:00	Feeding children and getting them ready to school	–
9.	9:15	Breakfast	–
10.	9:30	Go to work/fire wood collection (2 hours to come back)	–
11.	10:00	–	Take breakfast within work time
12.	13:00	–	Come back to home from work
13.	13:30	Come back home from work	Take bath at ponds/ sometimes go to market for shopping
14.	13:45	Take back livestock to home	–
15.	14:00	Bathing at ponds and washing clothes/collect water from pond for domestic uses/collect water from well for drinking	Lunch and some rest
16.	14:30	Lunch	–
17.	15:00	Take some rest	Go to work again
18.	15:30	Go to work again and take livestock to field again	–
19.	17:30	Come back to home and take livestock to home	–
20.	18:00	–	Come back to home
21.	18:15	Ablution to pond, collect water from well and pond	–
22.	18:30	–	Go to pond for ablution
23.	20:30	–	–
24.	21:00	–	Gossiping with family members
25.	22:00	Dinner	Dinner
26.	22:30	–	Go to sleep
27.	23:00	Prayer and go to sleep	–

Source: Field Survey 2017

17.4.2 Mobility Pattern

Mobility map depicts the variation of movements between female and male for different purposes. The mobility map shows that women visit much places than men to meet the household needs. Both women and men visit different places mostly on foot. Everyday women need to travel to collect drinking water, non-potable domestic water, visiting school, livestock grazing, visiting church for praying, etc. In every week they need to go market for buying necessary household items. Women need to go to union parishad office for collecting rice and work in the agricultural field once in a month. Based on the demand, they need to go outside the community for work, collect firewood, join in local fair, visit NGO offices for taking loan, visit community clinic, hospital, and church. Findings from mobility map suggest that women need to visit many places including office, clinic, market, agricultural field, and church. Women are accustomed to travel to a number of places due to family, social, and traditional responsibilities.

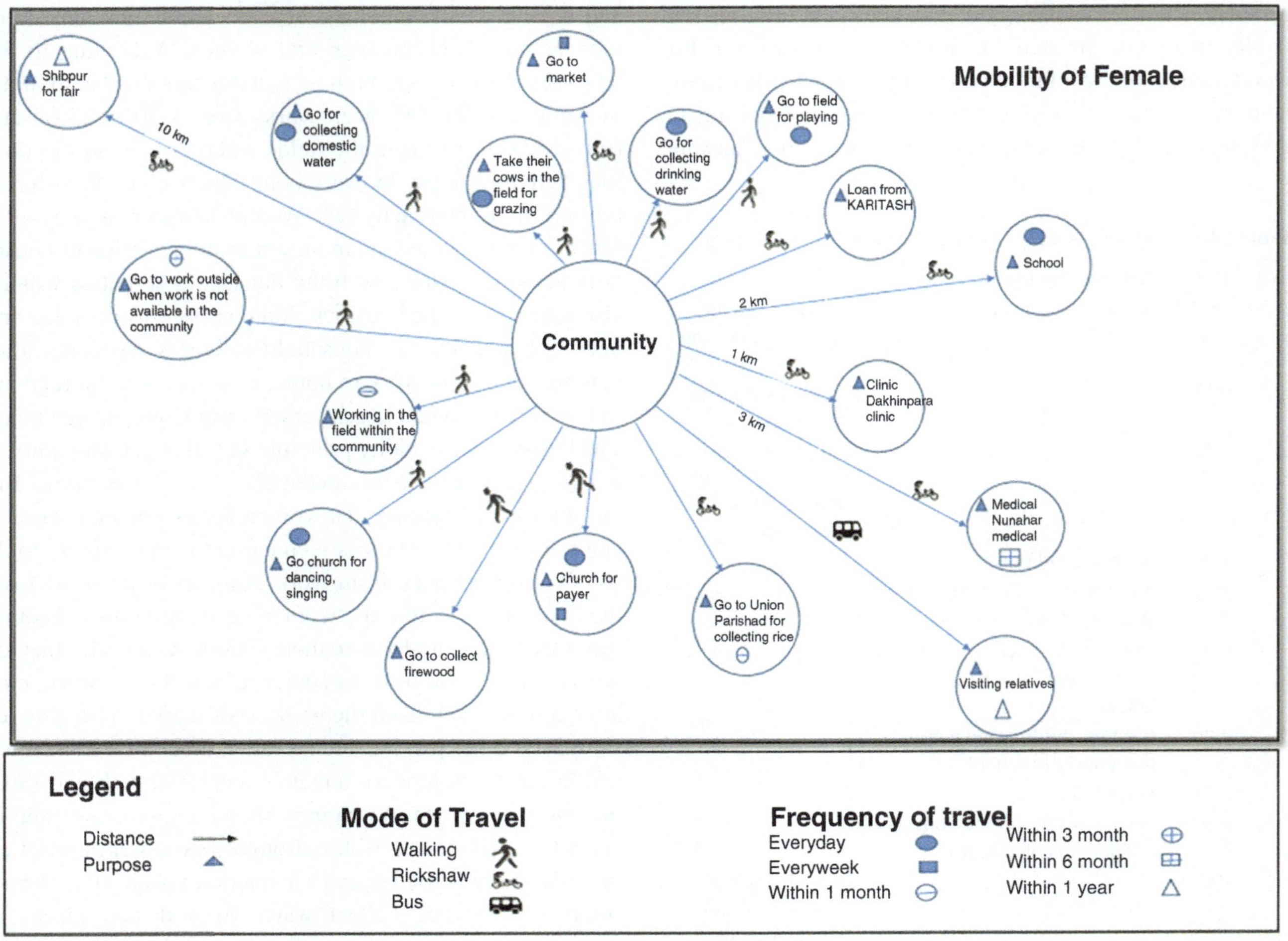

Fig. 17.1 Mobility map of female. (Source: Field Survey 2017)

On the other hand, mobility map of male affirm that men travel for two activities: working on agricultural field and visiting shops for gossiping. It gives a strong idea that apart from working, men enjoy leisure time everyday where women lack such opportunity. Every week men visit Bazar for household shopping. They need to visit church, school, market, and union parishad office for collecting seeds once in a month. Once in three months, men need to go to nearby city for job. The mobility map of female is more complex than male as the number of visiting places for women is more than the men. The traditional mobility map gets much complex in the summer time as women in drought-prone regions travel to multiple distant locations to collect water. Drought directly affects the mobility of women in these regions (Fig. 17.1).

17.4.3 Social Recognition and Respect

Although women contribute to multiple activities, their works are not acknowledged and referred as "work". Men's thoughts and opinions are considered for evaluation and the opposite is for women. It is socially established that working outside the home is job and indoor working is not job. It has a tremendous impact on the respect issue of women. As women's daylong activities are not recognized as work, their voice and opinion are not taken into account. According to Fraser's (1997) one of the principles that resembles gender equity is the "equality of respect principle" which says that women have to be acknowledged as person and their contribution to the social wellbeing must be valued. The study findings have come up with the significant aspect of gender disparity in the society and that is "inequality in respect" (Fig. 17.2).

17.4.4 Household Water Management and Water Crisis

Household water management can be discussed based on household water collection and storage. The daily activity schedule explores that a significant amount of time is spent by women on collecting water from ponds and other sources. Three times in a day women need to go to ponds and other

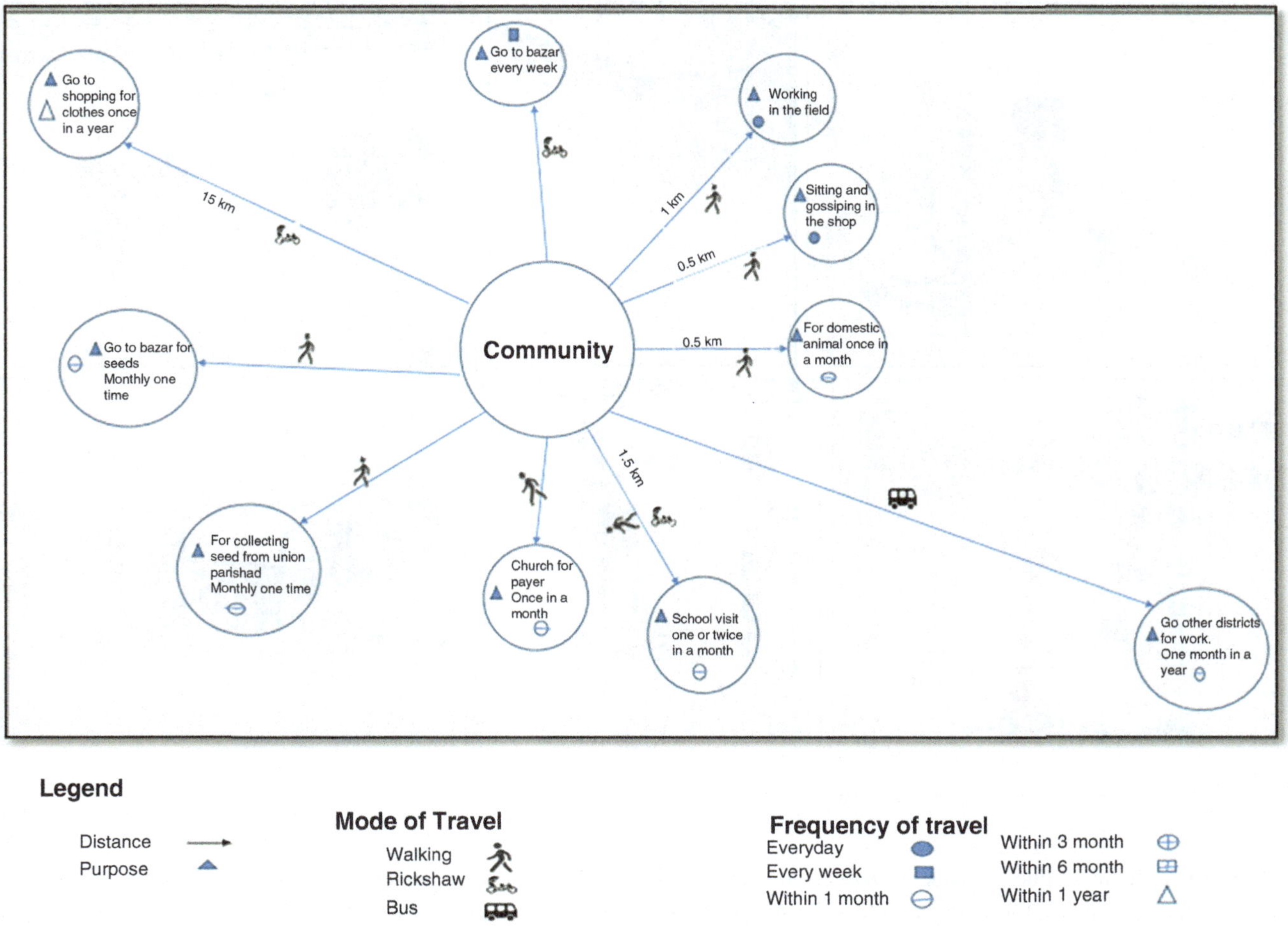

Fig. 17.2 Mobility map of male. (Source: Field Survey 2017)

sources to collect water. In the morning women go to the water source for collecting drinking water, at late noon they go to the pond for cleaning, washing, and bathing and at the evening again they need to go again to fetch drinking water. Women need to collect water three times a day due to storage problem and feed the daily demand (Fig. 17.3).

Water sources are far from the houses and it is really arduous to go and collect water repeatedly. During summer the water scarcity becomes more. Many shallow hand tube wells fail to supply drinking water. During the pick summer, water level of ponds falls down, dries up and the water scarcity goes beyond the survival limit. Women face extreme difficulties to collect water in such hot and dry climate. Moreover, the summer season in Brind tract stays for about 7 months. Each day women need to travel a long way carrying three pots with them to collect drinking water. Traditionally men of Bangladesh play a very passive role in household activities. They do not have the mindset to contribute in household activities such as cooking, dishwashing, cleaning, and collecting water. Men usually work outside. Sometimes they contribute in rearing domestic animals. On an average, they contribute little in household activity. Conventionally the mindset of people assumes that the performing household chores is solely female's task and working outside is men's job. Collecting water in such hostile weather puts them into a huge mental and physical stress. They suffer in psychological trauma, back and chest pain for doing this tricky laborious job daily. This scenario presents a caustic aspect of prevailing gender disparity of the region. The gender disparity becomes more prominent due to drought as drought-induced water crisis in this region forces women doing additional strenuous job of collecting domestic and drinking water from far sources to ensure the household water security.

17.4.5 Inequality in Job Sector

Women go to the agricultural fields after completing all household activities. So they get the chance to work for fewer hours. Due to working for fewer hours they earn less than the male. In addition to it, women have the opportunity to work only for 3 months in a year as the rice cultivation takes place within those 3 months. During the other season

Fig. 17.3 Different activities done by male, female and children. (Source: Field Survey 2017)

rice is not cultivated due to lack to irrigation water and drought condition makes the women jobless during this time. On the other hand, during this off job period of the year, men migrate temporarily to other cities for working. Women cannot migrate due to their family responsibilities and restricted to explore themselves for new job opportunities. The study ensures two eminent gender disparities in the form of income inequality and job opportunities.

17.5 Conclusion

The cumulative impact of water crisis and existing gender disparity cause a considerable negative impact of the mental and physical health of women. The research findings demonstrate that women are deprived in terms of respect, recognition, workload, income opportunity, socialization, and leisure opportunity. The women in the study communities lack social respect as their works are not recognized as works. Men and women are recognized differently in both social and economic terms and "equivalence"" in the communities is absent. The gender gap in the Barind region is more pronounced during drought, as women are more vulnerable to its physical and psychological impacts. Indeed, people are not aware of "gender disparity" despite the incident presents noticeably. This research suggests that drought implications on gender must be taken into account in making any policy agenda regarding drought risk management to establish sustainable development in disaster-affected regions.

References

BBS (Bangladesh Bureau of Statistics) (2011) Planning division, Ministry of Planning. Government of the people's republic of Bangladesh, Dhaka, Bangladesh

Bennett V et al (eds) (2005) "Front matter" opposing currents: the politics of water and gender in Latin America. University of Pittsburgh Press, Pittsburgh, PA, pp i–iv. www.jstor.org/stable/j.ctt9qh61c.1

Fraser N (1997) From redistribution to recognition. In: Justice interrupts: critical reflections on the "post-socialist" condition by. Routlege, Nancy Fraser; London, pp 68–93

Intergovernmental Panel on Climate Change (IPCC) (1998) In: Watson RT, Zinyowera MC, Moss RH (eds) The regional impacts of climate change: an assessment of vulnerability. Special report of IPCC working group II, vol 517. Cambridge University Press, Cambridge, United Kingdom

IWM (2006) Project brief on groundwater model study for deep Tubewell installation project on Barind area. Draft Final Report, Institute of Water Modeling, Dhaka

National Drought Mitigation Center (NDMC) (2006) What is drought understanding and defining drought. Retrieved from: http://www.drought.unl.edu/whatis/concept.htm

Shahid S, Chen X, Hazarika MK (2005) Assessment aridity of Bangladesh using geographic information, Center, Dhaka, Bangladesh

Women and Microcredit in Bangladesh

18

Shajeda Aktar and Rejaul Karim Bakshi

Abstract

Women's empowerment is indispensable for sustainable development. Microcredit initiatives are seen as critical in that adventure, enabling women to achieve economic, political, and psychological boosts that are otherwise absent for the destitute poor. However, microcredit is not a panacea that can cure all odds women face in a patriarchal society like ours. There are literature documenting positive effects of microcredit on the living and being of millions of poor women in the developing countries like Bangladesh. Microcredit enables destitute women to access financial services, facilitates employment-generating activities, boosts self-esteem, increases mobility, raises awareness, improves socio-political position, and helps form social and institutional capital. But such benefits are not sustainable and will not bring lasting effect until long-term challenges are addressed with due caution. This is because women's empowerment is a multidimensional phenomenon that requires a collective action for development. Microcredit programs have been subject to scrutiny and face harsh criticisms in recent time. There are often reports and anecdotes that are contradictory to the conventional claims of poverty alleviation of women empowerment. These controversies often led scholars to debate and divide across certain lines. Considering this ongoing debate, this paper critically explores the effect of microcredit on women's empowerment in the context of Bangladesh.

Keywords

Women empowerment · Microcredit · Microfinance · Poverty · Mobility · Bangladesh

18.1 Introduction

Microcredit is a critical instrument often advocated by scholars and development practitioners as an effective strategy empowering woman and to alleviate poverty across the globe. The current trend of microcredit movement is associated with the Grameen Bank model in Bangladesh. The Grameen Bank initiative has started originally as a pilot project "The Jobra" experiment by Professor Muhammad Yunus in the 1970s. Almost in the same time Central Bank of Bangladesh, *The Bangladesh Bank* initiated a micro lending project for the rural women entrepreneur under a project called "*The Dheki Rin Prokolpo*" in collaboration with another NGO "Swanirvar Bangladesh" (Aghion and Morduch 2007; Wright 2000). Over time, some other initiatives from various NGOs joined the tide, leading to microcredit movement in Bangladesh. Their collective effort and success took headlines in mainstream media not only in Bangladesh but also all over the world. Soon the microcredit movement became symbolized with Bangladesh. In the 1980s and 1990s, micro-credit has been argued unequivocally as key instrument to poverty reduction (Hashemi et al. 1996; Hulme and Mosley 1996; Khandker 1998; Pitt and Khandker 1999; Copestake et al. 1999). The proponents of microcredit urge that microcredit facilitates destitute women to access resources that will enable them to engage in productive employment. To their argument, such entrepreneurship eventually helps them achieve certain empowerment that was unthinkable before specially in terms of mobility outside home and participating in household's decision-making process (Bruntrup et al. 1997; Buckland 1996; Chowdhury et al. 2005). Microcredit was also argued as the

S. Aktar
Department of Public Administration, University of Rajshahi, Rajshahi, Bangladesh

R. K. Bakshi (✉)
Department of Economics, University of Rajshahi, Rajshahi, Bangladesh
e-mail: rkbakshi@ru.ac.bd

R. Ahmed et al. (eds.), *Transforming Bangladesh*, World Regional Geography Book Series, https://doi.org/10.1007/978-3-031-45093-8_18

cornerstone of poverty alleviation. To its advocates, the vicious circle of poverty is self-reinforcing unless you break the cycle with financial injection. According to Yunus, "the poor are not poor because they are untrained, or illiterate; they are poor because they cannot retain the genuine returns to their labor. The reason for this is obvious—they have no control over the capital that calls the tune. The poor work for the benefit of someone who controls the capital" (Yunus 2001: 26).

Microcredit operations provide access to financial services mostly to the poor and to destitute women often engaged self-employment and informal activities. Today, microcredit operations are also combined with savings, insurance, and often non-financial programs such as skill upgrading training and education to help small- and medium-scale producers' and business needs. The training programs are designed to foster skill, efficiency, marketing strategy, and awareness-raising to empower the destitute women. As Mayoux (1997) argued, "microcredit programs are currently being promoted as a key strategy for both poverty alleviation and women's empowerment on the basis that these programs have the impact of increasing women's income levels and control over income, which ultimately results in greater economic independence. The establishment of micro-credit programs enhances the perception of women's contribution to household income and family welfare and this increases women's participation in household decision-making about expenditure and invariably creates a greater expenditure on women's welfare. Finally, these programs tend to help greatly in changing the attitudes of men to the role of women in the household and the community in general".

Since the inception of the microcredit movement in the 1970s, women are the main focus of the microcredit programs. The argument is that there is a win-win situation for microcredit programs in focusing women. Microcredit helps women to boost confidence and self-esteem. It helps women play effective role in decision-making specially in the areas of household chores, family planning, children's education, marriage, and in sending children to school. Microcredit organization also helps to reduce domestic violence through collective action. In many instances, microcredit programs in Bangladesh also enabled them to achieve political offices including positions in the local governments (Kabeer 2002; Osmani 1998; Kabir and Amin 1995).

However, microcredit programs have been subject to scrutiny and face harsh criticisms in recent time specially in Bangladesh and India. There are often reports and anecdotes that are contradictory to the conventional claims of poverty alleviation of women empowerment. These controversies often led scholars to debate and divide across certain lines. Some literature criticizes the microcredit program as trap for rural poor women accusing that the NGOs are exploiting the poor women talking about opportunity of their poverty. According to this strand of literature, it is a clever trap that ties the poor rural women. In practice, NGO's primary concern is lending money and enforcing repayment of the loans (Kabeer 2002; Mahmud 2003; Rogaly 1996). As a result, the other issues of microcredit programs including participation in community activities get less attention due to their full focus on the credit programs. Some studies also express their frustration that microcredit programs cannot help borrowers to financial stability and welfare. Accordingly, offering women to credit-only programs might actually marginalize them (Mahmud 2003). Researchers also argue that microcredit is not an effective instrument for poor people who have no income-generating employment in day-one. These people do not require credit at day-one, rather needs skill improving training programs, access to market, improved transport, and access to financial services for their improvement. As Khandker (2005) opined, NGOs are operating microcredit programs for decades for empowerment of women and the alleviation of poverty; yet we still debate how microcredit operations actually help poverty alleviation and empower women. Considering this ongoing debate, this study explores the effect of microcredit on women's empowerment in Bangladesh.

18.2 Microcredit: A Conceptual Framework

Microcredit, also known as microfinance, literally means small loans. It also comes with small repayments as well as small savings. Microcredit programs are focused on low-income groups specially the poor women. Today, microcredit institutions are not confined to lending loans only, rather encompass other financial and non-financial services and in such refer to the arrangement of substantial kind of financial facilities to the poor. The small- and micro-entrepreneurs generally lack access to formal financial institutions. They are traditionally considered un-bankable due to lacking collateral for loan. Microcredit operations come in such situation lending the micro enterprises without formal physical or financial collateral (Aghion and Morduch 2007). Microcredit programs come up with certain characteristics:

- *Loan size*: The size of the loan is small.
- *No collateral*: There is no need of collateral to access the loan.
- *Loans are progressive*: The loan size increases with successful repayment of first loan.
- *Target group*: Microcredit focuses on low-income households and micro-entrepreneurs.
- *Utilization*: Loan proceeds are used for productive purposes to generate further income for the households.
- *Frequent repayment*: Loans are repaid with frequent installment in order to avoid heavy burden at the end.

18.3 Poverty, Women's Empowerment, and Microcredit

Women are instrumental in fighting against poverty (UN 2002). It is argued that a dollar is better spent by mothers for the health, nutrition, and education of the children than the same dollar earned and spent by men. Development activists and policymakers, therefore, are seen in no hesitation to tag women empowerment, and poverty alleviation with microcredit (Kulkarni 2011). Accordingly, "women's empowerment is indicated as a means that can produce extraordinary ends … their empowerment extolled as the solution to a host of entrenched social and economic problems. The predominant image of empowerment in development is that of women gaining (material) means to empower themselves as individuals, and putting this to the service of their families and communities" (Kulkarni 2011: 12–13). However, women's empowerment is not only related to material means and possessions, but also to socio-psychological issues (Kabeer 1998, Morduch 1999), self-esteem (Pitt and Khandker 1999), agency (Goetz 1995; Haque 2006), choice (Kabeer 2002, Chemin (2008), mobility and inter-personal relationships (Kabeer 2001; Chowdhury et al. 2005; Green et al. 2006). This perspective is well reflected by microcredit programs in all three paradigms of women's empowerment (Kulkarni 2011). The proponents of microcredit argue that microcredit helps achieving feminist agency paradigm, financial self-sustainability paradigm, and the poverty-alleviation paradigm (Kulkarni 2011; Mayoux 2005, 2006).

The poverty-microcredit-wellbeing paradigms, as argued by the proponents of the industry and supported by empirical literature, are presented in Fig. 18.1. Column 1 of the figure lists the multi-dimensional aspects of poverty that a rural poor woman usually faces in Bangladesh. This includes unemployment, unpaid employment, low income, food insecurity, illiteracy, ill health, violence against women, child marriage, dowry, stereotypes and prejudices against women, and so on. These multi-dimensional aspects of poverty are parts and parcel of the rural poor women's life that microcredit aims to address. The microcredit innovations consisting of small loan, savings, insurance, and non-financial measures including skill improving training, high-yield seeds, health and family planning kits, etc. are found effective in combatting poverty, malnourishment, ill health, and violence against women. Microcredit is credited to create employment specially for the destitute poor, enhances income-generating activities, boosts food security, children's health and education, women's agency, self-esteem, and well-being in the short and long run (Table 18.1).

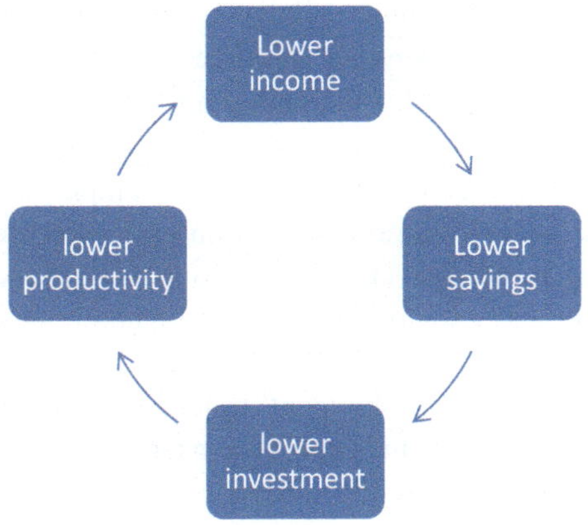

Fig. 18.1 The vicious cycle of poverty

The microcredit revolution started from the claim that credit is human right and is a catalyst in generating sustained increase of income and combating poverty (Yunus 2001). The proponents of microcredit revolution claimed that the injection of microcredit can fill the investment deficiency of the poor households to boost income through meaningful employment and investment. They see microcredit as a critical driver to breach the vicious cycle of poverty in the underdeveloped countries like Bangladesh. As Nurkse (1953) argued, a country is poor because it is poor. According to Nurkse (1953), the underdeveloped countries are in a vicious trap being poor and lower income. They cannot save and invest enough in health, education, and so on as their income is low. The lower level of investment is not helping but yielding lower level of productivity for their labor and capital, resulting in lower level of income further. The cycle of lower

Table 18.1 The poverty-microcredit-wellbeing nexus at a glance

(1)	(2)	(3)	(4)
Problems faced by the poor:	Microcredit interventions:	Short-term effects:	Long-term effects:
Unemployment Unpaid employment Low/ irregular income Illiteracy Hunger/ food insecurity Malnutrition Ill health Prejudices and stereotypes Violence against women Child marriage Dowry	Small loan without collateral Savings option Skill training Combined with social pledges (no dowry, no child marriage, family planning, child education, etc.) Higher loan with successful repayment Predominantly women clients Small enterprises Frequent (often weekly) loan repayment Closer monitoring	Small business Steady employment Steady income Food security Women's economic participation Women's agency within household Women's mobility outside home Women's visibility and position in the community Women's self-esteem	Social recognition of women Gender equity Women's life satisfaction Health for children Education for children Improved household Well-being Elimination of dowry Elimination of child marriage Elimination of family violence

Source: Authors

income, lower savings, lower investment, lower productivity, and lower income continues. The vicious cycle of poverty is summarized in Fig. 18.1.

Figure 18.1 shows that a country starting with lower level of income ends up with reproducing lower level of income again—as their low income enables them to save low, invest low, and productivity low. However, the proponents of microcredit argued to challenge the notion and to end poverty through microcredit intervention. The idea is that by acquiring microcredit, poor households can invest in meaningful investment project, find gainful employment, enhance productivity, and consequently earn better and higher income. Such a conjecture of the microcredit visionaries breaching the vicious cycle of poverty is drawn in Fig. 18.2.

Figure 18.2 shows that though the poorer households have lower income and lower savings, the added microcredit interventions in terms of small loan can increase their investment. The higher level of investment will enhance their productivity (and will increase further if the microcredit comes with non-financial packages such as skill-improving training, high-yielding seeds, etc.) and generate higher income. That is, the introduction of microcredit helps effectively breaching the lower income-lower savings-lower investment-lower productivity-lower income cycle.

Today, microcredit has been embraced not only in the developing countries like Bangladesh but also in countries across all continents worldwide. In Bangladesh, microcredit interventions are not only carried out by the pioneers like Grameen Bank and BRAC but also endured by commercial banks and government agencies. Table 18.2 portraits the evolving scenario of microcredit operations in Bangladesh across government and non-government agencies.

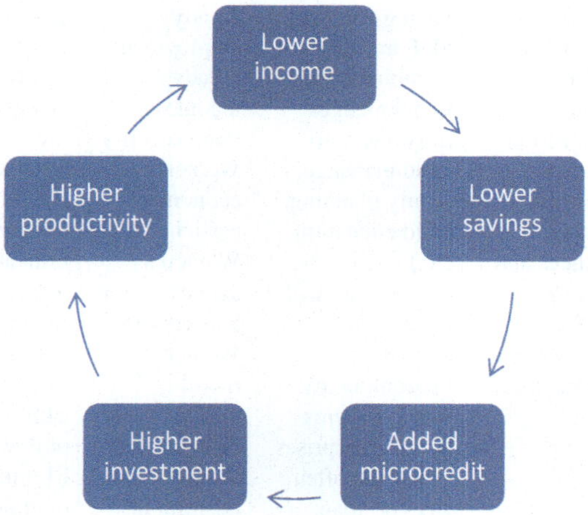

Fig. 18.2 Breaching the vicious cycle of poverty with microcredit. (Source: authors)

Table 18.1 shows that there are as much as 46.72 million people who are current members of microcredit associations (government or non-government) in a country of 165 million people. However, not all the members are joining the microcredit industry to borrow loan, as the third column of the table shows that nearly 36 million of the 47 million members borrowed money from the microcredit organizations. Others might have joined the industry for saving, insurance, and other financial and non-financial benefits. Column 5 of the table shows that the members of the microcredit groups also saved nearly 650 billion Taka compared to accumulated total borrowings of 1152 billion Taka. All these clearly show the depth and breadth of microcredit operations in Bangladesh.

18.4 Microcredit Facilitating Women's Empowerment: Empirical Evidences

Women are vulnerable and lagging behind men in the developing countries like Bangladesh. They lack in accessing formal employment and they own capital to initiate income-generating activities. Undoubtedly, microcredit programs in such can play substantial role in income-generating activities for women and in women's empowerment. Microcredit offers both direct and indirect benefits for women borrowers. The direct benefit is that women members get collateral-free loans, though micro in size, from microcredit organizations. At the same time they also get training services to improve their skill and productivity, legal aid, health advices, and education. Microcredit has documented that providing financial capital to those who are non-bankable can be productive and profitable (Mayoux 2000; Pitt and Khandker 2002). They proved that helping rural women with microcredit and necessary training can significantly increase their income, expenditure, health, nutrition, and self-esteem. Consequently, microcredit has gained the momentum to bring peace positive social change all over the world (Nago and Wahhaj 2011). There are plenty of studies citing evidences that credit for ultra-poor women has effect strengthening their bargaining position within household, and women are in better position than men in spending resources for the betterment of household specially for the children (Aghion and Morduch 2007; Pitt and Khandker 2002; Kabeer 2002; Khandker 1998; Mayoux 2000). Kabeer (2002) emphasizes microcredit as the key instrument of economic empowerment of women as it offers women access to financial services who were otherwise excluded from formal banking channels. To Nikkah and Redzuan (2010), microcredit program is an effective inclusive development strategy that correctly postulates women's economic contributions in order to bring overall financial efficiency of a country.

Table 18.2 Microcredit operations in Bangladesh (2020–2021)

Microcredit organizations	Members (million)	Borrowers (million)	Cumulated loan (billion Taka)	Cumulated savings (billion Taka)	Loan disbursed (billion Taka)
(1)	(2)	(3)	(4)	(5)	(6)
Grameen Bank	9.38	6.54	142.60	158.84	195.48
All other NGOs	35.19	27.80	949.85	422.39	1512.09
Government agencies	1.61	1.16	28.58	5.45	26.58
Commercial banks	0.54	0.41	31.58	62.31	12.55
Total	**46.72**	**35.91**	**1152.61**	**647.99**	**1746.69**

Source: Microcredit Regulatory Authority (2022)

18.4.1 Expanding Mobility Out of Home

Empirical research documents findings that microcredit helps women's mobility out of home in rural areas of Bangladesh where women are traditionally confined to their boundary wall. Microcredit borrowers report that their status has gone up with access to NGO membership because it allowed them to work as income-earners and provided opportunity to move out of their home (Hashemi et al. 1996, Pitt and Khandker 1999, Osmani 1998, Parveen and Chaudhury 2009). Hashemi et al. (1996) also report significant effects of microcredit on different dimensions of women's empowerment including mobility, self-esteem, financial security, etc.

18.4.2 Enabling Financial Services

Microcredit movement claims that credit is human rights (Yunus 2001). Accordingly, financial smoothing is essential to promote social welfare and alleviate poverty. It works with the clients who are very poor and are not considered creditworthy by the conventional banks (Aghion and Morduch 2007, Pitt and Khandker 1999). Furthermore, the principle of microcredit is that they go to the doorsteps of the poor women rather than asking them to go to the banks which essentially worked as a catalyst in providing financial services (Bruntrup et al. 1997; Buckland 1996; Chowdhury et al. 2005). Microfinance programs are, therefore, considered as the effective way to provide essential financial services to poor households at a minimum costs (Stephens and Tazi 2006; Miller and Martinez 2006; Elahi and Danapoulos 2004).

18.4.3 Generating Income

Literature evidenced findings in which households posted higher income through enterprises taken by microcredit programs (Hulme and Mosley 1996; Ghosh 1997). Literature also document notable improvement in the standard of living by the households participating in the microcredit programs. It is argued that microcredit helps poor people, especially women, to expand their sources of income and protect it during shocks which are important to combat poverty (Hashemi et al. 1996; Khandker 1998; Jalil 2005). The opportunity to get a small loan to pursue a business enterprise or to run a homestead garden appears to be the very first step in overcoming the vicious circle of poverty. Evidence suggests that majority of the borrowers in Bangladesh spent their microloan proceeds in productive purposes including poultry, nursery, grocery shop, medicine shop, rickshaw, van, cattle rearing, sewing machines, and so on. The success of these income-generating activities helped them to add to their household income, achieve financial solvency, and earn economic freedom.

18.4.4 Employment for Women

Microcredit provides opportunity for women to get involved with marketable enterprises. Women in Bangladesh are subject to stay home at large due to patriarchal norms (Chowdhury et al. 2005, Copestake et al. 1999; Coleman 1999). This often bars them to get involved with productive employment. Microcredit programs helped them become self-employed specially in non-agricultural sectors with higher income (Kabeer 2002; Kabir and Amin 1995; Khan 2001). Studies documented that a large number of microcredit borrowers run small shops and poultry farms generating productive employment (Ghosh 1997). All these helped to improve their living and being.

18.4.5 Human and Social Capital

Microcredit programs are designed in a way that facilitates building human resources and helps forming social and institutional capital. NGO operations require that women must form a group by their own in order to get financial services. The formation of small groups is conducted by the group leader (Osmani 1998; Parveen and Chaudhury 2009; Zaman 2001). By constructing and leading a group, woman members overcome age-old social stigma who are otherwise traditionally excluded from the public domain. This is revolutionary as the formation of group allows them to bring their social skill, help them get connected socially, and share their views (Kabeer 2002; Kabir and Amin 1995; Khan 2001). Furthermore, the formation of such groups is done by women themselves without assistance from the men and is critical to boost their self-confidence. This kind of social skill and social capital have lasting impact as it aware women of their identity, perceiving self-esteem and womanhood in economic as well as in social and political domain.

18.4.6 Strengthening Position Within Family

Microcredit programs enabled various income-generating activities for the poor women including poultry, gardening, nursery, silk and silk-worm production, embroidery, tailoring, etc. All these helped women to earn extra income and boost self-confidence for women that eventually help to improve women's position within the household (Madhura 2007; Osmani 2007; Parveen and Chaudhury 2009). Furthermore, microcredit programs are combined with various activities including family planning, awareness building, training, health and nutrition, etc. There were also programs to raise awareness to combat violence against women (Zaman 2001; Begum et al. 1998; Chowdhury et al. 2005). All these help to strengthen women's position within household.

18.4.7 Self-Esteem

Empirical studies also document that microcredit initiatives enabled poor women to become active agent and participate in decision-making process. It helps women's agency in various domains including household, economic, social, and political spheres. As a result women are seen with newfound respect, self-esteem, enhanced mobility, independence, and participation in household and community affairs (Amin et al. 2003).

18.4.8 Political Position

Women's participation in microcredit programs has not only enhanced their employment and income but also their position within and outside household. It has changed the way they contribute to the community and affiliation with the social setting. Microcredit borrowers are seen to form clubs, voluntary organization, and even join politics and get elected in the local government election in Bangladesh (Bruntrup et al. 1997; Buckland 1996; Coleman 1999; Ghosh 1997; Jalil 2005).

18.4.9 Social Position

Some literature indicates that women's position in the society has been increased as they garner higher income, productive enterprises, lead credit groups, and get awareness raising and skill improving trainings from NGOs (Kabeer 2002; Kabir and Amin 1995; Khan 2001; Khandker 1998). Women achieved certain level of power that enabled them to take part decision-making process and command social position, respect, and affiliation as their mobility, self-esteem, and skill improved.

18.4.10 Health and Education

In addition to financial services, microcredit programs also come up with non-financial services including health and education. BRAC, for example, run few thousand pre-primary schools for the children of the poor borrowers (Hulme and Mosley 1996; Khandker 1998; Copestake et al. 1999). Grameen Bank takes 16 decisions as obligatory principles including sending children to school, providing nutritious food, and maintaining hygienic practices for the borrowers (Osmani 2007, 1998; Zaman 2001). Empirical studies find that children of microcredit borrowers go to school more often than the similar cohorts (Osmani 2007; Parveen and Chaudhury 2009; Khandker 1998). The drop-outs from school are also fewer for the microcredit borrower households. Furthermore, NGOs like Grameen Bank and BRAC has introduced study loan for the children of the microcredit borrowers for their tertiary education (Parveen and Chaudhury 2009; Osmani 2007; Khan 2001).

18.4.11 Decision-Making

Microcredit programs help empowering women by increasing their agency and participation in the decision-making process. Studies documented women borrowers' enhanced participation in the decision-making in the areas of daily purchasing, buying and selling assets, family planning, sending children to school, marriage children, etc. (Hashemi et al. 1996; Hulme and Mosley 1996; Madhura 2007; Ghosh 1997).

18.4.12 Eradicate Poverty

Microcredit is mostly regarded for its ability to alleviate poverty through income-generating activities for the poor specially the poor women (Khandker 1998; Ghosh 1997). Self-employment and skill-upgrading training are the cornerstone for microcredit initiatives. Khandker (1998), and

Aghion and Morduch (2007) further argue that microcredit is the most cost-effective strategy among other alternatives. The essence of the microcredit program is that it is not charity and it operates with financial sustainability and hence its cost-effectiveness is unmatched compared to the traditional subsidy-based development programs. In addition, the effect of microcredit on household consumption and nutrition is found significant (Zaman 2001; Begum et al. 1998; Chowdhury et al. 2005; Bruntrup et al. 1997; Copestake et al. 1999; Coleman 1999).

18.4.13 Health Behavior and Hygiene

Literature documents that microcredit clients are more aware of health and nutrition status by joining microcredit program. They received training from NGOs about hygiene practices, sanitary latrine, drinking safe water, hand washing after defecation, etc. (Madhura 2007; Osmani 1998; Parveen and Chaudhury 2009). The higher income promoted by the income-generating activities also enabled borrowers to afford nutritious food, treatment while getting sick for household members (Zaman 2001; Begum et al. 1998; Chowdhury et al. 2005). Borrowers also report to receive training on protection from contagious diseases, including diarrhea and cholera, and information about immunization and vaccination.

18.4.14 Improving Relationships

There are also evidences that microcredit helps to improve relationships within the family, specially between husband and wife. The argument is that with increasing contribution from working women, husbands are seen to increasingly recognize their value and respect in such (Ghosh 1997; Jalil 2005). Husband often recognizes wife's entrepreneurship talent and feels proud of her success. They are also valued for their ability to manage family budget, and with helping hand when husbands are in crisis (Chemin 2008; Chowdhury et al. 2005).

18.5 Challenges and Caveats of Microcredit

Microcredit has been criticized on various grounds even in its birthplace Bangladesh. There are anecdotal evidences that microcredit organization often charges very interest rates and forces borrowers to repay even by distress sale of assets. Such assets work like insurance for the poor clients who are categorically excluded from formal insurance market and losing such assets due to microcredit repayment puts them in helpless condition during shocks.

18.5.1 Limited Effect on Poverty Reduction

There are some studies arguing the limitation of microcredit in reducing poverty. Accordingly, poor are getting poorer with the loan trap of microcredit. There are arguments that microcredit program is not suitable for the hardcore poor rather than the moderate poor and the effect of microcredit varies across income groups (Copestake et al. 2001; Morduch 1998; Dugger 2004). The empirical findings on the effect of microcredit eliminating poverty are mixed with success as well as failures (Bruntrup et al. 1997; Edgecomb and Barton 1998; Schrieder and Sharma 1999; Sebstad and Chen 1996).

18.5.2 Loan Trap

Studies found evidence that microcredit borrowers are caught in a debt trap in many instances specially in countries like Bangladesh and India. Accordingly, borrowers take new loans in order to repay their earlier loans as microcredit installments are weekly and NGO officials come to collect repayment at borrowers' doorsteps (Yuge 2011; Scully 2004; Khan 2001; Morduch 1998; Rahman 1999). Such debt-trap results in systemic risk of default. It also has negative impact on borrowers' empowerment. The scenario is very common in Bangladesh as there are no records of financial transactions shared between NGOs. Wright (2000) found that overlapping membership with NGOs in Bangladesh can be as high as 50%.

18.5.3 No Control Over Loan

There are literature citing that the loan proceeds are not used by the women borrowers rather than the husbands and male guardians of the households. Women borrowers in such lose control over credit and hence suffer badly when it's time to repay (Goetz 1995; Goetz and Gupta 1996; Jain 1996; Westover 2008; Zaman 1998; Kabeer 2001; Mahmud 2003). Some studies further noticed that women borrowers failed to use the loan amount according to their planned wishes (Goetz and Gupta 1996; Jain 1996; Zaman 1998; Kabeer 2001). The patriarchal nature of the society does not allow them to make strong decision over who will control the loan money and the business enterprise. The women borrowers also lack control over the income generating from the designated project. When husbands do not use loan effectively, family dispute arises (Jain 1996; Kabeer 2001; Nago and Wahhaj 2011).

18.5.4 Misuse of Credit by Borrowers

Microcredit loans are granted against some designated income-generating activities. However, use of loan money in unproductive purposes is reported in many literatures. However, very often loans are used for reasons other than stated in the loan application. Sometimes the loan proceeds are used for consumption smoothing, buying household durables, and even for social functions like marriage ceremony. This is a challenging issue undoubtedly where microcredit organizations are struggling increasingly (Rahman 1995; Haque 2006; Imai and Azam 2010; Rogaly 1996). Rahman (1995) investigated the use of loan proceeds in Gazipur district in Bangladesh and found that as much as 36% of the ASA loans are utilized in non-income generating activities. This is the key reason for loan default and multiple loan. Study also document that NGOs are not monitoring whether borrowers invested in the loan proceed in specified project or not. Rather they force them to repay the loan no matter what the borrowers are doing (Ebdon 1995; Amin et al. 2003).

18.5.5 Lack of Proper Training and Education

Empirical findings show that microcredit borrowers are largely uneducated. They do not have the necessary skill and training to operate profitable economic activities and to enhance productivity. But NGOs' key concern is loan disbursement and recovery rather than training and knowledge sharing (Dugger 2004; Ahmad 2007; Chemin 2008; Ebdon 1995; Mayoux 2000, 2005). Mayoux (2006) explains that the sustainability of the microcredit programs and the empowerment of the borrowers not only depend on credits but also on other critical issues surrounding it. There are various factors affecting women's empowerment such as education, training, social awareness, and social security. Education and training are important in enhancing skill and decision-making ability of the microcredit borrowers. Microcredit without skill-improving training and knowledge is ineffective in rendering desired outcome (Ebdon 1995; Mayoux 2000).

18.5.6 High-Interest Rates

A common criticism amidst all success of the microcredit program is it charges soaring interest rates putting huge toll on the shoulder of the already poor borrowers. The average interest rate of microcredit operations in Bangladesh is 25% (Aghion and Morduch 2007; Dugger 2004; Ahmad 2007; Fernando 2002; Gertler et al. 2003). It is true that NGOs are not charities rather are private entities and hence need financial sustainability to survive in the market in the long run. But in many cases they are charged with exorbitant interest rate on loan. In the 1970s and 1980s, NGOs were operating with external funding from donor agencies (Osmani 1998; Parveen and Chaudhury 2009). Though in paper the borrowers pay interest rates between 10–18%, the effective interest rates are as high as 27–31% for Grameen Bank, 39–41% for PROSHIKA, and 42–45% for BRAC and ASA (Ahmad 2007).

18.5.7 Excluding the Hardcore Poor

Another common fallacy about microcredit is that the hardcore poor are largely excluded from this noble initiative (Yuge 2011; Ebdon 1995; Coleman 1999; Buckland 1996). The Grameen Bank considers someone poor if he/she possesses farming land less than 1 acre whereas the hardcore poor are mainly landless. Evans et al. (1999) noticed that extreme poor people are excluded from microcredit operations due to various reasons including lack of education, lack of productive project, and awareness. Amin et al. (2003) mentioned that microcredit programs have failed to reaching out to the bottom poor though are successful in reaching the moderate poor. Kabeer (2001, 2002), on the other, argued that microcredit institutions always preferred people who are in better conditions in order to make sure repayment and financial sustainability. This is because NGOs consider the hardcore poor as risky clients and hence avoid them to minimize their risk of default.

18.5.8 Increasing Domestic Violence

There are anecdotal evidences that microcredit triggers domestic violence by increasing tension between husband and wife (Prothom Alo 2011). It was evidenced that in many instances the loan proceeds are used by men and often failed to generate income to repay the loan. This causes conflict within household and women become victim of domestic violence (Kabeer 2001; Amin et al. 2003; Yuge 2011; Scully 2004).

18.5.9 Increasing Dependency

The literature also argue that microcredit increases women's dependency on men to run income-generating enterprises, to sale in the market the items that they produce and often to pay the installment (Yuge 2011; Scully 2004; Goetz 1995). Such dependency is not helping women's empowerment rather aggravates the situation further where men dominate the subordinate women in a conservative society (Goetz and

Gupta 1996). The important concern is that once women lose control over resources, the empowerment dream specially in economic empowerment and in decision-making remains an illusion in a patriarchal society like Bangladesh (Kabeer 2001; Amin et al. 2003; Yuge 2011).

18.6 Conclusion

Women's empowerment is indispensable for sustainable development. The Beijing Platform for Action (UN 1995) declares that "empowerment of women and gender equality are prerequisites for achieving social, political, economic, cultural and environmental security among all peoples". Women's social, political, and economic status are at the core to the achievement of transparent and accountable government. Microcredit initiatives are seen as critical in that adventure enabling women to achieve economic, political, and psychological boost which is otherwise absent for the destitute poor. However, women's empowerment is a collective process and cannot be obtained just simply focusing on microcredit. Microcredit is not a panacea that can cure all odds women face in a patriarchal society like ours.

Undoubtedly, the literature documents certain positive effects of microcredit on the living and being of millions of poor women in Bangladesh. Microcredit enabled destitute women to get access to financial services, facilitates employment-generating activities, boosts self-esteem, increases mobility, raises awareness, improves socio-political position, and helps formation of social and institutional capital. But such benefits are not sustainable and will not bring lasting effect until long-term challenges are addressed with due caution. There are certain caveats of the microcredit program including high-interest rates, excluding the extreme poor, conflict over controlling resources, incidences of domestic violence, mismanagement of loan proceeds by borrowers, lack of appropriate training and education, etc. these are important in order to facilitate income generating activities and for the empowerment of women through microcredit.

It is imperative to note that women's empowerment is a multidimensional phenomenon that requires a collective action for development. As OECD (2012) correctly highlights, "achieving women's empowerment is not a quick fix. It will take sound public policies, a holistic approach and long-term commitment from all development actors. Innovative approaches and partnerships are needed to scale up women's empowerment". One should, therefore, consider microcredit as a critical weapon to empower women in a developing country like Bangladesh but must not think it as a panacea and only instrument to empower women. Therefore, the policymakers must formulate an integrated approach for the empowerment of women and for the destitute women in particular. Furthermore, women's empowerment is a combination of diverse aspects which are closely related. Therefore, initiatives that only target one aspect of empowerment may not succeed at all (Kabeer 2001). All actors and stakeholders in such should work together in order for the empowerment of women.

References

Aghion A, Morduch J (2007) The economics of microfinance. MIT Press

Ahmad Q (2007) Socio-economic and indebtedness-related impact of micro-credit in Bangladesh. The University Press Limited, Dhaka

Amin S, Rai A, Topa G (2003) Does microcredit reach the poor and vulnerable? Evidence from northern Bangladesh. J Dev Econ 70:59–82

Begum S, Ahmed F, Rahman M (1998) The impact of RDRS programmes on income and women's status. J Rural Develop 28(1):107–108

Bruntrup M, Huda A, Alauddin S, Rahman M (1997) Impact assessment of ASA. Association for Social Advancement (ASA), Dhaka

Buckland J (1996) The distributional impact of income-generation programs in Bangladesh. J Dev Stud 3(17):385–404

Chemin M (2008) The benefits and costs of microfinance: evidence from Bangladesh. J Dev Stud 44(4):463–484

Chowdhury M, Ghosh D, Wring R (2005) The impact of micro-credit on poverty: evidence from Bangladesh. Prog Dev Stud 5(4):298–309

Coleman B (1999) The impact of group lending in Northeast Thailand. J Dev Econ 45:105–141

Copestake J, Bhalotra S, Johnson, Susan (2001) Assessing the impact of micro-credit: a Zambian case study. J Dev Stud 37(4):81–100

Dugger C (2004) Debate stirs over tiny loans for World's poorest. New York Times

Ebdon R (1995) NGO expansion and the light of reach the poor: gender implication of NGO scaling up in Bangladesh. IDS-Bulletin 26(3):19–55

Edgecomb E, Barton L (1998) Social intermediation and microfinance programs: a literature review. Development Alternatives Inc, Bethesda

Elahi K, Danapoulos C (2004) Microfinance and third world development: a critical analysis. J Polit Military Sociol 32(1):61–77

Evans et al (1999) Demystification nonparticipation in microcredit: a population-based analysis, World Dev 27(2): 426–427

Fernando N (2002) Microfinance industry in Asia: current status, recent trends and some thoughts on the future. Available on: http://www.adb.org/sds/doc/micNFernado.pdf

Gertler P, Levine D, Moreti E (2003) Do microfinance programs help families insure consumption against illness? Working paper no. CO3-129. Center for International Development Economics Research (CIDER), University of California, Berkeley

Ghosh A (1997) Impact of homestead farming on income and women's development in PROSHIKA programme area in Gabtali Upazilla of Bogra District. Unpublished M.S thesis, Department of Agricultural Economics, BAU, Mymensingh

Goetz A (1995) Credit for poor women in Bangladesh: who controls the loan use? Develop Corporat 6:11–15

Goetz A, Gupta R (1996) Who takes the credit? Gender, power, and control over loan use in rural credit programs in Bangladesh. World Dev 24:No.1

Green C, Kirkpatrick C, Murinde V (2006) Finance for small enterprise growth and poverty reduction in developing countries. J Int Dev 18(7):1017–1030

Haque K (2006) Microfinance in Bangladesh: major achievements and new challenges. Paper proceedings of the SAARC

FINANCE Governors' Symposium on Microfinance held in Dhaka, Bangladesh, 15–16 February

Hashemi S, Schuler S, Riley A (1996) Rural credit programs and Women's empowerment in Bangladesh. World Dev 24(April):635–653

Hashemi S et al (1996) Rural credit program and women empowerment in Bangladesh. World Dev 24(4):635–653

Hulme D, Mosley P (1996) Finance against poverty. Routledge Publishing, New York

Imai K, Azam M (2010) Does microfinance reduce poverty in Bangladesh? New evidence from household panel data. Research Institute for Economics and Business Administration, Kobe University, Rokkodai, Nada

Jain P (1996) Managing credit for the rural poor: lessons from the Grameen Bank. World Dev 24(1):79–89

Jalil MA (2005) Impact of micro-credit program of RDRS on rural women in some selected Aress of Kurigram District. Unpublished M.S. thesis, Department of Agricultural Economics (Finance). Bangladesh Agricultural University, Mymensingh

Kabeer N (1998) Money Can't buy me love? Re- reevaluating gender, credit, and empowerment in rural Bangladesh, IDS, discussion paper no. 363, Brighton: IDS

Kabeer N (2001) Conflicts over credit: re-evaluating the empowerment potential of loans to women in rural Bangladesh. World Dev 29(1):63–84

Kabeer N (2002) Reflections on the measurement of women's empowerment, SIDA studies no 3: discussing women's empowerment- theory and practice. Elanders Novum AB, Göteborg

Kabir M, Amin R (1995) The impact of poor Women's participation in village based development programme of facility. J Rural Develop 25(1):25–45

Khan A (2001) Bangladesh NGOs and poverty alleviation. In: The Bangladesh observer on 2 September 2001, Dhaka, Bangladesh

Khandker SR (1998) Fighting poverty with microcredit, experience in Bangladesh. Oxford University Press, New York

Khandker SR (2005) Microfinance and poverty: evidence using panel data from Bangladesh. World Bank Econ Rev 19(2):263–286

Kulkarni V (2011) Women's empowerment and microfinance. An Asian Perspective Study, Occasional paper, IFAD

Madhura S (2007) The microcredit alternative? Econ Polit Wkly 42(13):1171–1175

Mahmud S (2003) Actually how empowering is microcredit? Dev Chang 34(4):577–604

Mayoux L (1997) Microfinance programs and Women's empowerment: towards a framework for participatory planning. Action Aid United Kingdom

Mayoux L (2000) Microfinance and the women empowerment, a review of the key issues., Working Paper,. International Labor Office

Mayoux L (2005) Women's Empowerment through Sustainable Microfinance: Rethinking Best Practice sustainable micro-finance for Women Empowerment. www.genfinance.info/

Mayoux L (2006) Women's Empowerment and Microfinance: A 'think piece' for the Microfinance field, www.genfinance .info/

Microcredit Regulatory Authority (2022) Annual Report 2021. MRA, Dhaka

Miller J, Martinez R (2006) Championship league: an overview of 80 leading Latin American providers of microfinance. Microbanking Bulletin 2006

Morduch J (1998) Does microfinance really help the poor? New evidence from flagship programs in Bangladesh. https://ideas.repec.org/p/pri/rpdevs/morduch_microfinance_poor.pdf.html

Morduch J (1999) The microfinance promise. J Econ Lit XXXVII:1569–1614

Nago TM-P, Wahhaj Z (2011) Microfinance and gender empowerment. J Dev Econ:1–12

Nikkah H, Redzuan M (2010) The role of NGOs in promoting empowerment for sustainable community development. J Hum Ecol 30(2):85–92

Nurkse R (1953) Problems of capital formation in underdeveloped countries. Oxford University Press, London

OECD (2012) Women's economic empowerment, the OECD DAC network on gender equality, http://www.oecd.org/dac/povertyreduction/50157530.pdf, 10.05.2014

Osmani L (1998) Impact of credit on the relative Well-being of women: evidence from the Grameen Bank, IDS bulletin, 29(4). IDS, Brighton

Osmani L (2007) A breakthrough in Women's bargaining power: the impact of microcredit. Internat J Develop 19:695–716

Parveen S, Chaudhury M (2009) Micro-credit. Intervention and effects on empowerment of rural women: the BRAC experience. Bangladesh Res Publicat J 2(3):641–647

Pitt M, Khandker SR (1999) The impact of group-based credit program on poor households in Bangladesh: does the gender of participants matter? J Polit Econ 106:5

Pitt M, Khandker SR (2002) The impact of group-based credit on poor households: an analysis of panel data from Bangladesh (mimeo)

Prothom-alo (2011) "Microcredit and Poverty Alleviation", A round table discussion organized by *Prothom-Alo* on 13 March, 2011 (Published in 27 March, 2011)

Rahman M (1995) A socio-economic study of credit programme of the ASA in some selected areas of Gazipur District. M.S. thesis, Department of Agricultural Finance, BAU, Mymensingh

Rahman A (1999) Microcredit initiatives for equitable and sustainable development: who pays? World Dev 27(1):262–270

Rogaly B (1996) Microfinance evangelism, destitute women and the hard selling of a new anti-poverty formula. Dev Pract 6(2):100–112

Schrieder G, Sharma M (1999) Impact of finance on poverty reduction and social capital formation: a review and synthesis of empirical evidence. Savings Develop 23(1):67–93

Scully N (2004) Microcredit no panacea for poor women. Global Development Research Centre, Washington, DC. http://www.gdrc.org/icm/wind/micro

Sebstad J, Chen G (1996) Overview of studies of the impact of microenterprise credit. Manag Syst Internat, Washington

Stephens B, Tazi H (2006) Performance and transparency: a survey of microfinance in South Asia. Microbanking Bulletin 2006

UN (1995) The Beijing platform for action. The Fourth World Congress on Women, Beijing, p 1995

UN (2002) World development report, UN

Westover J (2008) The record of microfinance: the effectiveness/ineffectiveness of microfinance programs as a means of alleviating poverty. Elect J Sociol 1198:3655

Wright G (2000) Micro finance systems: designing quality financial services for the poor. The University Press Limited, Dhaka

Yuge Y (2011) The current situation of microfinance in Bangladesh: a growing concern about overlapping loan problems – from a field visit to Rajshahi and Camilla Center for Emergency Markets Enterprises, Student Research Series. The Fletcher School, Tufts University

Yunus M (2001) Banker to the poor: the story of the Grameen Bank. Aurum Press, New York, London

Zaman H (1998) Who benefits and to what extent? An evaluation of BRAC's micro-credit program. University of Sussex, D.Phil thesis

Zaman H (2001) Assessing the poverty and vulnerability impact of micro-credit in Bangladesh: a case study of BRAC. Background paper for the WDR 2000/2001

Index

R. Ahmed et al. (eds.), *Transforming Bangladesh*, World Regional Geography Book Series,
https://doi.org/10.1007/978-3-031-45093-8

SPRINGER NATURE

GPSR Compliance

The European Union's (EU) General Product Safety Regulation (GPSR) is a set of rules that requires consumer products to be safe and our obligations to ensure this.

If you have any concerns about our products, you can contact us on ProductSafety@springernature.com

In case Publisher is established outside the EU, the EU authorized representative is:

Springer Nature Customer Service Center GmbH
Europaplatz 3
69115 Heidelberg, Germany

The manufacturer's authorised representative in the EU is Springer
Nature Customer Service Centre GmbH, Europaplatz 3, 69115 Heidelberg,
Germany. If you have any concerns regarding our products, please
contact ProductSafety@springernature.com

Printed and bound by CPI Group (UK) Ltd, Croydon, CR0 4YY
01/10/2026
02225884-0005